WATER QUALITY IN NORTH AMERICAN RIVER SYSTEMS

EDITED BY

C. DALE BECKER AND DUANE A. NEITZEL

Environmental Sciences Department
U.S. Department of Energy
Pacific Northwest Laboratory
Operated by Battelle

River, take me along
In your sunshine
Sing me a song
Ever mornin'
Ever windin' and free
You rollin' old river
You changin' old river
Let's you and me river
On down to the sea.

Michael Staines
Mineral River Records

BATTELLE PRESS
Columbus • Richland

Library of Congress Cataloging-in-Publication Data

Water quality in North American river systems / edited by C. Dale
  Becker and Duane A. Neitzel.
      p.   cm.
    Includes bibliographical references.
    ISBN 0-935470-50-6 : $44.95
    1. Stream ecology—North America—Congresses. 2. Water quality—
  North America—Congresses. 3. Water quality management—North
  America—Congresses. 4. Water—Pollution—North America—
  Congresses. I. Becker, C. Dale (Clarence Dale) II. Neitzel, D. A.
  QH102.W38   1992
  574.5′26323′097—dc20   92-4935
                                                                     CIP

Printed in the United States of America.

Copyright © 1992, Battelle Memorial Institute

All rights reserved. No part of this book may be reproduced or transmitted in any form or by any means, electronic or mechanical, including photocopying, recording or by any information storage and retrieval system, without permission from the publisher.

Battelle Press
505 King Avenue
Columbus, Ohio 43201-2693
614-424-6393
Toll Free: 1-800-451-3543
Fax: 614-424-5263

Acknowledgments

The effort, first, of holding the original symposium and, second, of coordinating and preparing this publication was supported by the Water Quality Section (WQS) of the American Fisheries Society over the terms of three presidents—Ronald H. Preston (1984–1986) of the U. S. Environmental Protection Agency, Charles C. Coutant (1986–1988) of Oak Ridge National Laboratory, and James W. Wiener (1988–1990) of the U. S. Fish and Wildlife Service. The coordinating editors received encouragement and logistical backing from Pacific Northwest Laboratory, which is operated by Battelle Memorial Institute for the U. S. Department of Energy under Contract DE-AC06–76RLO 1830.

Coordinators and Editors
    C. Dale Becker and Duane A. Neitzel
    Pacific Northwest Laboratory
    Richland, Washington 99352 USA

Contents

Preface	ix
Introduction	1
Water Quality and Ecology of the Chena River, Alaska	5
Physical Features	5
Biota and Ecosystem Properties	10
Water Quality and Resource Management	17
Land Use and Management	20
Future of the Chena River	23
Acknowledgments	23
References	24
Longitudinal Zonation of the Biota and Water Quality of the Bow River System in Alberta, Canada	29
Morphology, Hydrology, and Water Quality	32
Longitudinal Zonation of Biota	35
Municipal Effluent Impact on Mainstem Water Quality	38
Tertiary Sewage Treatment of Effluents to the Bow River	43
Conceptual Model of River Nutrient Enrichment	45
Water Quality Control Efforts	46
Acknowledgments	47
References	48
Water Quality and Biota of the Columbia River System	51
Morphometry	54
Hydrology	55
Water Quality and Resource Management	59
Biota and Ecosystem Properties	70
Water Quality control and Fishery Mitigation	75
Acknowledgments	76
References	77
Water Quality and Water Management Sacramento-San Joaquin River System	85
Morphology and Hydrology	90
Water Quality and Resource Management	92

Biota and Ecosystem Properties	104
Water Quality Control Efforts	111
Future Prospects	112
References	114

Some Geohydrological Features of the Santo Domingo Basin, Sierra San Pedro Mártir, Baja California Norte, Mexico ... 117

Physiography	121
Water Quality Characteristics	123
Climate	124
Ecology	126
Human Influences	128
Future Prospects	131
Acknowledgments	131
References	132

The Missouri River—Great Plains Thread of Life ... 135

Morphometry and Hydrology	138
Physical and Chemical Characteristics	141
Biota and Abiotic Environment of the Missouri River	144
Industrial and Municipal Pollution	146
Problems with Heavy Metals	148
Future Prospects	154
References	155

Water Quality Changes and Their Relation to Fishery Resources in the Upper Mississippi River ... 159

River Basin Characteristics	163
River Features	163
Water Quality Issues	166
Trends in Fish Populations	173
Water Quality Control Efforts	175
Future Prospects	177
Acknowledgments	178
References	178

La Grande Rivière (Northern Québec) ... 181

Morphology	186
Climate	188
Hydrology	190
Water Quality and Resource Management	191
Properties of the Natural Ecosystem	196
Changes from Hydroelectric Development	198

Control of Water Quality	203
Future Projects	203
Acknowledgments	203
References	204

Historical Changes in Water Quality and Fishes of the Ohio River ... 207

Morphology and Hydrology	210
Water Quality	214
Changes in Fish and Other Biota	221
Water Quality Control Effects	227
Future Prospects and Hopes	228
Acknowledgments	229
References	229

Water Quality in the Cumberland River Basin ... 233

History	235
Morphology and Hydrology	236
Properties of the Ecosystem	239
Future Prospects	244
References	246

Seasonal Patterns of Water Quality in Blackwater Rivers of the Coastal Plain, Southeastern United States ... 249

Morphology and Hydrology	252
Water Quality Characteristics	256
Biota and Ecosystem Characteristics	269
Water Quality Control Efforts	271
Future Prospects	272
Acknowledgments	273
References	273

Water Quality and Biological Communities of the Mobile River Drainage, Eastern Gulf of Mexico Region ... 277

Eastern Gulf of Mexico Region	282
Prehistoric Perspectives	287
Historical Perspectives and Current Status	287
Conservation Efforts and Future of Area	297
Conclusions	300
Acknowledgments	300
References	300

Preface

This book is about water quality and other characteristics of selected ecosystems in North America. It is also about changes that have occurred in these ecosystems as a result of recent human activities—changes that result primarily from development and exploitation to sustain the needs of an ever-increasing population and the technical innovations that sustain it.

Even to the casual observer, it should be evident that major physical and chemical changes have occurred in North American river ecosystems during the last 100 years. In terms of the millennia during which rivers have evolved, covering thousands of generations for most aquatic organisms, the last 100 years is a relatively brief span. But this span has been ecologically significant because of the rate and extent of changes that have taken place. In terms of the human race, 100 years represents only four generations. Yet modification of river ecosystems in North America (indeed, throughout the world) continues at a rapid pace in step with the unrelenting growth in population and technology. What will these rivers look like 500 years from now? Will any resemblance of these rivers remain, as the aboriginal people and first European settlers once knew them?

A ray of hope lies in the present social and political atmosphere in North America: the philosophy of many regulatory agencies has partially shifted from development and exploitation of natural resources to their protection and enhancement. However, these agencies feel justified in taking action to protect and enhance a river ecosystem, or to retrieve and restore a severely deteriorated river ecosystem, only when they have extensively gathered and analyzed data on hand.

It is no longer sufficient simply to observe and describe in broad terms the major physical alterations of river ecosystems and changes in the biota's abundance and distribution. Identifying and quantifying the many minor alterations in physical, biological, and ecological attributes have become more important to maintain or to initiate any change in the status quo. This effort must be undertaken at considerable financial cost, extracted from the taxpayer and funneled from federal, state, and local authorities to engineers, surveyors, hydrologists, ecologists, biologists, and other experts working together to provide the necessary quantitative data. The concerned public now also provides considerable input to decision making. More often than not, these extensive investigations are mandatory under environmental laws.

This book, a compilation of information on twelve modified river ecosystems in North America, contributes to our understanding and to decision making. The contents of each chapter are revealing.

The Chena River (Chapter 1), a clearwater river system in the subarctic interior Alaska near Fairbanks, exists under a severe thermal regime with long periods of winter ice cover. The river supports 13 species of fish (7 salmonids) and the largest fishery for Arctic grayling (*Thymallus arcticus*) in the state. Stream food webs receive and store less organic carbon, species of fish and benthic macroinvertebrates are less diverse, and individual fish grow more slowly than in similar streams of more temperate regions. Water quality has been affected by mining (primarily for placer gold), limited agriculture, flood plain development, and municipal wastewater discharges. The most important instream development is the Fairbanks Flood Control Project, which provides reasonable control of floods without the ecological changes of a large reservoir but raises potential problems for fish migration and flood plain management. Clearing of land in the watershed poses the greatest threat to the integrity of the Chena River.

The Bow River (Chapter 2) originates in the subalpine forests of the Rocky Mountains in Canada and flows eastward through the foothills and mixed grass prairies of southern Alberta. The harsh climate, combined with rapidly changing riparian habitat, leads to marked zonation in physical and chemical environments and, hence, in biota along the river's length. Cold water species of fish such as rainbow trout (*Oncorhynchus mykiss*) and mountain whitefish (*Prosopium williamsoni*) live in the upper, forested reaches of the Bow River; cool water species such as northern pike (*Esox lucius*) and walleye (*Stizostedion vitreum*) live in the lower, prairie reaches. Discharges from the city of Calgary contribute phosphorus and nitrogen to the lower reaches during all seasons, which stimulate the production of attached macrophytes and algae. Revised water-management practices include tertiary treatment of sewage to improve both water quality and fisheries in reaches of the Bow River negatively affected by human activities.

The Columbia River (Chapter 3) originates in the Rocky Mountains of southwestern Canada, enters the United States, receives the Snake River, and flows into the Pacific Ocean as the boundary between Washington and Oregon. It ranks fifth in drainage area and third in discharge volume among North American rivers. Water quality and fish habitat have been degraded since the 1930s, largely from installation of main-stem dams and reservoirs to produce hydroelectric power, but also from irrigation withdrawals, logging, mining, stream channeling, and urbanization. Although water quality in the main-stem Columbia River generally remains high, production of anadromous salmonids (genus *Oncorhynchus*) has been greatly reduced from historical levels. As a result, water-quality issues take second in importance to restoration of salmon runs. Strong conflicts have emerged between consumptive and nonconsumptive users of Columbia River water. Possible large-scale transfers to areas deficient in water in the southwestern United States threaten.

The Sacramento and San Joaquin rivers (Chapter 4) drain the Central Valley of California and the surrounding mountains, join in the Sacramento-San Joaquin Delta, and enter the Pacific Ocean through the San Francisco Bay. Combined, these rivers provide more than half the water used in California. Irrigated agriculture receives most of this water, and it is carried throughout the state in some of the world's largest engineered distribution systems. As a result, morphology and hydrology of the watershed have been greatly modified during the last 200 years. Introduced aquatic species and water transfers make maintenance of a static ecosystem impossible. Major issues facing resource managers include drinking water quality, agricultural return flows, and relationships between stream flows and fish abundance. Purchase from agricultural interests has been proposed to obtain water for California's increasing population, thus circumventing the need for additional storage projects.

The Santo Domingo Basin (Chapter 5) is a major water district in the hot and arid Baja California Norte, Mexico. The Rio Santo Domingo drains San Pedro Mártir Park, and is the only river in the basin that has sufficient precipitation and water storage in its headwaters to flow to the Pacific Ocean throughout the year. Overgrazing, improper management of rangeland, and natural catastrophes such as fires have made the basin susceptible to erosion and other destructive forces of nature. Water quality in the lower reaches of the basin has been altered by tenants along the border of the park, overstocking of cattle, and use of fertilizers and pesticides. The endemic San Pedro Mártir trout (*Oncorhynchus mykiss nelsoni*) occurs in the upper reaches, where relative isolation provides limited protection. Storage dams proposed for streams of the Santo Domingo Basin to meet the region's urgent need for more water may adversely affect this salmonid and other natural resources.

The Missouri River (Chapter 6), the longest river in the United States, begins in Montana and flows generally south and east to join the Mississippi River near St. Louis, Missouri. The main stem now consists of channelized, impounded, and flowing reaches of nearly equal overall lengths. Although the Missouri River was once called the "Big Muddy," much of its load of suspended sediments is now deposited in reservoirs. However, sediment-free water released at dams renews its load downstream by scouring the river bed. Influx of allochthonous organic material to the river is reduced by use of the reservoirs to control flooding, erosion, and channel meandering. Consumptive water losses, primarily from irrigation and evaporation, account for more than 8.6% of the river's flow. Sediments in tributaries of the middle Missouri River still contain large amounts of mercury from early gold mining activities. Fish inhabiting the lower Missouri River accumulate insecticides and polychlorinated biphenyls (PCBs) at levels that require surveillance. Despite the river's size, continued development and use will one day result in an insufficient water supply.

The upper Mississippi River (Chapter 7) is a 1370-km stretch between north-central Minnesota and the Ohio River at Cairo, Illinois. The most notable change occurred in the 1930s with construction of a lock and dam system to facilitate commercial transport. In 1988, traffic at Lock and Dam 26 near Alton, Illinois, included 7500 barge tows. While the low-head dams created diverse lentic habitats, they also changed the stage and sediment transport features of the river. The main fishery-related issues concern water quality and the effects of sediments and toxic contaminants from non-point sources. Between 42 and 99% of the streams in the upper Mississippi River Basin, encompassing five states, now fail to support all uses designated by water-quality criteria because of pollution. Sediment input results in a significant loss of fishery habitat, primarily in productive side channels and backwaters. An additional loss of 22 to 49% of existing open water areas is predicted within the next 50 years. Toxic contaminants transported along with fine sediments become more available to stream biota. Today, the upper Mississippi River is a multiuse system with a difficult future.

The La Grande Rivière (Chapter 8) is the largest river entering the east side of James Bay, Northern Québec, Canada. The area lies in the cold continental subarctic with short, mild summers and long, harsh winters. The river's drainage basin was greatly expanded when water was diverted from the Eastmain and Caniapiscau rivers after the La Grande hydroelectric project began in 1972. The project called for five large reservoirs that flooded a 11,350-km^2 area at maximum levels, of which 85% was previously land. Before the reservoirs were filled, water in the La Grande Rivière was clear, almost colorless, slightly acid, and low in dissolved salts and minerals. After the reservoirs first filled, nutrient levels increased because of leaching and, especially, the decomposition of flooded organic matter. However, water quality returned to its former state in a few years. The paucity of nutrients, limited organic contribution from terrestrial vegetation, and rigors of the northern climate keep aquatic production low. Flooding altered the composition of 24 resident fish species, of which two species (and occasionally a third) increased in abundance. Some game fish now carry high levels of mercury.

The Ohio River (Chapter 9) begins at Pittsburgh, Pennsylvania, at the junction of two large tributaries and flows 1578 km southwesterly to join the Mississippi River at Cairo, Illinois. It drains one of the great coal-producing regions of the world, which supports an extensive human population. The entire river was affected by siltation after forests were cleared in the 19th century and by low-head navigation dams built between 1900 and 1927. The navigation dam system converted the Ohio river from a free-flowing form, with depths less than 1 m at many locations during dry seasons, to a canalized form in which depths are maintained above 2.7 m to benefit heavy barge traffic. Water pollution became severe by 1940, but abatement efforts have now raised dissolved oxygen levels to the point where they again support aquatic life. Between 1819 and 1988, 159 species of fish were reported from the Ohio River. Of these, 13 species present before 1970 have not been found since. Historical changes in fish abundance are associated with the effects of siltation and channelization on substrates. Pollution control requires continued vigilance and monitoring.

The Cumberland River (Chapter 10) drains a large area adjacent to the Tennessee River in the southeastern United States; the two rivers enter the Ohio River near one another. The Cumberland River has been extensively developed by the U.S. Army Corps of Engineers, beginning in the 1800s with a series of navigation impoundments. Today, the system encompasses ten multipurpose tributary and mainstem dams that allow control of flows to enhance navigation, water quality, and recreation. Another upstream dam provides hydroelectric power. The upper Cumberland Basin is affected by acid mine drainage, siltation, and residues from oil and gas extraction. The middle and lower areas are affected by expanding urbanization and agriculture. Water quality in the main-stem Cumberland River generally is good but it is only fair near Nashville because of nutrient enrichment and changes by upstream reservoirs. Composition and distribution of the diverse fish and unionid mussel fauna have changed with the altered hydrologic conditions, especially by hypolimnetic releases. Eight species of mussels are endangered. Strict enforcement of state and federal environmental regulations is needed to ensure the integrity of the Cumberland River Basin.

The blackwater rivers of the Coastal Plain (Chapter 11) in the southeastern United States are named after the dark color acquired from leached humic substances and a low load of suspended sediments. These rivers have low gradients and drain extensive flood plain swamps. The Ogeechee and Satilla rivers, representative of other streams in the area, are warm water rivers that regularly flood during winter. Water quality is generally low because of their high color and dissolved organic matter, and their low alkalinity and pH. Dissolved oxygen levels decline in the summer because rates of ecosystem respiration become high as organic materials decay in the flood plain swamps. Nutrients enter the Ogeechee River from agricultural activity and the Satilla River from sewage treatment effluent below the watershed's largest city. Water quality has changed little in the last decade, but the unregulated withdrawal of water for agriculture is a threat. Removal of woody debris, a hazard to navigation, has affected aquatic biota more than has any change in water quality. The integrity of blackwater rivers can be protected by maintaining the extensive flood plain forests.

The Mobile River (Chapter 12) is the largest system east of the Mississippi River that flows to the Gulf of Mexico. It drains most of Alabama, and parts of Mississippi and Georgia, and includes six large rivers: the Alabama, Tallapoosa, Coosa, Cahaba, Black Warrior, and Tombigbee. The Mobile River system differs from other river systems in North America by its geologic diversity, which contributes to the biological richness of its rivers and streams. The four major physiographic provinces drained by the system have distinct underlying materials that influence water chemistry as well as the distribution and abundance of aquatic organisms. The large rivers were dredged, channelized, and cleared of snags in the 1800s to aid transportation. Construction of impoundments in the Mobile River drainage began in the early 1900s and now all the large rivers, except the Cahaba River, are a series of reservoirs. Many aquatic species are now restricted to smaller tributaries. At least 15 species of fish were eliminated by construction of the Tennessee-Tombigbee Waterway, and at least 99 species of gastropods are now either extinct or threatened and endangered. Agriculture, coal mining, and urbanization have changed water quality in many small river and streams, but those in forested watershed retain historical features.

Introduction

This volume originated with a symposium on water quality at the 117th Annual Meeting of the American Fisheries Society (AFS) at Winston-Salem, North Carolina, in September 1987. The symposium's purpose was to promote an understanding of water quality conditions and ecological interactions in some diverse river ecosystems throughout North America. Through synthesis and interpretation of information for a particular river system, each presentation was expected to provide a reasonable portrayal of other rivers in the same geographical area.

The number of rivers that could be covered by presentations at the AFS symposium, however, was limited. More papers were contributed during subsequent efforts to publish the symposium presentations in a single volume. The 12 papers appearing here were the result.

It was neither possible nor desirable to obtain informative papers on water quality and ecological interactions for all large rivers (on the basis of size) in North America in this volume. For some large rivers, relevant information can be found in volumes focused on fish communities and fishery resources. For example, recent articles containing information on water quality and ecology of the Colorado River, Frazier River, Mackenzie River, and other North American ecosystems can be found in *The Ecology of River Systems*, 1986 (B. R. Davies and K. F. Walker, eds., Dr. W. Junk Publishers, Dordrecht, The Netherlands), *Community and Evolutionary Ecology of North American Stream Fishes*, 1987 (W. J. Matthews and D. C. Helms, eds., University of Oklahoma Press, Norman), and *Proceedings of the International Large Rivers Symposium (LARS)*, 1989 (D. P. Dodge, ed., Can. Spec. Publ. Fish. Aquat. Sci. 106, Dept. of Fisheries and Oceans, Ottawa).

Contributors to this volume had various academic and research backgrounds. They were provided a general outline that incorporated the theme of the symposium—water quality and ecological interactions—but were allowed the freedom to develop papers in their areas of strength.

The twelve contributions in this volume, then, may be viewed as supplements to existing information on river ecosystems in North America. On the basis of geographic distribution (Figure 1), morphology, and habitat type, the rivers in this volume represent a wide and disparate spectra. North to south, coverage extends from the Chena River (1), a tributary of the Yukon River in interior Alaska and the La Grande Rivière (8) in Northern Québec, Canada, to the Santo Domingo River Basin (5), a small watershed in arid Baja California, Mexico, and the Mobile River drainage (11), eastern Gulf of Mexico. West to east, coverage extends from the Columbia (3) and Sacramento-San Joaquin river (4) systems, two major watersheds draining to the Pacific Ocean, to the blackwater rivers draining the southeastern coastal plain (12). From the interior heartland are contributions for the Bow River system (2) in Alberta, Canada, and the Missouri (6), upper Mississippi (7), Ohio

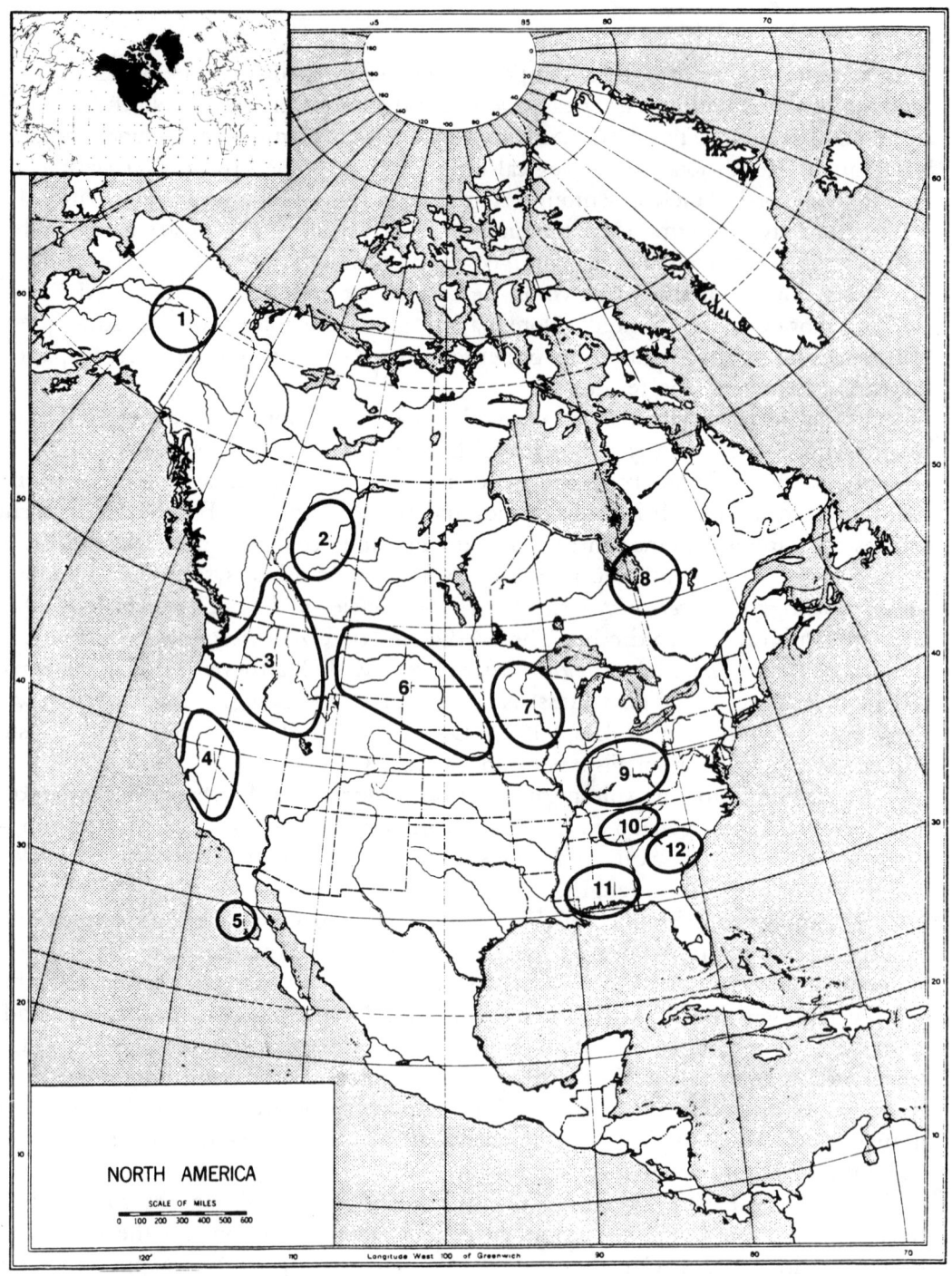

(9), and Cumberland (10) rivers of the United States. Information from these widely distributed areas provides a great deal of insight on water quality conditions and environmental disparities in various regions of North America.

It is safe to say that river ecosystems are complex. Their complexity arises from distinctive hydrologic, morphologic, and biological features that encompass entire drainage basins and, today, are greatly influenced by human activities (e.g., "modified river ecosystems"). We are fortunate that many users and abusers of rivers and other aquatic ecosystems now realize, or have begun to realize, that maintaining the quality of water and the functional integrity of aquatic ecosystems is essential to the health, economic status, and long-term survival of the human race.

It is also safe to say that every river ecosystem in the world has been altered significantly, in some way, by human activities. Rivers in North America are no exception.

Those of us familiar with rivers, and who have examined their individual characteristics and ecological processes, have come to view them as living things. Rivers resemble living things because they undergo constant changes that enable them to display different modes of behavior and to develop different personalities. Such changes are due to both natural (abiotic and biotic) processes and exploitative (human use) processes.

Natural processes that cause changes in river ecosystems include battering storms and winds, torrential rains and floods, spells of hot or cold weather, forest and marsh fires, shifting of river channels, and annual cycles of biotic reproduction and abundance. Changes resulting from the interactions of nature and river ecosystems have occurred since the primordial rains and are responsible for the historical evolution of each river. Natural processes tend to rejuvenate and may be beneficial to, and even necessary for, the long-term existence of a viable river capable of sustaining health life forms.

On the other hand, changes associated with exploitative processes in rivers are of recent origin. Some are desirable because they contribute to the benefit of mankind while maintaining the integrity of a life-supporting ecosystem. Some are undesirable in that they lead to a gradual, often subtle, deterioration that, sooner or later, detracts from the quality of human life. And, depending on viewpoint, some exploitative processes have both desirable and undesirable features.

Exploitative processes as a whole tend to cut more deeply than natural processes into ecosystem functions, apply more widely in space, and persist longer over time. They often result from application of new technologies. Many exploitative processes eventually lead to degraded water quality and reduced stream flows. But that is not all. Evidence now indicates that a river system with degraded water quality and reduced flow has a lower capacity to compensate for, or recover from, further human-induced stress.

Exploitative processes in river ecosystems include 1) the harvesting of renewable resources (hunting, fishing, trapping, and withdrawal of water); 2) addition of physical and chemical substances (inert solids, silt, nutrients, sewage, pesticides, herbicides, industrial toxicants, agricultural and urban runoff, and heat); 3) restructuring of physical features related to the river bed, flow, and morphology, and to riparian and estuarine areas (dam building, dredging, diking and filling of flood plains, flow regulation, channelization, logging, overgrazing, mining, consumptive use, and irrigation); and 4) intentional and accidental introduction of exotic organisms, some of which become unwanted, competitive, or destructive pests.

Today, it's the interaction of natural and exploitative factors that make the study of river ecosystems so necessary, controversial, and, to a large degree, interesting. Natural factors must be accepted because they are not avoidable, but sometimes they can be dealt with. Accepting and dealing with exploitative factors that are known or believed to degrade rivers in considerably more difficult. While implementation of environmental laws during the past two decades has provided tools to limit many activities leading to degradation, lower water quality, and reduced or altered aquatic populations, confrontations between protective and exploitative users of rivers remain inevitable.

Where do we go from here? One need not be clairvoyant to see continued conflicts between the integrity of river ecosystems in North America and the increasing number of people using them. This statement also applies on a global scale, where the United Nations Population Fund projected in 1990 that the current world population of 5.2 billion people will increase in the next decade by nearly one billion, the fastest population expansion in history. Not only will people added to this planet require more water, but development of third-world countries will require further exploitation of their often limited aquatic resources. Unfortunately, all too many countries are poorly equipped to resolve the water quality issues they face. There are no lasting solutions in sight, in North America or elsewhere, as long as human populations continue to expand.

As coordinators and editors, we are proud of the articles presented in this volume. They provide another step in understanding water quality conditions and ecological interactions in North America river systems. Perhaps they will also help to moderate conflicts in resource use.

C. Dale Becker and Duane A. Neitzel
May 1992

Water Quality and Ecology of the Chena River, Alaska

MARK W. OSWOOD
Institute of Arctic Biology, University of Alaska
Fairbanks, Alaska 99775, USA

JAMES B. REYNOLDS AND JACQUELINE D. LAPERRIERE
U.S. Fish and Wildlife Service, Alaska Cooperative Fishery Research Unit[1],
University of Alaska, Fairbanks 99775, USA

ROLLAND HOLMES AND JEROME HALLBERG
Alaska Department of Fish and Game, Sport Fish Division
300 College Road, Fairbanks, Alaska 99701, USA

JULIA H. TRIPLEHORN
Geophysical Institute, University of Alaska, Fairbanks 99775, USA

---

1. The Unit is jointly sponsored by the U.S. Fish and Wildlife Service, Alaska Department of Fish and Game, and the University of Alaska.

ABSTRACT. *The Chena River, a clearwater river system located in subarctic interior Alaska near Fairbanks, is characterized by a severe thermal regime with long periods of winter ice cover. The river supports 13 species of fish (7 salmonids) and the largest fishery for Arctic grayling (Thymallus arcticus) in the state. In the Chena River, stream food webs receive and store less organic carbon, the benthic macroinvertebrate community is taxonomically less diverse, and fish growth rates are slower than in otherwise similar streams in temperate regions. Water quality has been affected by mining (primarily placer gold mining), limited agriculture, floodplain development, and municipal wastewater discharges. Diversion of wastewater in the early 1970s improved the water quality of the lower Chena River, especially by decreasing numbers of coliform bacteria. The most important instream development is the Fairbanks Flood Control Project, which has provided reasonable flood control without the financial cost and ecological changes imposed by a large permanent reservoir. However the project holds potential problems in regard to fish migration and flood plain management. Land clearing and human development pose the greatest threat to the integrity of the watershed. The Chena River is of great value to the community of Fairbanks and protecting the vegetation of the watershed is the key to its future.*

Extensive research in midlatitude regions of North America and Europe has led to several new advances in stream ecology (Barnes and Minshall 1983). However, there have been far fewer studies of streams at high and low latitudes (Harper 1981). Only wide-ranging comparative studies can establish the geographic limitations of the emerging generalities and hypotheses in stream ecology. The growing literature in stream ecology represents a valuable database for the examination of spatial patterns beyond the scope of individual investigators. Stream and river management schemes developed in temperate latitudes must be tested and possibly modified for use at high latitudes. For example, it is not possible to use macroinvertebrate "indicator" species or communities derived from temperate studies to assess the "health" of northern streams because many temperate taxa are not present in high latitude streams (Oswood 1989).

We here provide a selective summary of the ecology of the Chena River, emphasizing water quality and fisheries. Comparisons with lotic systems in lower latitudes are made when possible. The Chena River is a typical river in the subarctic boreal forest of interior Alaska. Unlike most Alaskan rivers, but like rivers in more developed regions, the lower Chena River has withstood a variety of human perturbations, including mining, sewage disposal, land development, intensive recreational fishing, and flood control.

Physical Features

Physiography

The Chena River (Figure 1) is a clear subarctic river that originates in east-central Alaska at 64°47'N, 147°54'W. It flows in a meandering westerly course 241 km to

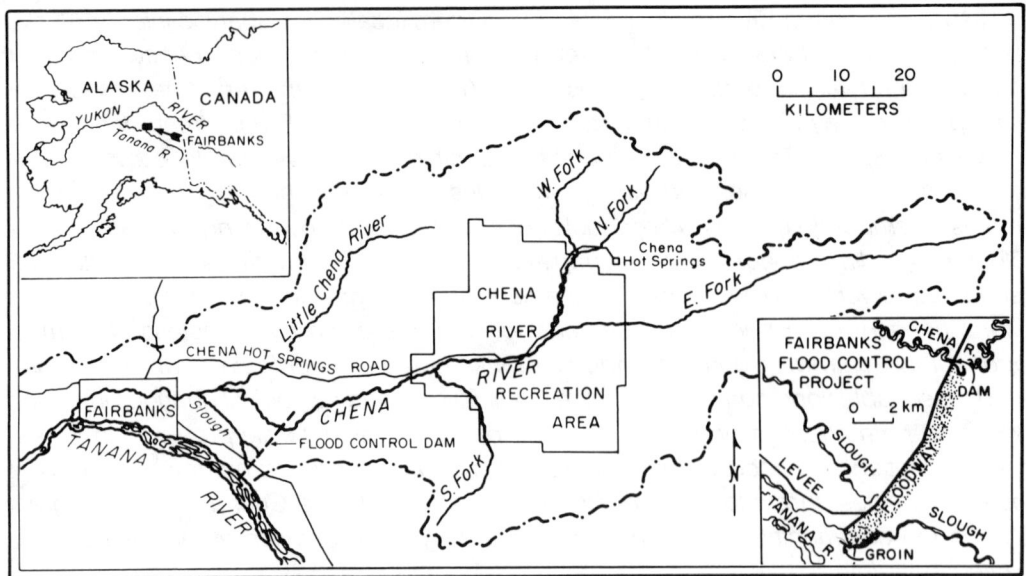

FIGURE 1. The Chena River drainage showing the locations of major tributaries and physical features.

discharge into the glacially turbid Tanana River about 17 km southwest of downtown Fairbanks. The Tanana River, in turn, joins the Yukon River, which flows to the Bering Sea. The Chena River drains a basin of 762 km². Its major tributaries are the Little Chena, North Fork, South Fork, Munson Creek, West Fork, and East Fork. Monument Creek is a frequently studied headwater stream.

The Chena basin has a continental climate influenced by polar air masses from the north. The Alaska Range to the south of the Chena basin forms an effective barrier against the maritime influences of warm moist air from the North Pacific Ocean because the mountains receive most of the precipitation. The extremely low temperatures in winter are caused by polar air as it moves across the state generally untempered by maritime influences. Ice forms on the river as early as September and breakup occurs in May. The growing season in the Chena basin lasts about 100 days from the last frost in spring to the first frost in fall and the average annual temperature is $-3.3°C$. Precipitation is highest in July or August, and lowest from February to April.

The Chena basin lies in the physiographic province of the Yukon-Tanana Upland, a central plateau in interior Alaska between the Yukon and Tanana Rivers, and consists mainly of rolling hills (Wahrhaftig 1965). There are no continuous mountain ranges, only rounded mature ridges trending northeast. These ridges are dissected by relatively flat alluvial-filled valleys resulting from uninterrupted deposition. Elevations range from 1219 m above mean sea level at the headwaters of the Chena River to 131 m at the mouth. The river valley has numerous meanders, old ponds, bogs, and oxbow lakes.

The northern boreal forest is primarily open, slow-growing spruce, interspersed with occasional dense well-developed forest stands and treeless bogs. This type of regional vegetation is referred to as "taiga", a Russian term, and the Chena basin is an example. Black spruce (*Picea mariana*) characterize the north-facing slopes, which are poorly drained and underlain by permafrost. Willow (*Salix* spp.) and alder (*Alnus* spp.) are typical riparian trees. On the south-facing slopes, which are better drained and have no permafrost, the principal trees are white spruce (*Picea glauca*), cottonwood (*Populus tacamahacca*), birch (*Betula papyrifera*), and aspen (*Populus tremuloides*).

Hydrology

The Chena River is a "clearwater" river, fed by precipitation and subsurface flows rather than by glaciers or bogs. The river's annual discharge cycle (Figure 2) is typical of such systems in the subarctic. Freeze-up is generally in late October or early November. Discharge near the mouth remains at a stable minimum (20–30 m^3/s) during winter when ice cover is 1–2 m thick and only ground water flows freely. Rapid spring warming results in ice breakup during late April to early May. Snowmelt creates an abrupt peak in runoff during May that often exceeds 100 m^3/s; this peak

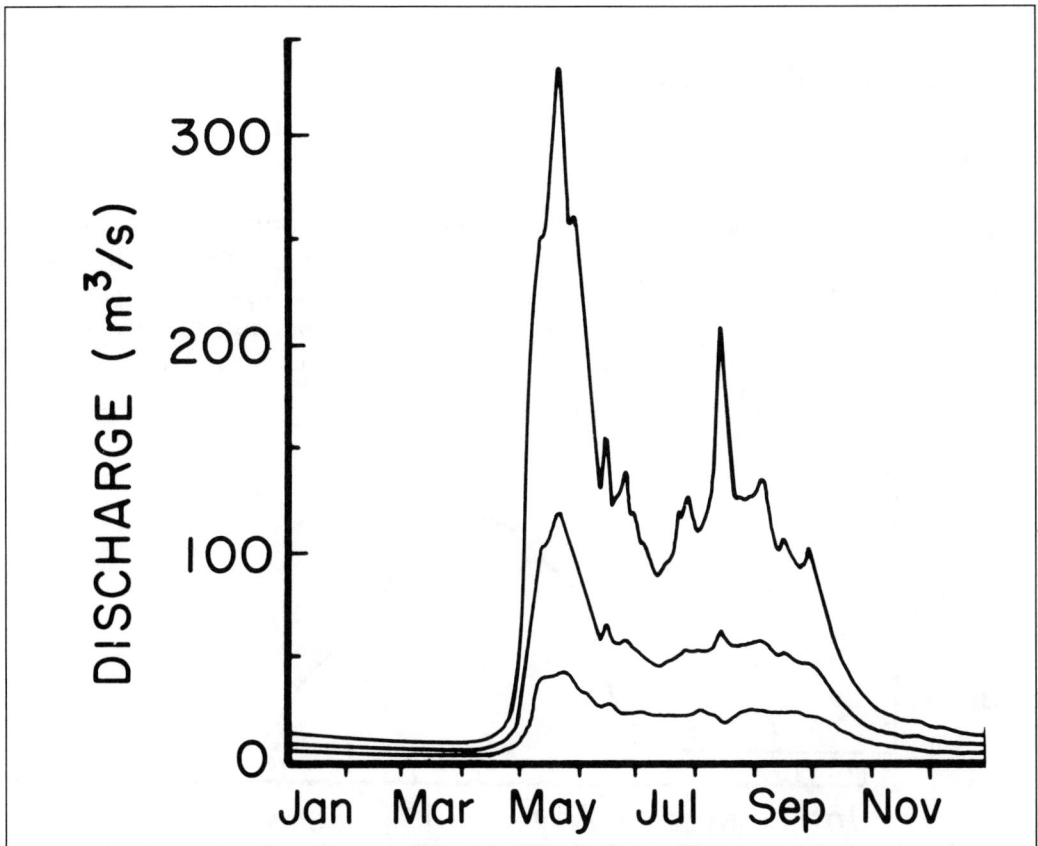

FIGURE 2. Average monthly discharge of the Chena River at Fairbanks (1948–1979). Upper, middle, and lower curves represent, respectively, calculated flows that were exceeded 10, 50 and 90% of the time. (Redrawn from Chapman 1982.)

typically declines to about 50 m³/s during June and early July, a dry period. A season of spates occurs from mid-July to mid-September, causing a secondary, more variable period of peak discharges. The Fairbanks flood of 1967, in which discharge reached 2,107 m³/s on August 15, was the result of an unusually long and heavy basin-wide rainfall. From 1949 to 1978, the average annual discharge (average of yearly averages) was 40 m³/s (USGS 1979).

Water Temperature

Water temperature in the Chena River at Fairbanks remains near 0°C from October to March and increases rapidly in April as a result of spring breakup (Figure 3). Temperatures reach about 18°C in July, fluctuate near this maximum, then decline during August and September as solar radiation rapidly decreases. Farther upstream, the temperature also peaks in July but at a lower temperature (about 14°C) and with a narrower curve of warming and cooling from mid-May to mid-September. Low winter temperatures and accompanying ice-cover occur for 6 to 8 months each year, depending on river size and stream elevation.

Biota and Ecosystem Properties

Lower Food Chains

The two sources of organic carbon for stream food webs are primary production by algae and macrophytes within the stream (autochthonous input), and particulate and dissolved organic material from streamside terrestrial areas (allochthonous input). The importance of leaf litter from riparian vegetation in the economics of stream food webs has become increasingly evident. However, in habitats with sparse riparian

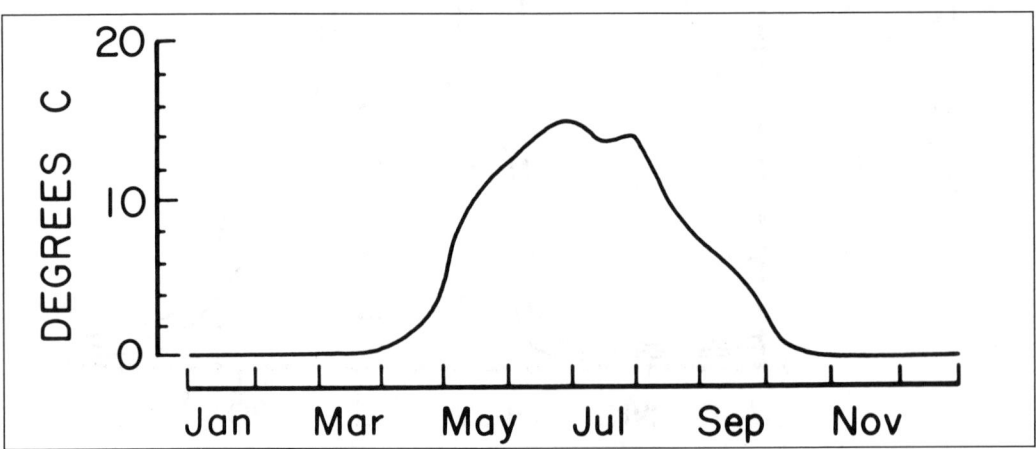

FIGURE 3. Average water temperatures (1951–1952, 1967–1968, 1977–1978) at river km 7.2 (Mile 4.5) of the Chena River. (Redrawn from Tilsworth and Bateman 1982.)

vegetation (e.g., desert, high altitude, and high latitude areas), decreased leaf litter input and increased light availability to streambed primary producers may raise the relative importance of autochthonous input (Minshall 1978). Low total carbon input may limit secondary production of aquatic consumers (e.g., insects), and production of fish and other animals at higher trophic levels.

Seasonal (spring, summer, fall, winter) average chlorophyll *a* densities (mg/m^2) in Monument Creek and West Fork (Anderson 1984) ranged from 0.053 (spring, in Monument Creek) to 0.741 (fall, in West Fork). Chlorophyll *a* values for the Chena River reported earlier (McCoy 1974) ranged from 0.05 to 5.70. By comparison, seasonal averages from the four sites studied by Bott et al. (1985) ranged from 0.63 (Michigan, first order stream, in spring) to 136.24 (Michigan, fifth order river, in winter). Chlorophyll *a* values from a review of published data (Bott et al. 1985: Table 16) range from 0.1 to 29250 mg/m^2. Thus, the highest chlorophyll *a* value for the Chena River was at the low end of the range for temperate streams and rivers. These comparisons confirm the visual impression that periphyton biomass in subarctic streams is much less than periphyton biomass in streams of temperate regions.

Only a few estimates of primary productivity are available for the Chena River; consequently, regional comparisons are tentative. Gross primary production estimates (n = 5) for the Chena River range from no measurable production (winter, in Monument Creek) to 11.0g O_2/m^2·day (summer, in Chena River) calculated from unpublished data (collected by P. A. Anderson and available from J. D. La Perriere) and McCoy (1974). Seasonal means from the four sites studied by Bott et al. (1985) ranged from 0.06 to 6.39g O_2/m^2·day compared with a range of 0.01 to 48 g O_2/m^2·day for streams throughout North America. Summer gross primary production in the Chena River (based on limited data), thus, appears to be comparable with that in temperate streams. Decreased shading by the relatively sparse riparian vegetation and longer summer days may facilitate primary production in subarctic rivers.

Input of leaf litter for Monument Creek totaled 62.5 g AFDW (ash-free dry weight)/m^2·yr (Cowan 1983). This value is low compared to a range of 286 g/m^2·yr to 1719 g/m^2·yr in published values for temperate streams (Cowan and Oswood 1983), but is similar to that reported for three streams in Southeast Alaska (52.1, 89.2 and 295.4 g AFDW/m2·yr) (Duncan and Brusven 1985). Monument Creek is a low-order headwater stream. No information on allochthonous input to other streams or rivers of the Chena River system is available; obtaining such data is critical to testing one of the major tenets of the "river continuum concept" (Vannote et al. 1980) e.g., allochthonous input decreases downstream from headwaters to outlet in large rivers.

Although primary production of periphyton may be high in the Chena River system during summer, production during the extended long winter (ice and snow cover) and spring (breakup) may be very low. Our two estimates of primary production during winter showed no detectable production. Therefore, even though leaf litter input is low, the relative contributions of autochthonous and allochthonous energy sources are still unknown for the Chena River system.

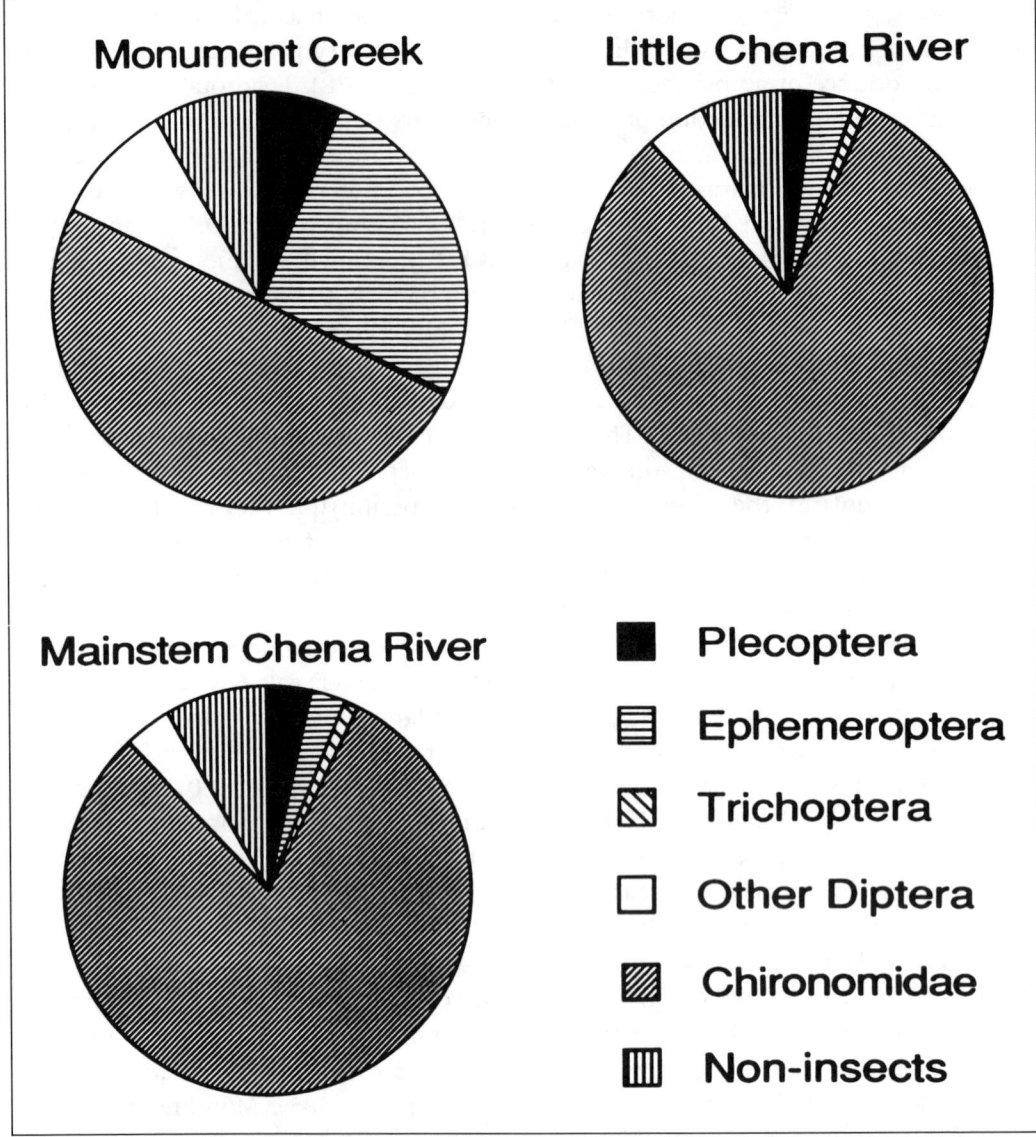

FIGURE 4. Composition (based upon numerical abundance) of major macroinvertebrate taxa in the Chena River and its tributaries. (Monument Creek data calculated from Cowan 1983, Little Chena River and mainstem data calculated from McCoy 1974.) Non-insects largely include oligochaetes, nematodes, and mites. Mainstem site in vicinity of Flood Control Dam (see Figure 1).

Benthic Macroinvertebrates

Data on taxonomic composition of benthic macroinvertebrates are available for a small tributary, a major tributary, and the main-stem Chena River. The macroinvertebrate fauna is dominated by insects. Less than 10% of the macroinvertebrates are non-insect taxa such as oligochaetes and nematodes, mollusks, and mites. The relative contribution of the insect orders to the benthic insect fauna varies somewhat between sites (Figure 4). Clearly, the true flies (Diptera) predominate, while the Plecoptera (stoneflies) and Ephemeroptera (mayflies) run a poor second. The caddisflies (Trichoptera) make up only a small proportion of the total fauna. Diptera typically comprise a larger proportion of stream fauna in Alaska than in temperate regions (Oswood 1989).

The invertebrate fauna of stony streams throughout the world is generally uniform (with some notable exceptions) and certain taxa "almost invariably provide the major constituents of the fauna" of such streams (Hynes 1970). Among the major taxa missing from the Chena River are amphipods, isopods, Odonata (dragonflies and damselflies), Megaloptera (dobsonflies, alderflies, and fishflies), and net-spinning caddisflies (Trichoptera: Hydropsychoidea). There is one reference each to an unidentified Hydropsychidae, Lepidoptera (aquatic moths), and Coleoptera (aquatic beetles).

Fishes

Taxonomic Richness. Of the 13 species of fish that inhabit the Chena River, seven are salmonids: round whitefish *Prosopium cylindraceum*, humpback whitefish *Coregonus pidschian*, least cisco *Coregonus sardinella*, inconnu (sheefish) *Stenodus leucichthys*, chinook salmon *Oncorhynchus tshawytscha*, chum salmon *Oncorhynchus keta*, and the Arctic grayling *Thymallus arcticus*. Two of the more abundant non-salmonids are the slimy sculpin *Cottus cognatus* and the longnose sucker *Catostomus catostomus*. Additional species in the lower Chena River include northern pike *Esox lucius*, burbot *Lota lota*, Arctic lamprey *Lampetra japonica*, and (occasionally) lake chub *Couesius plumbeus*.

The limited number of fish species from the Chena River (13) conforms with the general trend for lower numbers of fish species at higher latitudes (Table 1). This is not a smooth continuum, however, and considerable regional variability exists. For example, the number of fish species is greater in rivers of Western Alaska than in interior rivers at similar latitudes, and rivers on Alaska's North Slope have species counts only slightly lower than those in the more southerly interior rivers (Table 1).

Distribution and Abundance. The upper reaches of the Chena River may be characterized as "a succession of deep, wide, slow-moving pools, long shallow runs and many short steep riffles" (Frey et al. 1970). Five of the 13 fish species in the Chena River inhabit headwater tributaries: Arctic grayling, round whitefish, slimy sculpin, longnose sucker, and chinook salmon. Of these, the Arctic grayling and slimy sculpin penetrate farthest upstream to the headwaters (Hallberg 1977; Tack 1972). The headwater tributaries of the Chena River provide little overwintering habitat for fish and most summer residents generally move downstream in the fall to overwintering areas in the main-stem of the lower Chena River (Tack 1975; Holmes et al. 1986).

In the lower 160 km of the Chena River, the morphology of the channel changes considerably. As it flows through the broad Chena Valley, the channel meanders and becomes wider, deeper, and slower, with many pools and only occasional riffles. A wider variety of habitat is available in this area, enabling occupation by a larger

TABLE 1. Comparison of the number of fish species in the Chena River, Alaska, with those in other areas. (Data are derived from Bishop 1975; Holden and Stalnaker 1975; Stauffer et al. 1975; Maughan 1976; Alt 1977; Birch and Scalet 1977; Johnson and Beadles 1977; Bendock 1979; Bendock and Burr 1984).

Location	River	Number of species
Interior Alaska	Chena	13
	Salcha	10
	Chatanika	15
Northern Alaska	Chandler	10
	Ikpikpuk	10
	Itkillik	9
Western Alaska	Aniak	22
	Noatak	16
	Unalakleet	17
Alberta, Canada	Peace	22
Idaho	Clearwater	19
Colorado	Yampa	22
South Dakota	Little Missouri	22
West Virginia	East	49
Arkansas	Eleven Point	90

number of fish species. Arctic grayling, round whitefish, longnose sucker, and slimy sculpin are year-round residents throughout the lower Chena River. Each of these species spawn, rear, feed, and overwinter in the lower river. Chinook and chum salmon also spawn in the Chena River. Adult chinook and chum salmon first enter the Chena River from the Bering Sea in late June. Chinook salmon usually finish spawning by mid-August, but a few chum salmon continue spawning until fall. Fry of both species emerge from the gravel in the spring. The newly hatched chum salmon begin out-migration after hatching but chinook salmon fry spend at least 1 year in the Chena River and out-migrate the second or third spring. Three whitefish species, the least cisco, humpback whitefish, and inconnu, are all relatively scarce and occur only in about the lower 75 km of the Chena River (Van Hulle 1968; Tack 1971). A few northern pike are distributed throughout the lower 75 km and in Badger Slough. Burbot, like northern pike, occur in the lower reaches of the stream (up to 100 km). Arctic lamprey occur in the lower 160 km. Very little is known about the life history and distribution of arctic lampreys, but they occur in both the large anadromous and the dwarf freshwater form. Lake chubs are occasionally found in the lower 5 km of the Chena River.

Counts or population estimates of Chena River fish are available only for Arctic grayling in specific index sections. Grayling estimates on the upper Chena River (above km 100) have been consistently higher than those obtained from the lower and middle sections. In 1985, Arctic grayling in the Chena River averaged 479/km (Holmes et al. 1986). This estimate is similar to densities estimated from some interior Alaskan rivers (Salcha, Goodpaster, and Susitna tributaries) where the Arctic grayling is a predominant species (Figure 5). Estimates of Arctic grayling abundance from other areas of Alaska and from Montana (Upper Big Hole River) are generally lower than those from the Chena River (Figure 6).

Growth and Production. Growth data are available for Arctic grayling, round whitefish, slimy sculpin, and burbot in the Chena River. In general, all of these species grow slower and live longer than they do in rivers in the contiguous United States (Figure 6). These are common traits of northern fish populations, probably associated with harsh conditions and low productivity in northern systems (Hobbie 1973).

Slimy sculpin in the Chena River take nearly 5 years to reach the same length that those in Valley Creek, Minnesota reach in 2 years (Figure 6). We found no growth rates for burbot and round whitefish in temperate rivers; however, the same species in Minnesota and Michigan lakes grow much faster than those in the Chena River. Round whitefish grow somewhat slower, however, in the Koksoak River, Ungava, Quebec (59° N. lat.), than those in the Chena River (65° N. lat.). Arctic grayling in the Chena River take 5 to 7 years to reach the same length that those in Montana streams reach in 2 to 3 years (Figure 6). After 1970, the average growth rate of Chena River grayling slowed somewhat, perhaps because those in some heavily fished sections grew more slowly (Grabacki 1981). This indicates that the overall decrease

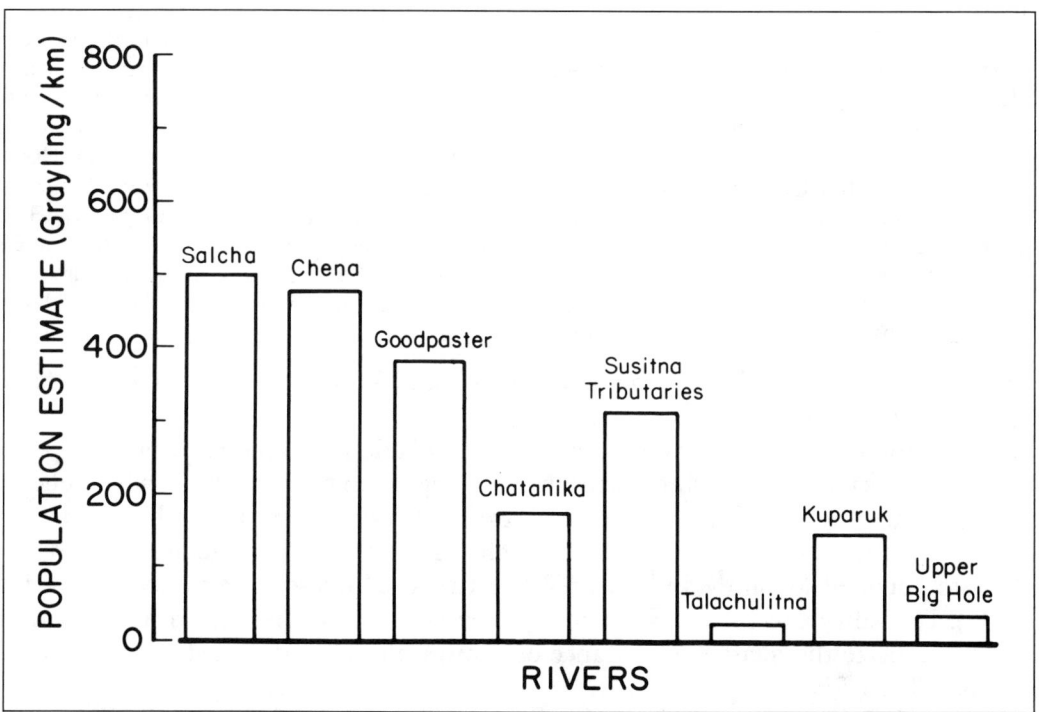

FIGURE 5. Estimates of Arctic grayling populations (grayling/km) for the Chena River and other rivers in Alaska and Montana. (Data for the Salcha, Chena, and Chatanika rivers are from Holmes et al. 1986; Goodpaster River are from Peckham 1979; Susitna tributaries are from ADFG 1981; Talachulitna River are from Kubik and Chlupack 1976; Kuparuk River are from Alt 1976; and Upper Big Hole River, Montana are from Liknes 1981.)

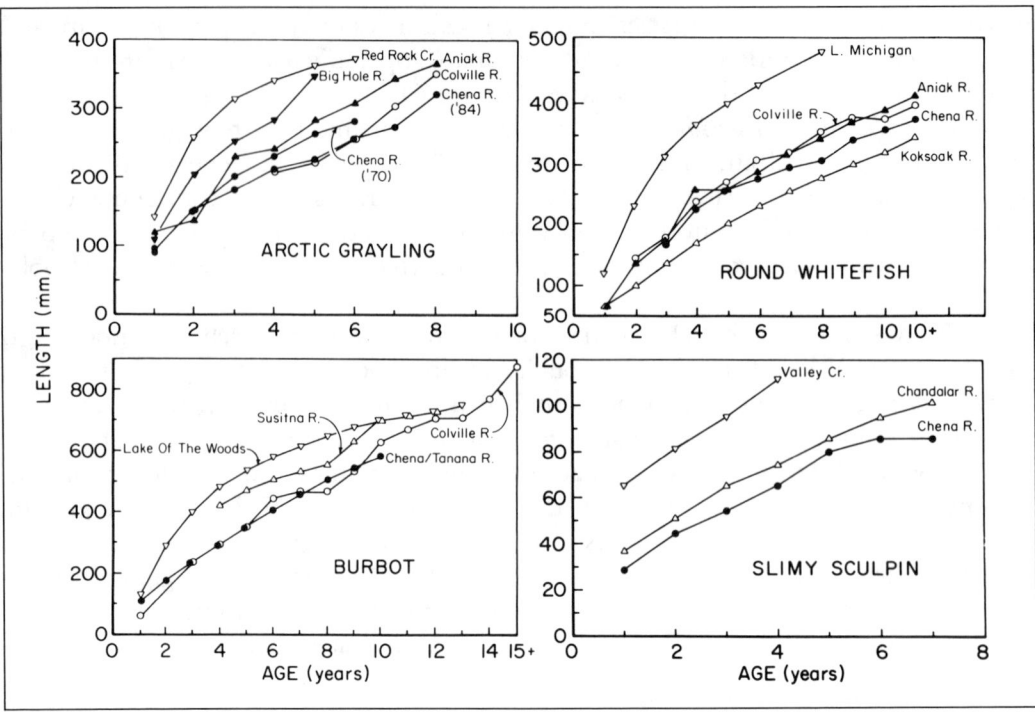

FIGURE 6. Growth of Arctic grayling, round whitefish, burbot, and slimy sculpin in the Chena River compared with growth rates of these species in other areas within and outside of Alaska. (Data for Chena River Arctic grayling are from Tack 1971 and Holmes 1985; Aniak River Arctic grayling and round whitefish are from Alt 1977; Colville River Arctic grayling, burbot, and round whitefish are from Bendock 1978; Red Rock Creek Arctic grayling are from Nelson 1954; Upper Big Hole River, Montana, Arctic grayling are from Liknes 1981; Lake of the Woods, Minnesota, burbot are from Muth 1973; Chena and Tanana Rivers burbot are from Chen 1969; Susitna River burbot are from ADFG 1983; Koksoak River round whitefish are from Mackay and Power 1968; Lake Michigan, Wisconsin, round whitefish are from Mraz 1964; Chena River slimy sculpin are from Sonnichsen 1981; Chandalar River slimy sculpin are from Craig and Wells 1976; and Valley Creek, Minnesota, slimy sculpin are from Petrosky and Waters 1975.

in growth rate of Arctic grayling since 1970 could be due to the harvest of faster growing fish by the sport fishery.

The only estimate of annual production available for a Chena River fish species is 8.5 kg/hectare for the slimy sculpin made by Sonnichsen (1981), who believed this estimate might be high for subarctic streams because of sampling bias. However, this production estimate is still substantially lower than the estimate of 59.4 kg/hectare obtained for slimy sculpin in Valley Creek, Minnesota (Petrosky and Waters 1975). For Valley Creek, this estimate represented 32% of the total fish production; it may indicate the relative importance of slimy sculpins in the total fish production of the Chena River.

Food Habits. Aquatic insects provide the energy base for fish in the Chena River (Figure 7). Arctic grayling, least cisco, slimy sculpin, round whitefish, humpback whitefish, juvenile chinook salmon, and longnose sucker feed directly on aquatic insects. Burbot, northern pike, and sheefish feed on the above fish species. Only Arctic lamprey ammocoetes, which burrow in the river bottom and feed by filtering

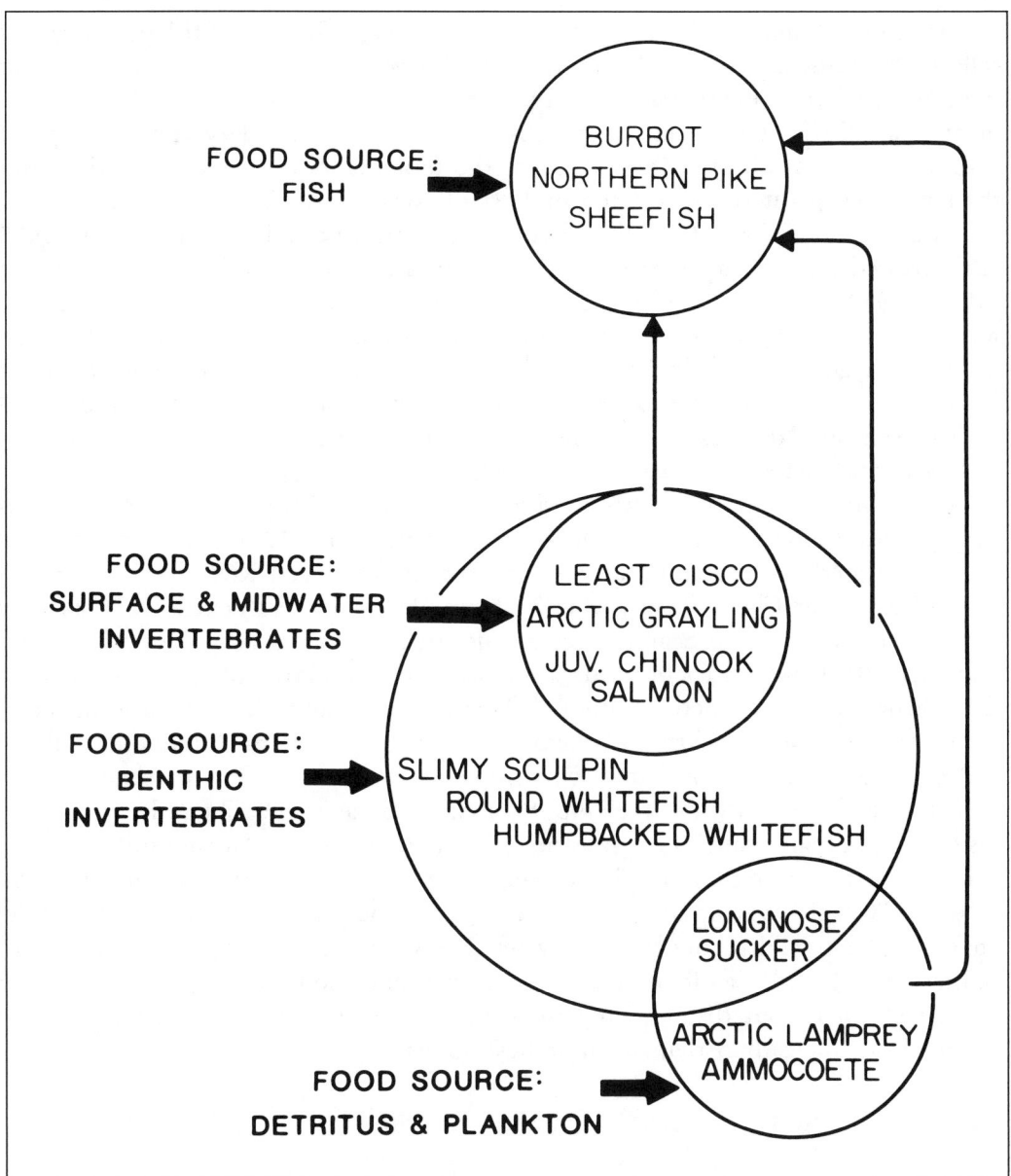

FIGURE 7. Major food sources of fishes that live in the Chena River.

seston (Scott and Crossman 1973), do not rely on aquatic insects or other fish directly for food.

Water Quality and Resource Management

Water Quality

Water quality in the Chena River basin has been influenced primarily by mining and sewage discharges. Lesser impacts occur from land development (urbanization and agriculture) and flood control. Before 1963, sewage from the Fairbanks area was collected and discharged directly into the Chena River. From 1963 to 1971, wastewater from four treatment plants (three with primary treatment, one with secondary treatment) was discharged to the Chena River. By 1976 a secondary treatment plant discharged wastes collected from three of the four plants to the Tanana River and the remaining plant was connected in 1981 (Tilsworth and Bateman 1982).

Comparison of water quality features from the Chena River for 1977 to 1978 (after treated sewage was diverted to the Tanana River) with similar information for 1951 to 1952 (untreated sewage was discharged) and 1967 to 1978 (treated sewage was discharged) indicated no differences in temperature or dissolved oxygen, although dissolved oxygen occasionally became low (minimum near 2.0 mg/L) beneath winter ice cover (Tilsworth and Bateman 1982). Likewise, values for chemical oxygen demand (COD) were similar across these time periods. Conversely, coliform bacteria counts and orthophosphate levels were substantially higher in 1967–1968 than in 1977–1978. The higher values were attributed to the discharge of treated sewage. Decreases in suspended solids from 1954 to 1965–1966, and then to 1977–1978, were attributed to reduced gold placer mining activities in the upper Chena River basin.

Water in the Chena River (Table 2) shows pH values near neutrality. Alkalinity and perhaps hardness appear to increase downstream, as is typical of most rivers (Hynes 1970). However, chemical data for the upper, middle, and lower sections of the Chena River were derived from different sources and the comparisons were, therefore, tentative. Calcium + magnesium concentrations are high, but sodium and chloride concentrations are low, compared to average river waters in North American and the world (Wetzel 1983), probably because the Chena River basin is distant from oceanic sources of airborne sodium and chloride. Average sulfate, silica, and nitrate concentrations in the Chena River are close to the North American averages (Wetzel 1983). Total phosphorous concentrations (values for upper river only) appear to be typical of oligotrophic waters. Very few data on organic carbon concentrations are available. Total organic carbon concentrations are typical of river water (Thurman 1985), but dissolved organic carbon appears to be a smaller fraction of the total organic carbon than in most freshwaters.

Fisheries

The Chena River, because it is located near Fairbanks, the population center of interior Alaska, and because much of the river is easily accessible by road, supports the largest fishery for Arctic grayling in the state (Holmes 1981). On the average, over 90% of the fish harvested from the Chena River are Arctic grayling (Table 3). Average catch (grayling/h) by anglers in successive decades was 0.84 in the 1950's, 0.61 in the 1960's, 0.78 in the 1970's, and 0.52 in the 1980's (Warner 1958; Van Hulle 1968; Roguski and Tack 1970; and Holmes et al. 1986). Fishing effort increased on the Chena River through the early 1980's, peaking at more than 41,000 angler days in 1984 but dropping to only 27,000 by 1985. Harvest of Arctic grayling peaked at 41,825 fish in 1980 and then dropped sharply to a low of 8,038 in 1985 (Mills

TABLE 2. Water chemistry data for the Chena River. (Sample sizes are in parentheses.)

Feature	River section		
	Upper[a]	Middle[b]	Lower[c]
Specific conductance (μmhos)	97.0 (26)	144.0 (74)	182.0 (56)
pH	7.13 (26)	7.34 (44)	7.38 (37)
Alkalinity (mg/L as CaCO3)	22.5 (23)	52.1 (43)	78.6 (38)
Hardness (mg/L Ca + Mg)	—	68.0 (25)	93.0 (33)
Sodium (mg/L Na)	—	1.6 (25)	3.0 (24)
Potassium (mg/L K)	—	0.9 (25)	1.8 (24)
Chloride (mg/L Cl)	—	1.0 (23)	1.2 (24)
Sulfate (mg/L SO4)	—	17.0 (23)	16.0 (24)
Silica (mg/L SiO2)	7.9 (26)	6.1 (25)	12.0 (24)
Nitrate (mg/L N)	0.24 (26)	0.20 (20)	0.31 (21)
Total phosphorus (mg/L P)	0.004 (10)	—	—
Dissolved organic carbon (mg/L C)	—	4.5 (2)	—
Total organic carbon (mg/L C)	—	8.2 (2)	—

[a]Average of data from Monument Creek and West Fork Chena River (Anderson 1984) sampled monthly from May 1982 to May 1983. Total phosphorus (LaPerriere 1981) is an average of five samples taken over the ice-free season, 1979.

[b]USGS Water Resources Data for Alaska Series. Average of data collected at various intervals from May 1972 to May 1975, Chena River near North Pole.

[c]USGS Water Resources Data for Alaska Series. Average of data collected at various intervals from January 1969 to December 1972, Chena River at Fairbanks.

1986). The declines in effort and harvest are attributable to a decline in the Arctic grayling population throughout the Chena River (Holmes et al. 1986).

Estimates of the number of Arctic grayling per kilometer in the Chena River index sections have also declined over time. Declines were especially rapid in the early 1970's. High population's of Arctic grayling in the early 1970's may have represented a recovery response from low populations after the 1967 flood (Tack 1971). Nutrient enrichment from the Fairbanks and Fort Wainwright sewage facilities could have increased invertebrate densities and, thereby, accounted for high populations of Arctic grayling in the lower river (Tack 1971; Hallberg 1979). Holmes et

TABLE 3. Harvest by species for the Chena River sport fishery, 1977–1985. (From Mills 1979-1986.)

Species	Number of fish harvested									
	1977	1978	1979	1980	1981	1982	1983	1984	1985	Average
Grayling	27,723	33,330	27,977	41,825	27,548	29,318	21,866	30,400	8,038	27,558
Burbot	642	389	807	1,127	1,317	1,457	1,055	1,233	2,065	1,121
Northern pike	871	452	437	458	333	377	780	1,064	787	618
Whitefish	538	187	622	1,032	683	462	1,064	883	3,780	1,028
Inconnu (sheefish)	37	18	26	21	50	10	0	156	210	59
Chinook salmon	29	23	10	0	39	31	31	0	37	22
Chum salmon	43	20	9	21	0	0	10	39		16

al. (1986) correlated low populations of Arctic grayling in the Chena River with a series a high-water years that resulted in poor recruitment of year classes.

Overfishing has been suggested as a partial cause of the Arctic grayling decline. Fishing mortality on the Chena River in 1983 and 1984 averaged about 22% and natural mortality averaged about 31% (Holmes 1985). Grabacki (1981), who compared areas of high and low fishing pressure on the upper Chena River, concluded that abundance of the Arctic grayling declined due to fishing. In a dynamic pool model of Arctic grayling population in the Chena River, Holmes et al. (1986) determined that significant improvements in population abundance and in age and size structure could be obtained in 5 years by changes in fishing regulations. Several new regulations were imposed in 1987: fishing closure during the spawning season, a minimum length limit of 30.5 cm (12 in), and artificial lures only.

Most sport fishing for burbot takes place in the lower 10 km of the Chena River. The annual harvest of burbot in 1977–1984 averaged about 1,000 fish (Table 3). The average annual harvest increased 19% from 1977 to 1982 and thereafter leveled off. The Alaska Trophy Fish Program gives special recognition to anglers catching large fish. For burbot, the minimum weight to qualify as a "trophy" is 3.63 kg (8 lb). From 1967 to 1983, more trophy burbot were caught in the Chena River than in any other body of water in Alaska. No other species qualifying as a trophy has ever been certified from the Chena River.

Land Use and Management

Intense use of the Chena River watershed began in the early 1900's. Fairbanks was built primarily with white spruce logs floated downriver from the surrounding hills. Gold was discovered near Fairbanks, but there has been only limited mining in the Chena River area, sparing it from extensive watershed impacts. Since World War II, most land use has involved urban, residential, and agricultural development. Statehood, followed by the Trans-Alaska Pipeline, accelerated this trend in the 1960's and 1970's. Instream impacts and developments have been minor, considering the size of the present-day human population. Direct discharge of sewage into the Chena River at Fairbanks affected water quality until 1976, when treated sewage was first discharged to the Tanana River.

The most important instream development is the Fairbanks Flood Control Project. It began in 1945 when the head of the Chena Slough was blocked off to prevent turbid, glacial water from the Tanana River from entering the Chena River. The lower part of old Chena Slough is now called Badger Slough; its confluence with the Chena River, about 16 km upstream from Fairbanks, was considered to be the mouth of the Chena River until 1945. Badger Slough became a clear, productive, ground water-fed stream that supports an early spring fishery for Arctic grayling by Fairbanks anglers. Blockage of other Tanana River sloughs near Fairbanks has also produced clear streams with stable flows and enhanced fisheries. Slough blockage may be a useful mitigation technique in interior Alaska if it is not overdone. The sloughs also absorb floodwaters from the Tanana River, and provide natural protection for low-lying areas of South Fairbanks.

The Fairbanks Flood Control Project was substantially expanded from 1969 to 1981 with the construction of Moose Creek Dam (a gated dike) on the Chena River, a floodway connecting the Chena and Tanana rivers, and 32 km levee along the north shore of the Tanana River from Moose Creek Dam to the present mouth of the Chena River at the southwest corner of Fairbanks (Figure 1). The dam is closed only during flood stages ($330 \text{ m}^3/\text{s}$ or more), creating a temporary reservoir that drains along the floodway into the Tanana River. The dam was first closed during spring breakup in 1981, and was closed on several later occasions for a few days each time.

The Fairbanks Flood Control Project has been a useful compromise: Fairbanks enjoys reasonable flood protection while avoiding the financial cost and ecological changes imposed by a large, permanent reservoir. However, the project poses a serious threat to the migratory fish species of the Chena River. Closure of flood gates in the spring could block the out-migration of chinook and chum salmon smolts (Williamson 1984). Depending on the severity of the flood event, the salmon smolts could become entrapped in the impounded area or receive false imprinting (location of spawning grounds by chemical cues) if forced to reach the Tanana River by way of the floodway. Significant fish mortality from entrapment at the point where the reservoir floodway waters return to the main Chena River has occurred during past dam operations (Hallberg 1982; J. Burns, U.S. Army Corps of Engineers; and R. Simmons, U. S. Fish and Wildlife Service, personal communications 1985). The lag between the time the flood gates are closed and the fish ladder begins operation (Hallberg 1981) could, during a spring or summer flood, block Arctic grayling or adult salmon migrating upstream to spawn. Poor stock recruitment might result. As construction in the flood basin of the Chena River increases, the Army Corps of Engineers will probably be pressured to operate the dam at lower and lower water levels, thus, increasing the possibility of harm to fishes of the Chena River. Furthermore, human development has been permitted well within the 100 year flood zone (an area somewhat less than that covered by the 1967 flood) and the gates have been forced to close at discharges of less than $300 \text{ m}^3/\text{s}$ to offer protection. These sub-standard closures reduce scour and accelerate siltation, further exacerbating the flooding problem by reducing channel capacity.

Other instream developments on the Chena River are minor in comparison to the flood control project, and most are related to road-building. The infrequent bridges over the length of the main-stem Chena River and its North Fork have little effect on water quality and flow. Culverts, mostly on tributaries of the lower river, may create velocity barriers for fish during peak flows. Channelization along the North Fork to protect Chena Hot Springs Road was kept to a minimum to protect natural, instream habitat.

Extensive development of the floodplain and surrounding hills has been largely limited to the lower watershed, downstream from the confluence of the main channel and the Little Chena River. Most land use has involved the clearing of vegetation for construction projects aimed at urbanization. As a result, erosion and runoff have increased turbidity and sediment deposits in the lower Chena River. Development in mid-reaches of the watershed, from the Little Chena River upstream to the Chena River Recreation Area, have occurred primarily as small agricultural projects along Chena Hot Springs Road. Erosion and sedimentation, therefore, have been minor. The upper watershed is largely underdeveloped because of protection afforded by the Chena River Recreation Area (ADNR 1984). However, private land holdings and leases in the headwaters of the West, North, and East forks have been mined and developed for tourism. Some sedimentation has resulted in the headwaters but it is not a serious problem now.

The greatest possibility for altering the water quality and ecology of the Chena River is loss of habitat. As the human population increases, development of the nearby land and river continues and the likelihood of problems increases. Agriculture, mining, roads, housing, dams, and recreational development have all had some effect on the Chena River. Siltation and riparian damage due to mining and agriculture have the potential for major effects on fish species. Siltation has both direct and indirect effects on fishes (Simmons 1984), resulting in avoidance of affected waters, reduced feeding by sight-feeding fishes, reduced availability of aquatic insects for food, reduced growth rates, decreased reproductive success, lowered ability to avoid predators, increased stress-related diseases, egg mortality from decreased subsurface water flow, impediment of fry movement, and decreased respiratory function from abrasive damage to gill surfaces. Mining currently occurs on the East Fork of the Chena River and tributaries of the Little Chena River. The Little Chena River drainage carries a heavy load of placer mining silt. In all areas where water moves slowly, the gravel bottom in the Little Chena River is now covered with a layer of silt (Tack 1972). Arctic grayling grow at a slower rate in the Little Chena River than in the main-stem Chena River (Holmes 1985).

In addition to effects on the fish themselves, siltation of a stream also reduces sport fishing potential. Reduced fishing success in areas of the upper Chena River during July and August 1983 was due to deposits of silt from mines along the East Fork (Holmes 1984). Although only limited loss of recreational fishing has occurred on the main Chena River, the Little Chena River has became virtually unfishable due to silt from mining (Tack 1972; Holmes 1985).

In general, low-lying areas of interior Alaska contain permafrost and standing water. Thus, low-lying areas are poorly suited for human development but, ecologically, they absorb runoff and dampen the highs and lows of river discharge. Development has proceeded at a high rate in low-lying areas along the lower Chena River. Warmer, well-drained, and vegetated soils on higher grounds are more attractive for development but they erode rapidly in face of indiscriminate clearing. The key to preventing further degradation of the Chena River is protective zoning of wetlands with minimal and selective removal of vegetation throughout the watershed.

Future of the Chena River

Land development in the Chena River watershed has been reasonably well planned from an ecological perspective. Instream development has been minor, and the success of the Fairbanks Flood Control Project will largely depend on future building trends in low-lying areas subject to flooding. Local government may be faced with a choice of enforcing flood-zone ordinances or losing national flood insurance for area residents. Development in the watershed has been largely limited to the lower Chena River, but recent land lotteries in headwater areas raise the specter of extensive land clearing in areas now well-protected by taiga forests. These forests not only provide the physical and chemical basis of water quality, but also the detrital food for much of the biological production. Vegetation removal in the upper watershed will probably decrease or destroy the values that make the river attractive and useful in the first place.

The Chena River is a renewable resource of inestimable value to residents of Fairbanks and the surrounding area. The value of the river lies in the character it lends to the community. Aesthetics, fishing, and water-borne recreation are long-term, low-cost benefits that accrue from good water quality. The quality and ecological integrity of the Chena River are primarily preserved by the forests and riparian vegetation of the watershed. The future of the Chena River rests squarely on this vegetation. The upper watershed must be well protected through careful land use and stringent requirements for developers. Greenbelts must be encouraged throughout the watershed where development occurs to preserve the vegetation and the river it protects.

Acknowledgments

This summary of the Chena River had its origins in a symposium (Ecology and Fisheries of the Chena River) held in October 1985 in Fairbanks. In addition to the listed authors, the following people participated in preparing this paper: P. Bateman, R. Carlson, A. Jubenville, D. Snarski, and T. Tilsworth. We thank D. Borchert for assistance with graphics.

References

ADFG (Alaska Department of Fish and Game). 1981. Resident fish investigation on the upper Susitna River. Susitna Hydro aquatic studies, Subtask 7.10. Anchorage.

ADFG (Alaska Department of Fish and Game). 1983. Resident fish investigation on the upper Susitna River. Winter aquatic studies, 1982–1983, Phase II data report. Anchorage.

ADNR (Alaska Department of Natural Resources). 1984. Chena River State Recreation Area master plan. Division of Parks and Outdoor Recreation, Juneau.

Alt, K. T. 1976. Inventory and cataloging of North Slope waters. Federal Aid in Fish Restoration, Annual report of progress, 1975–1976, Project F-9-8,17 (G-I-O). Alaska Department of Fish and Game, Juneau.

Alt, K. T. 1977. Inventory and cataloging of sport fish and sport fish waters of western Alaska. Federal Aid in Fish Restoration, Annual performance report, 1976-1977, Project F-9-9,18 (G-I-P). Alaska Department of Fish and Game, Juneau.

Anderson, P. R. 1984. Seasonal changes of attached algae in two Alaskan subarctic streams. Master's thesis. University of Alaska, Fairbanks.

Barnes, J. R., and G. W. Minshall. 1983. Stream ecology: an historical and current perspective. Pages 1–5 *in* J. R. Barnes and G. W. Minshall, editors. Stream ecology: application and testing of ecological theory. Plenum, New York.

Bendock, T. N. 1978. Fisheries survey of the major watersheds of NPR-A. Unpublished data summary report, Alaska Department of Fish and Game, Juneau.

Bendock, T. N. 1979. Inventorying and cataloging of arctic area waters. Federal Aid in Fish Restoration, Annual performance report, 1978–1979 Project F-9-11,20 (G-I-I). Alaska Department of Fish and Game, Juneau.

Bendock, T. N., and J. Burr. 1984. Inventory and cataloging of arctic area waters. Federal Aid in Fish Restoration, Annual performance report, 1983–1984 Project F-9-16,25 (G-I-I). Alaska Department of Fish and Game, Juneau.

Birch, J. P., and C. G. Scalet. 1977. Fishes of the Little Missouri River, South Dakota. Proceedings of the South Dakota Academy of Science 56:163–177.

Bishop, F. G. 1975. Observations of the fish fauna of the Peace River in Alberta. Canadian Field-Naturalist 89:423–430.

Bott, T. L., J. T. Brock, C. S. Dunn, R. J. Naiman, R. W. Ovink, and R. C. Petersen. 1985. Benthic community metabolism in four temperate stream systems: an interbiome comparison and evaluation of the river continuum concept. Hydrobiologia 123:3–45.

Chapman, D. L. 1982. Daily flow statistics of Alaskan streams. NOAA (National Oceanic and Atmospheric Administration) Technical Memorandum, National Weather Service AR-35, Anchorage.

Chen, L. 1969. The biology and taxonomy of the burbot, *Lota lota leptura*, in interior Alaska. Biological Papers of the University of Alaska 11, Fairbanks.

Cowan, C. A. 1983. Phenology of benthic detritus input, storage and processing in an Alaskan subarctic stream. Master's thesis. University of Alaska, Fairbanks.

Cowan, C. A., and M. W. Oswood. 1983. Input and storage of benthic detritus in an Alaskan subarctic stream. Polar Biology 2:35–40.

Craig, P. C., and J. Wells. 1976. Life history notes for a population of slimy sculpin (*Cottus cognatus*) in an Alaskan arctic stream. Journal of the Fisheries Research Board of Canada 33:1639–1642.

Duncan, W. F. A., and M. A. Brusven. 1985. Energy dynamics of three low-order southeast Alaska streams: allochthonous processes. Journal of Freshwater Ecology 3:233–248.

Frey, P. J., E. W. Mueller, and E. C. Berry. 1970. The Chena River: the study of a subarctic stream. Federal Water Quality Administration, Alaska Water Laboratory, Project No. 1610—10/70, College, Alaska.

Grabacki, S. T. 1981. Effects of exploitation on the population dynamics of Arctic grayling in the Chena River, Alaska. Master's thesis. University of Alaska, Fairbanks.

Hallberg, J. E. 1977. Distribution, abundance and natural history of the Arctic grayling in the Tanana River drainage. Annual report of progress, 1976–1977, Project F9-9, 18(R-I). Alaska Department of Fish and Game, Juneau.

Hallberg, J. E. 1979. Distribution, abundance and natural history of the Arctic grayling in the Tanana River drainage. Annual report of progress, 1978–1979, Project F-9-9, 20 (R-I). Alaska Department of Fish and Game, Juneau.

Hallberg, J. E. 1981. Distribution, abundance and natural history of the Arctic grayling in the Tanana River drainage. Annual report of progress, 1980–1981. Project F-9-13, 22(R-I). Alaska Department of Fish and Game, Juneau.

Hallberg, J. E. 1982. Distribution, abundance and natural history of the Arctic grayling in the Tanana River drainage. Annual report of progress, 1981–1982, Project F-9-13,23 (F-I). Alaska Department of Fish and Game, Juneau.

Harper, P. P. 1981. Ecology of streams at high latitudes. Pages 313–337 *in* M. A. Lock and D. D. Williams, editors. Perspectives in running water ecology. Plenum, New York.

Hobbie, J. E. 1973. Arctic limnology: a review. Pages 127–169 *in* M. E. Britton, editor. Alaskan arctic tundra. Arctic Institute of North America, Technical Paper 25.

Holden, P. B., and C. B. Stalnaker. 1975. Distribution of fishes in the Delores and Yampa River systems of the upper Colorado basin. Southwestern Naturalist 19:403–412.

Holmes, R. A. 1981. Angler effort, expectations, and values on the upper Chena River, Alaska. Master's thesis. University of Alaska, Fairbanks.

Holmes, R. A. 1984. Distribution, abundance and natural history of the Arctic grayling in the Tanana River drainage. Annual report of progress, 1983–1984, Project F-9-15, 25 (R-I). Alaska Department of Fish and Game, Juneau.

Holmes, R. A. 1985. Population structure and dynamics of the Arctic grayling, with emphasis on heavily fished stocks. Federal Aid in Fish Restoration, Annual report of progress, 1984–1985, Project F-9-17, 26 (R-I). Alaska Department of Fish and Game, Juneau.

Holmes, R. A., W. P. Ridder, and R. A. Clark. 1986. Distribution, abundance, and natural history of the Arctic grayling in the Tanana drainage. Federal Aid in Fish Restoration, Annual report of progress, 1985–1986, Project F-9-18,27 (R-I). Alaska Department of Fish and Game, Juneau.

Hynes, H. B. N. 1970. The ecology of running waters. University of Toronto Press, Toronto.

Johnson, B. M., and B. K. Beadles. 1977. Fishes of the Eleven Point River within Arkansas. Proceedings of the Arkansas Academy of Science 31:58–61.

Kubik, S., and R. Chlupach. 1975. Inventory and cataloging of sport fish and sport fish waters of the lower Susitna and central Cook Inlet drainages. Federal Aid in Fish Restoration, Annual Report of Progress, 1974–1975., Project F-9-7, 16 (G-I-H). Alaska Department of Fish and Game, Juneau.

La Perriere, J. D. 1981. Chemical and physical influences on invertebrate drift in subarctic Alaskan streams. Doctoral Dissertation. Iowa State University of Science and Technology, Ames.

Liknes, G. A. 1981. The fluvial Arctic grayling, *Thymallus arcticus*, of the upper Big Hole River drainage, Montana. Master's thesis. Montana State University, Bozeman.

Mackay, I., and G. Power. 1968. Age and growth of round whitefish (*Prosopium cylindraceum*) from Ungava. Journal of the Fisheries Research Board of Canada 25:657–666.

Maughan, O. E. 1976. A survey of the fishes of the Clearwater River. Northwest Science 50:76–86.

McCoy, G. A. 1974. Preconstruction assessment of biological quality of the Chena and Little Chena rivers in the vicinity of the Chena Lakes Flood Control Project near Fairbanks. Water-Resources Investigations 29-74. U. S. Geological Survey, Anchorage.

Mills, M. J. 1979–1986. Alaska statewide sport fish harvest studies. Federal Aid in Fish Restoration, Annual Reports of Progress from 1979 to 1986. Alaska Department of Fish and Game, Juneau.

Minshall, G. W. 1978. Autotrophy in stream ecosystems. BioScience 28:767–771.

Mraz, D. 1964. Age and growth of the round whitefish in Lake Michigan. Transactions of the American Fisheries Society 93:46–53.

Muth, K. M. 1973. Population dynamics and life history of burbot, *Lota lota* (Linneaus), in Lake of the Woods, Minnesota. Doctoral dissertation. University of Minnesota, Minneapolis.

Nelson, P. H. 1954. Life history and management of the American grayling (*Thymallus signifer tricolor*) in Montana. Journal of Wildlife Management 18:324–342.

Oswood, M. W. 1989. Community structure of benthic invertebrates in interior Alaskan (USA) streams and rivers. Hydrobiologia 172:97–110.

Peckham, R. 1979. Evaluation of interior Alaska waters and sport fish with emphasis on managed waters-Delta district. Federal Aid in Fish Restoration, Annual Report of Progress, 1978–1979, Project F-9-12, 19 (G-III-G). Alaska Department of Fish and Game, Juneau.

Petrosky, C. E., and T. F. Waters. 1975. Annual production by the slimy sculpin population in a small Minnesota trout stream. Transactions of the American Fisheries Society 104:237–244.

Roguski, E. A., and S. L. Tack. 1970. Investigations of the Tanana River and Tangle Lakes grayling fisheries: migration and population study. Federal Aid in Fish Restoration, Annual Report of Progress, 1969–1970., Project F-9-2, 11 (16-B). Alaska Department of Fish and Game, Juneau.

Scott, W. B., and E. J. Crossman 1973. Freshwater fishes of Canada. Fisheries Research Board of Canada, Bulletin 184.

Simmons, R. C. 1984. Effects of placer mining sedimentation on Arctic grayling of interior Alaska. Master's thesis. University of Alaska, Fairbanks.

Sonnichsen, S. K. 1981. Ecology of slimy sculpin, *Cottus cognatus*, in the Chena River, Alaska. Master's thesis. University of Alaska, Fairbanks.

Stauffer, J. R., C. H. Hacutt, M. T. Masnik, and J. E. Reed. 1975. The longitudinal distribution of the fishes of the East River, West Virginia-Virginia. Virginia Journal of Science 26:121–125.

Tack, S. L. 1971. Distribution, abundance and natural history of the Arctic grayling in the Tanana River drainage. Federal Aid in Fish Restoration, Annual Report of Progress, 1970–1971, Project F-9-3, 12 (R-I). Alaska Department of Fish and Game, Juneau.

Tack, S. L. 1972. Distribution, abundance, and natural history of the Arctic Grayling in the Tanana River drainage. Federal Aid in Fish Restoration, Annual Report of Progress, 1971–1972, Project F-9-4, 13 (R-I). Alaska Department of Fish and Game, Juneau.

Tack, S. L. 1975. Distribution, abundance and natural history of the Arctic grayling in the Tanana River drainage. Federal Aid in Fish Restoration, Annual Report of Progress, 1974–1975, Project F-9-7, 16(R-I). Alaska Department of Fish and Game, Juneau.

Tilsworth, T. and P. L. Bateman 1982. Changes in Chena River Water Quality. Northern Engineer 14:29–37.

USGS (United States Geological Survey). 1979. Water resources data for Alaska, water year 1978, water-data report AK-78-1. U.S.G.S. Anchorage, Alaska. (Not seen; cited in Tilsworth and Bateman 1982.)

Van Hulle, F. D. 1968. Investigations of the fish populations in the Chena River. Federal Aid in Fish Restoration, Annual Report of Progress, 1967–1968., Project F-5-R-9, 9 (15-B). Alaska Department of Fish and Game, Juneau.

Vannote, R. L., G. W. Minshall, K. W. Cummins, J. R. Sedell, and C. E. Cushing. 1980. The river continuum concept. Canadian Journal of Fisheries and Aquatic Sciences 37:130–137.

Wahrhaftig, C. 1965. Physiographic divisions of Alaska. U. S. Geological Survey Professional Papers 482, USGS, Washington, D.C.

Warner, G. W. 1958. Catch distribution, composition, and size structure of sport fish in the Fairbanks area. Federal Aid in Fisheries Restoration, Project F-1-R-8, Volume 8, Work Plan A, Job 3C. Job Completion Report. Alaska Game Commission, Juneau.

Wetzel, R. G. 1983. Limnology. Saunders Company, Philadelphia.

Williamson, D. 1984. Chena River salmon out migration studies, 1981–1983, Final Draft. Northern Alaska Ecological Services, U. S. Fish and Wildlife Service, Fairbanks, Alaska.

Longitudinal Zonation of the Biota and Water Quality of the Bow River System in Alberta, Canada

JOSEPH M. CULP[1]
Ecology Division (Aquatic Group)
Department of Biological Sciences
University of Calgary, Calgary, Alberta T2N 1N4, Canada

HAL R. HAMILTON
Hydroqual Consultants Incorporated, Suite 380
4500-16th Avenue Northwest
Calgary, Alberta T3B OM6, Canada

AL J. SOSIAK[2]
Alberta Fish and Wildlife, Suite 200
5920-1A Street Southwest
Calgary, Alberta T2H OG3, Canada

RONALD W. DAVIES
Ecology Division (Aquatic Group)
Department of Biological Sciences
University of Calgary, Calgary, Alberta T2N 1N4, Canada

---

1. Present address: National Hydrological Research Institute, 11 Innovation Boulevard, Saskatoon, Saskatchewan, S7N 3H5.
2. Present address: Environmental Assessment Division, Alberta Environment, 2938–11 Street Northwest, Calgary, Alberta, T2E 7L7.

ABSTRACT. *The Bow River system originates in the subalpine forests of the Rocky Mountains and flows east through the foothills and mixed grass prairies of southern Alberta. The hydrologic regime is dominated by peaks in discharge and suspended sediments during spring from local and mountain runoff, and during summer from intense rainfall. Ice cover extends from November to March and water temperatures rise from 0°C in winter to 20–30°C in summer. This template of a harsh climate combined with a rapidly changing riparian vegetation leads to a marked longitudinal zonation in the physical, chemical, and biotic characteristics of the river. Upstream of major municipalities, the river is saturated with dissolved oxygen and concentrations of plant nutrients are so low that they appear to limit production of primary producers during summer. In upper reaches of forested streams, the macroinvertebrate community consists largely of shredders and collectors, and the fish communities contain coldwater species like rainbow trout* Oncorhynchus mykiss *and mountain whitefish* Prosopium williamsoni. *The lower stream gradients in the prairies and the major input of phosphorus and nitrogen from the City of Calgary significantly increase plant nutrients during all seasons, stimulating the production of attached macrophytes and algae. In river sections with extensive beds of macrophytes, dissolved oxygen concentrations exhibit diel fluctuations and levels often approach 4–5 mg/L at dawn. In the lower river, fish communities are dominated by coolwater species such as northern pike* Esox lucius *and walleye* Stizostedion vitreum, *and the macroinvertebrate fauna is comprised of scrapers, filter-feeders, and deposit-feeders. Currently, revised water management practices that include tertiary sewage treatment have been implemented to improve the water quality and fisheries potential of river reaches negatively affected by human activities.*

As the Bow River flows southeasterly from its origin in the subalpine forests of the Canadian Rocky Mountains and joins the Oldman River in the mixed prairie to form

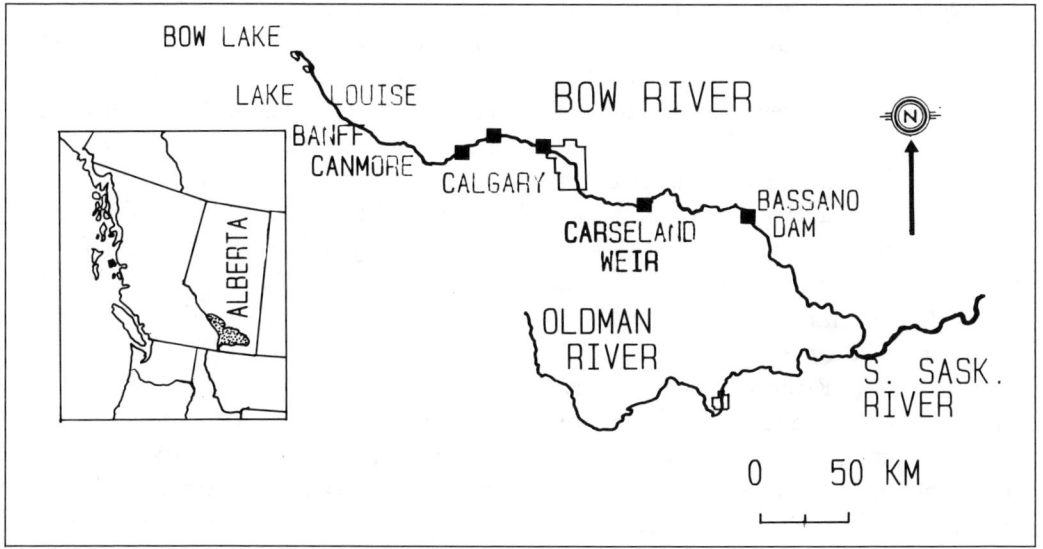

FIGURE 1.—Catchment area and location of the Bow River in southern Alberta. Dams and weirs are indicated by squares.

the South Saskatchewan River (Figure 1), the physicochemical environment changes dramatically. These changes affect the distribution of aquatic biota. For example, the fish communities are dominated by coldwater species (e.g., mountain whitefish *Prosopium williamsoni*) at the higher elevations and by coolwater species (e.g., walleye *Stizostedion vitreum*) in the lower reaches flowing through the prairies (Longmore and Stenton 1981). Similarly, the distribution of stonefly species along the Bow River are related to the vegetational zones through which the river flows (Donald and Mutch 1980), a distribution also shown by macroinvertebrates along the Oldman River (Culp and Davies 1982). Thus, natural longitudinal shifts are evident in temperature and hydrological regimes, concentration of dissolved substances and nutrients, and biota.

Hydroelectric dams and municipal sewage input clearly have altered the natural longitudinal distribution of aquatic biota in the Bow River. Because the severe impact of sewage discharge from the City of Calgary was recognized by the Alberta Department of Public Health as early as 1944, much research on the Bow River has focused on the effects of this effluent on water quality downstream of Calgary. Few data examine the effects of river regulation on water quality of the Bow River.

Our objectives are to provide a brief overview of the longitudinal changes in the biotic and abiotic components of the Bow River, followed by a more detailed account of the effects of municipal sewage. Although hydroelectric impoundments have caused a variety of downstream modifications of rivers, including changes in sediment transport and water temperature (Ward and Stanford 1979), we will discuss these features only in terms of their potential interaction with sewage effluent.

Morphology, Hydrology, and Water Quality

The Bow River provides a classic illustration of a concave longitudinal profile as it passes 625 km through the subalpine, montane, boreal, and mixed prairie grassland vegetation zones (Coupland 1961; Rowe 1972). Average gradients along this profile are 6.4 m/km between Bow Lake and the town of Banff, 2.0 m/km from Banff to the Carseland Dam, and 0.8 m/km between the Carseland Dam and the confluence with the Oldman River. The river channel is wide and shallow. Long-term means are 25 m wide and 0.5 m deep near Lake Louise, 101 m wide and 0.9 m deep at Calgary, and 143 m wide and 1.1 m deep near its confluence with the Oldman River (Kellerhals et al. 1972). The Bow River valley gradually increases from a 0.4 km wide mountain rift at Lake Louise to a 6.5 km wide, 40 m deep, stream-cut valley at the town of Bassano.

The Bow River was a glacial spillway during the late Pleistocene. Thus, its substrate is derived from glacial debris of the parent bedrock formations that largely consist of marine deposits of limestone, semiconsolidated sandstones, shales, and unconsolidated clay. Generally, the substrate is very coarse throughout the river's length and is mostly composed of gravel, cobble, and boulders (Longmore and Stenton 1981; Blachford and Ongley 1984). Silt and sand are lodged in the bottom sediments as a result of tractive force gradients within the streambed, and silt deposition over the substrate is particularly apparent in the lower reaches from Carseland to the Oldman confluence (Longmore and Stenton 1981). River bank erosion is often the main source of fine silt and sand particles.

The Bow River drains a 25,000 km² catchment, and most of the tributary flow is accumulated upstream of Calgary. Long-term mean discharges along the river are 11 m³/s at Lake Louise, 93 m³/s at Calgary, and 125 m³/s near the Oldman River confluence (Kellerhals et al. 1972). Peak discharge generally occurs in June as a result of mountain snowmelt, although runoff from foothills and prairie occurs during April and May (Figure 2). Minimum discharge is during the winter when the river is covered with ice. Daily fluctuations in discharge occur throughout the year in the reach between Canmore and Calgary due to irregular releases from hydroelectric operations, while two re-regulating dams moderate the daily fluctuations from Calgary to the confluence of the Oldman River. Because discharge has a direct effect on water quality by diluting the concentration of dissolved substances, poor water quality often appears downstream of Calgary during drought or during the late-summer and fall (Hamilton and North 1986).

Mean maximum summer water temperatures increase rapidly from 11°C at Lake Louise to 19°C at Carseland, and then gradually rise to 20°C at the river's mouth (Figure 3a). However, water temperatures up to 30°C are often reached during hot and dry summers below the dam at Bassano (Longmore and Stenton 1981). Maximum temperatures occur in August. The Bow River is covered with ice from November to March (Figure 3b), except in reaches downstream of heated effluents or reservoirs.

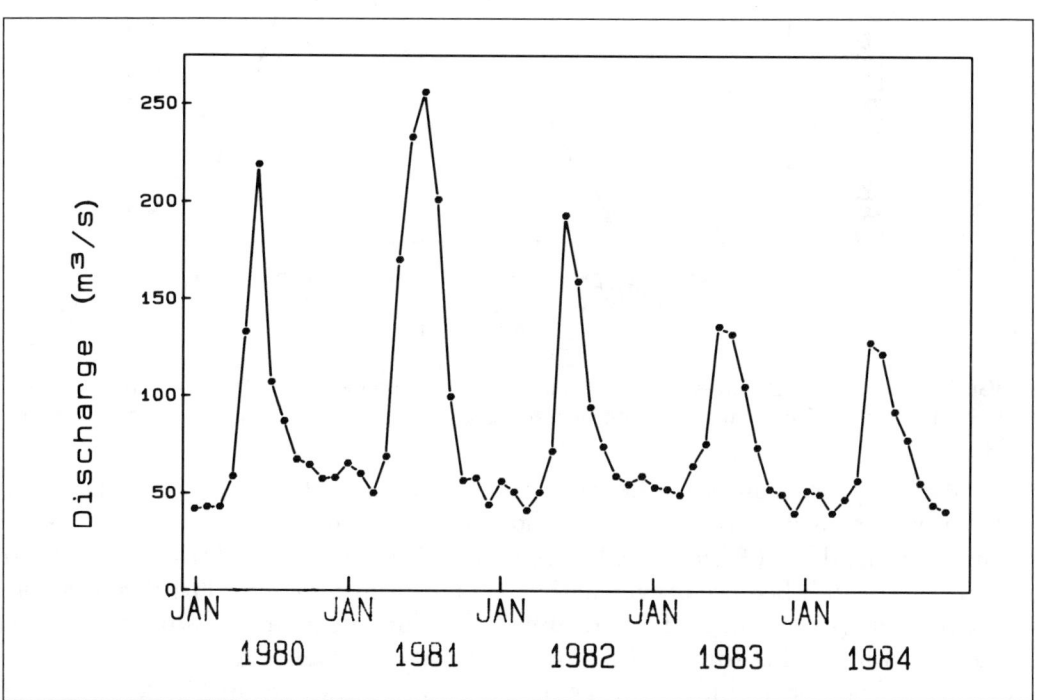

FIGURE 2.—Mean monthly discharge of the Bow River at Calgary, Alberta, 1980–1985. (From Environment Canada 1987.)

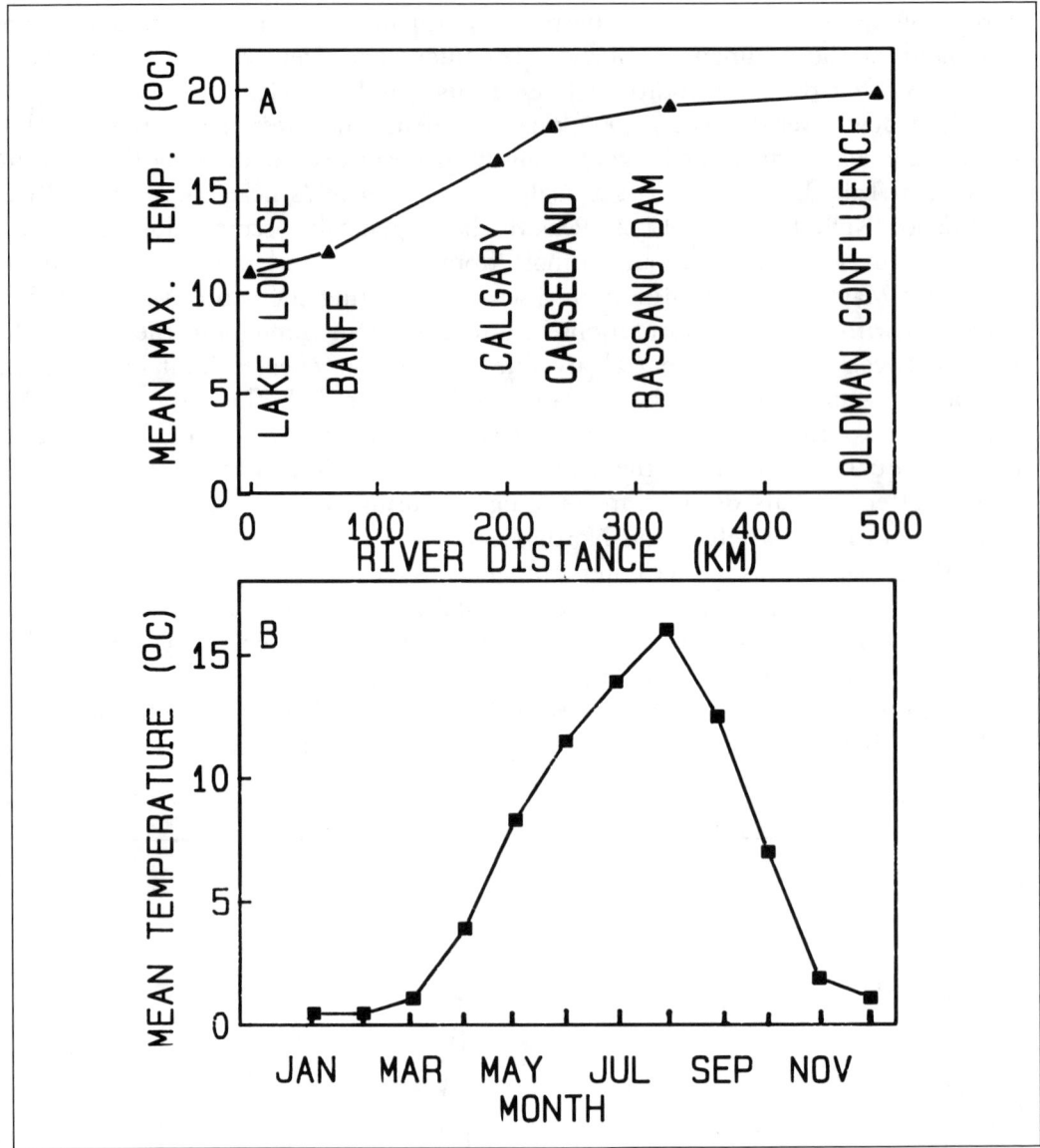

FIGURE 3.—For the Bow River, Alberta, (A) the longitudinal profile of mean maximum temperature (1958–1975) and (B) the mean monthly temperature near Calgary (1975). (From Environment Canada 1976.)

Because the bedrock of the basin is a limestone-shale complex, the Bow River is a hardwater stream and has high specific conductance, total alkalinity, total dissolved solids, and hardness (Table 1). All of these variables increase with distance from the source presumably because of evaporation, erosion from agricultural land, and reduced precipitation (Hamilton and North 1986). During the year, minimum values are reached during spring runoff while maxima occur in winter (Cross et al. 1986). Total organic carbon and turbidity also gradually increase downstream (Table 1). The pH generally remains between 7.0–8.5 over the entire length of the river, except downstream of Calgary in the summer when high levels of photosynthesis often raise the

TABLE 1. —Long-term (1970–1975) means (maximum values in brackets) for selected water quality variables at five sites on the Bow River, Alberta. Sample sizes (n) range from 19 to 98. (From Hamilton and North 1986.)

	Longitudinal site location								
	Below Banf		**Above Calgary**		**Carseland**		**Bassano**		**Near mouth**
Silica (mg/L)	3.8	(4.4)	3.5	(10.9)	2.7	(5.8)	2.7	(5.4)	2.4 (12.5)
Specific conductance (μS/cm at 20C)	281	(376)	295	(350)	333	(431)	357	(515)	357 (444)
Total alkalinity (mg CaCO3/L)	111	(126)	118	(148)	129	(189)	131	(193)	127 (162)
Total organic carbon (mg/L)	2	(22)	4	(23)	6	(28)	7	(19)	6 (20)
Total dissolved solids (mg/L)	163	(199)	168	(191)	183	(235)	202	(292)	200 (246)
Total hardness (mg CaCO$_3$/L)	150	(182)	155	(208)	166	(260)	172	(241)	163 (218)
Turbidity (JTU[a])	3	(17)	4	(25)	12	(300)	11	(105)	18 (120)

[a]Jackson Turbidity units

pH to 9.4 (Cross et al. 1986). During 1970, 22 metals were routinely monitored in the Bow River but most metals were at concentrations below analytical limits of detection and only iron ever exceeded the Alberta Surface Water Quality Objectives (A.S.W.Q.O.) (Hamilton and North 1986). Similarly, of 23 organic compounds monitored—including pesticides, herbicides, and polychlorinated biphenyls—only very low concentrations of the pesticide alpha-BHC and the herbicide 2,4-D were regularly detected.

Because of the addition of municipal sewage effluent, concentrations of ammonia, total nitrogen, and total phosphorus increased downstream of Calgary between 1970 and 1980 (Table 2). These variables were highest in the winter during periods of low primary productivity and heterotrophic decomposition, often exceeding the recommended A.S.W.Q.O. in the Bow River below Calgary. During the ice-free period, nutrient uptake by macrophytes and algae leads to a substantial decrease in concentrations of ammonia, total nitrogen, and phosphorus in the Bow River between Calgary and the Bassano Dam.

Longitudinal Zonation of Biota

The attached algae community is numerically dominated by the Cyanophyta (blue-green algae) and Bacillariophyceae (diatoms) over the entire course of the Bow River (Charlton et al. 1986). Abundance of algae fluctuates markedly among years and seasons, and sharp longitudinal zonation of taxa is not evident at this taxonomic level. However, the macrophytes (e.g., *Potamogeton* spp.) show a distinct longitudinal zonation; their biomasses increase dramatically in the low gradient reach downstream of the ·sewage effluent outfall at Calgary (Charlton et al. 1986). Macroalgae like

TABLE 2. —Long term (1970-1980) means (maximum values in brackets) for ammonia, total nitrogen, and total phosphorus during the winter and summer season at five sites along the Bow River, Alberta. Sample sizes (n) range from 8 to 21. (From Hamilton and North 1986.)

	Immediately above Calgary		Immediately below Calgary		Near mouth	
	Winter	Summer	Winter	Summer	Winter	Summer
Ammonia (mg/L)	<0.10 (<0.10)	<0.10 (<0.10)	1.30 (2.80)	0.20 (0.50)	0.60 (1.40)	<0.10 (<0.10)
Total nitrogen (mg/L)	0.17 (0.37)	0.14 (0.23)	2.19 (3.90)	1.14 (1.82)	1.86 (2.88)	0.40 (0.69)
Total phosphorus (mg/L)	0.01 (0.02)	0.01 (0.02)	0.33 (0.58)	0.13 (0.26)	0.19 (0.28)	0.05 (0.17)

Cladophora are also very abundant in the Bow River downstream of Calgary during years with low discharge (Charlton et al. 1986), and the macroalgae and drifting macrophytes can clog irrigation pumps when biomass peaks during the summer.

Although aquatic vegetation likely serves as a food source for benthic macroinvertebrates either directly or through decomposer trophic pathways, longitudinal distribution of benthic macroinvertebrates appears to relate best to the various zones of terrestrial vegetation. Relatively distinct faunal associations of adult Plecoptera are recognized in the sub-alpine, montane, boreal, and mixed prairie zones (Donald and Mutch 1980). The macroinvertebrate communities of the lower Bow River further substantiate this relationship in the montane-boreal and mixed prairie zones (Reynoldson 1973). A district longitudinal zonation in macroinvertebrate communities is also evident in unregulated streams, such as the Oldman River, that flow through the same terrestrial zones of southern Alberta. These biotic changes appear to be related to underlying longitudinal shifts in hydraulic, geomorphic, and climatic conditions (Culp and Davies 1982).

Sport fish in the main-stream Bow River have a longitudinal distribution that closely resembles the macroinvertebrate pattern (Longmore and Stenton 1981). The forested reaches upstream of Calgary are dominated by coldwater trout and mountain whitefish, with the prairie reaches below Carseland occupied mostly by coolwater species like walleye (Table 3). The reach between Carseland and Bassano is thought to represent a transition zone where both warm and coldwater species of fish can be found. The introduced rainbow trout, *Oncorhynchus mykiss*, has displaced the endemic cutthroat trout, *Oncorhynchus clarki*, in most reaches of the Bow River. Longitudinal changes in fish distribution are likely affected by shifts in the various abiotic variables recognized for macroinvertebrates but, especially for fish, increases in water temperature (Figure 3a and 3b) and reduced dissolved oxygen levels downstream of Calgary may be most the important factors. Currently, the Bow River is not stocked with sport fish. The last recorded introduction of rainbow trout was in 1947; brown trout (*Salmo trutta*) were introduced in 1925 near Banff (Paetz and Nelson 1970).

TABLE 3. —Presence (+) of fish species along the Bow River, Alberta. (From Henderson and Peter 1969).

Scientific Name	Common Name	Bow River above Calgary	Bow River below Calgary
Sport Fish			
Oncorhynchus clarki	Cutthroat trout	+	+
Oncorhynchus mykiss	Rainbow trout	+	+
Salvelinus fontinalis	Brook trout	+	+
Salvelinus confluentus	Bull trout	+	+
Prosopium williamsoni	Mountain whitefish	+	+
Salvelinus namaycush	Lake trout	+	
Salmo trutta	Brown trout	+	+
Esox lucius	Northern pike	+	+
Lota lota	Burbot	+	+
Coregonus clupeaformis	Lake whitefish		+
Perca flavescens	Yellow perch		+
Stizostedion vitreum	Walleye		+
Hiodon alosoides	Goldeye		+
Non-Sport Fish			
Catostomus catostomus	Longnose sucker	+	+
Catostomus platyrhynchus	Mountain sucker	+	+
Catostomus commersoni	White sucker	+	+
Culea inconstans	Brook stickleback	+	+
Couesius plumbeus	Lake chub	+	+
Rhinichthys cataractae	Longnose dace	+	+
Cottus ricei	Spoonhead sculpin	+	+
Percopsis omiscomaycus	Trout-perch	+	+
Semotilis margarita	Pearl dace	+	
Pimephales promelas	Fathead minnow	+	+
Notropis atherinoides	Emerald shiner		+
Notropis hudsonius	Spottail shiner		+
Carpiodes cyprinus	Quillback		+
Moxostoma macrolepidotum	Shorthead redhorse		+

The best known and most heavily angled section of the Bow River is the 50 km reach from Calgary to Carseland where about 52,000 angler-days per year were spent in search of mountain whitefish, rainbow trout, and brown trout during the summer of 1985 (Sosiak 1986). In Calgary and immediately upstream, the fishery is mainly catch-and-keep for mountain whitefish. Between Calgary and Carseland, it is largely a catch-and-release fishery for trout. Regulations on this section impose a maximum size limit of 40 cm (total length) for all trout, allow only unbaited artificial lures, and permit the retention each day of two trout smaller than the size limit. This reach is internationally known by dry fly-fishing enthusiasts for its large trout. The Bow River now has a rapidly growing, fishing-guide industry that supported 1869 guided angler-days in 1985 (Sosiak 1986), largely in the reach between Calgary and Carseland.

Municipal Effluent Impact on Mainstem Water Quality

Primary Producers

There were 18 point-source discharges of sewage to the Bow River downstream of Banff in 1980 (Hamilton and North 1986). The proportions by volume of this discharge contributed by the various effluent categories were about 7% urban, 33% industrial, and 60% from municipal sewage plants. The industrial inputs were largely cooling waters from fertilizer plants, and more than 97% of the treated sewage input to the river occurred at Calgary. Loadings to the Bow River from the two Calgary sewage treatment plants were greater than loading estimates from other sources for almost all measured variables, including biochemical oxygen demand (BOD), total salts, phenols, trace metals, and plant nutrients (Figure 4). Thus, most wastewater management activities have focused on the Calgary area, in particular, evaluating the effect of the city's sewage effluent on the river.

The features commonly examined to assess the impact of municipal discharges on the receiving stream include BOD, suspended solids, coliform bacteria, metals, nitrogen compounds, and phosphorus. All of these variables have been implicated in the deterioration of water quality downstream of Calgary. After secondary treatment was installed at the Bonnybrook sewer plant during 1970, water clarity increased markedly because loadings of suspended solids were reduced and the maximum BOD load declined by 75% to 4,010 kg/d. Thus, secondary sewage treatment greatly reduced the oxygen depletion so evident downstream of Calgary in fall and winter during the late 1960s. Despite the reductions in suspended solids, municipal effluent discharges remain the primary loading source for copper, zinc, and nutrients. Furthermore, slightly elevated levels of the metals copper, cadmium, chromium, lead, and nickel occur on river particulates below Calgary (Blachford and Ongley 1984). Fecal coliform densities also increase significantly downstream of the effluents relative to consistently low levels upstream of the city. Because the wastewater discharge is not disinfected, fecal and total coliform densities during the summer months can exceed contact recreation guidelines for 70–75 km below Calgary before returning to levels observed upstream (Figure 5a and 5b).

However, the primary effects on the Bow River that remain after secondary sewage treatment started are the additions of the plant nutrients nitrogen and phosphorus. Prior to the fall of 1982, mean annual concentrations of phosphorus increased from near analytical detection limits (i.e. 3 μg/L) downstream (Table 4). Similar increases in nitrate-nitrite concentrations have been recorded (Hamilton and North 1986; Cross et al. 1986). Coincident with these high nutrient conditions was the development of dense beds of aquatic macrophytes, the dominant species being *Potamogeton vaginatus*, *P. pectinatus*, and *P. crispus*. Macrophyte densities in these beds during summer were great enough to affect (1) water extraction for municipal supply and irrigation, (2) recreational use of the system for boating and fishing, and (3) fisheries habitat through wide fluctuations in the diel regime of dissolved oxygen (4–5 mg/L to supersaturation).

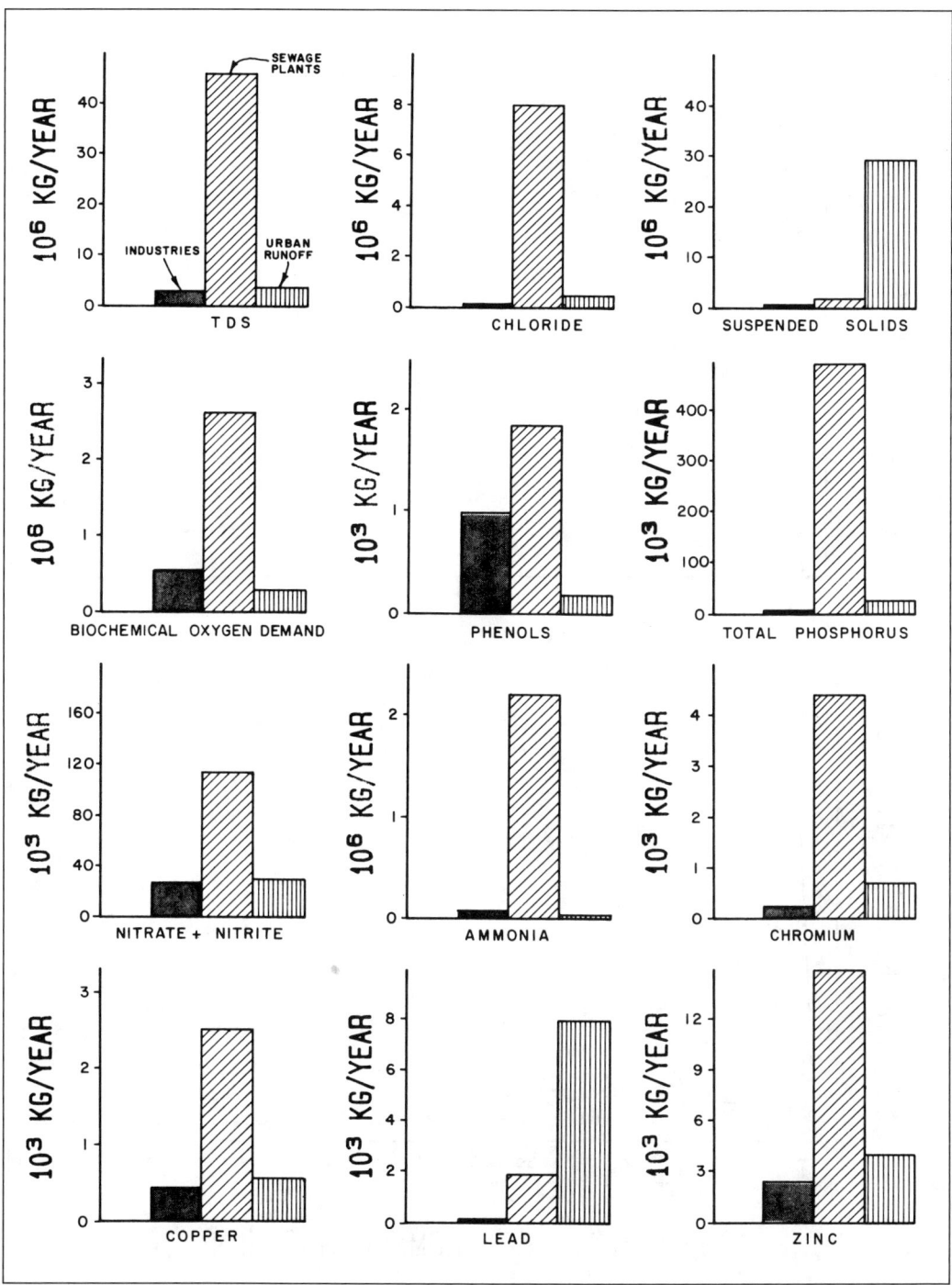

FIGURE 4.—Net annual loadings of 12 water quality variables to the Bow River during 1980 from industrial discharges, sewage effluents, and urban runoff originating at Calgary, Alberta. TDS is total dissolved solids. (From Hamilton and North 1986.)

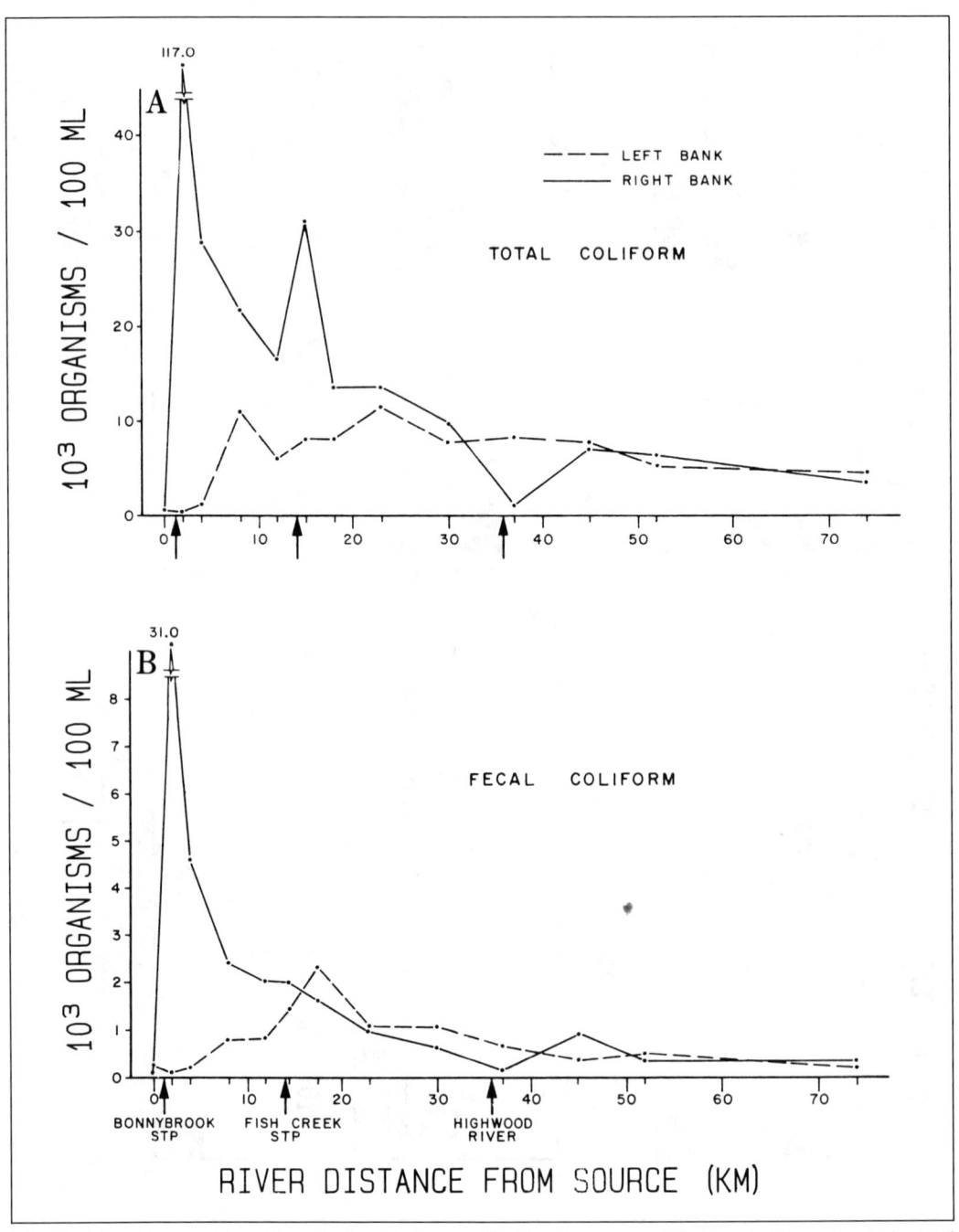

FIGURE 5.—For the Bow River below Calgary, Alberta, the mean (A) total coliform bacteria and (B) fecal coliform bacteria on the left and right river banks during autumn 1981 (n = 4). STP is sewage treatment plant. (From Hamilton and North 1986.)

TABLE 4. —Mean (range in brackets) nitrogen and phosphorus concentrations in the Bow River above and below sewage treatment effluents at Calgary and Carseland, Alberta before (1980–1982) and after (1983–1985) phosphorus removal. (From Charlton and Bayne 1986.)

	Above Calgary effluent		Below Calgary effluent		Coarseland Dam	
	Before	After	Before	After	Before	After
Dissolved phosphorus (μg/L)	4 (3–23)	6 (3–40)	169 (10–370)	22 (30–70)	147 (30–290)	22 (3–83)
Nitrogen (μg/L)	54 (10–160)	27 (10–100)	1191 (170–2400)	917 (310–1800)	816 (260–1500)	927 (70–1500)

Because the problems of high phosphorus and nitrogen concentrations and dense macrophyte growths persisted even after secondary sewage treatment was implemented, the Alberta Department of Environment began a study in 1979 to assess the longitudinal and temporal patterns of nutrient concentrations and the distribution and growth of aquatic primary producers in the Bow, Oldman and South Saskatchewan rivers (Charlton et al. 1986; Cross et al. 1986). Because the city of Calgary implemented advanced phosphorus removal at both of its sewage plants in 1982, this research also provided a baseline for comparing conditions before and after phosphorus reduction.

These studies (Charlton et al. 1986; Cross et al. 1986) indicated that, although the particulate phosphorus form was often elevated throughout the river system, much of it was biologically inert. Conversely, dissolved phosphorus (of which 91% was from the sewage plants) appeared to be the form most available to aquatic primary producers. With few exceptions, phosphorus rather than nitrogen was the nutrient in short supply.

Standing crops of macrophytes and epilithic algae responded to increased nutrients, and were significantly higher downstream of Calgary relative to upstream. Minimum values of benthic algal biomass occurred in the winter and early summer, and peak algal biomass and production coincided with late-summer senescence of the macrophyte community. Throughout much of the growing season, macrophytes predominated in the aquatic plant community for about 70 km downstream of Calgary. Epilithic primary producers predominated from that point downstream to the confluence with the Oldman River.

Macrophyte growth was also strongly affected by discharge and several related variables, as suggested for other lotic systems (Westlake 1973; Holmes and Whitton 1977; Haslam 1978). Accumulation of plant growth approached zero at mean current velocities greater than 1 m/s and water depths greater than 1.2 m (Figure 6a and 6b). Consequently, in high discharge years like 1981, the increased tractive forces tend to scour the substrate, thereby reducing the accumulations of macrophyte biomass to below nuisance levels throughout the summer (Cross et al. 1984). Currently, extensive flow regulation by dams and weirs on the Bow River above Calgary is believed to dampen the effects of scouring by natural floods each spring and, thus, allow perennial macrophytes to become well established. In contrast, benthic algal

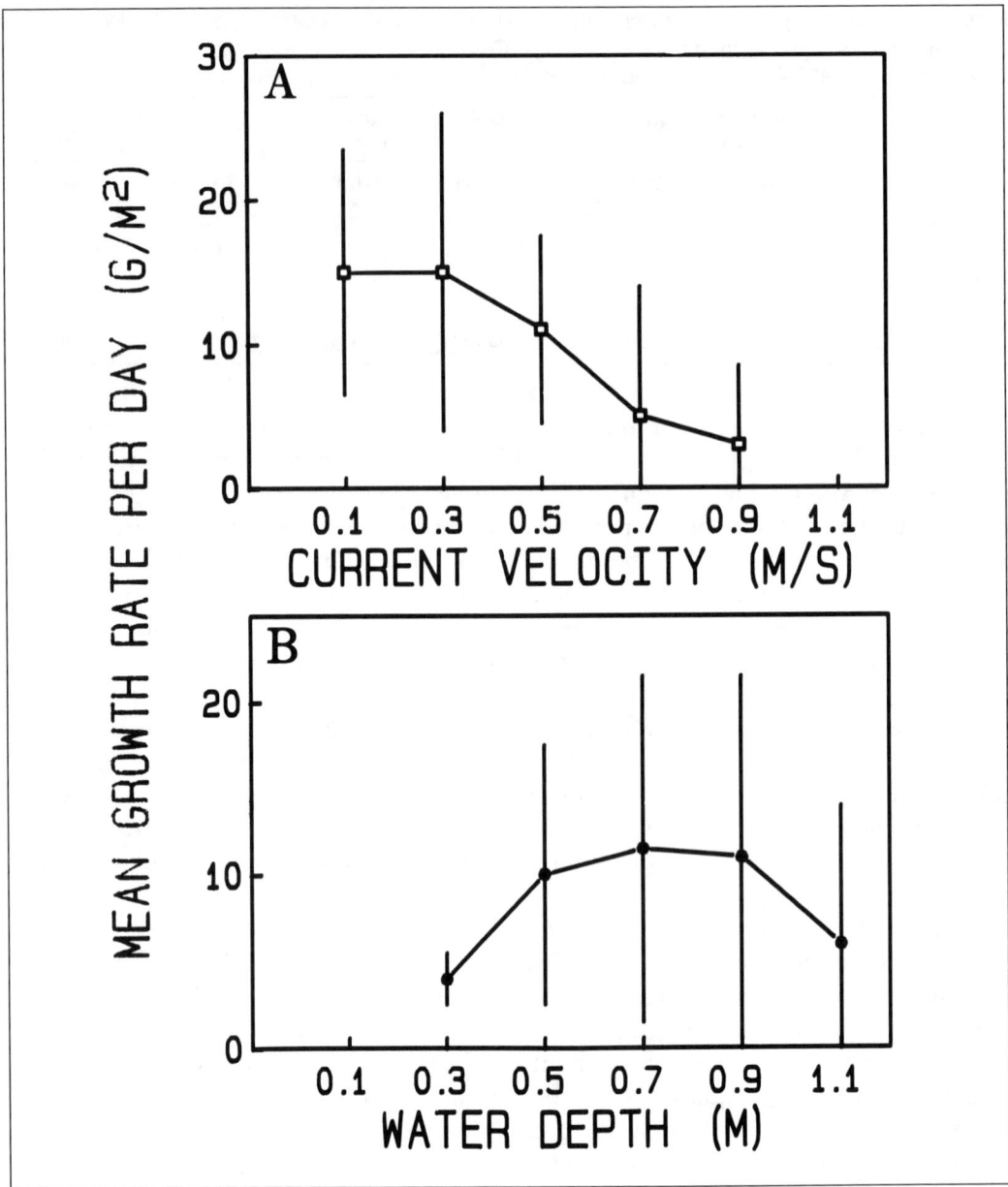

FIGURE 6.—The relationship of mean daily growth rate of macrophyte dry mass to (A) current velocity and (B) water depth for the Bow River at Calgary, Alberta. (From Charlton et al. 1986.)

and macrophyte biomass in the Oldman River never reach the levels observed in the Bow River (Charlton et al. 1986), perhaps because the flow in the Oldman River is presently unregulated and the stream bed is annually scoured (Warner 1973).

Secondary Producers

Enrichment of the Bow River by treated sewage effluent decreases species richness and diversity between Calgary and Carseland, but increases macroinvertebrate density

(Reynoldson 1973). The higher density of invertebrates, and presumably an increase in production, in the reach between Calgary and Carseland may enhance the production of trout. But the regulated flow of this reach and the associated reduction in streambed scour could also play an important role in this relationship. Although detailed studies are not available, rainbow trout downstream of the effluent have high growth rates that more closely resemble lentic rather than lotic populations (Figure 7). Growth rates and densities of young trout may also be related to the extensive macrophyte beds, which could provide a refuge for fish from predators and high current velocities. However, the role of the macrophyte beds as fish habitat is clearly complex because dense beds along braided side channels also restrict the flow, and can create oxygen supersaturation in the day and near hypoxia (4–5 mg/L) at night. These extreme fluctuations of dissolved oxygen as well as high ammonia levels may have caused fish mortality prior to 1982 (Longmore and Stenton 1981).

Tertiary Sewage Treatment of Effluents to the Bow River

Advanced phosphorus removal by alum precipitation began at the two Calgary sewage treatment plants in late 1982. Phosphorus concentration in the effluent was reduced from near 4 mg/L to less than 1 mg/L, and dissolved phosphorus concentrations in the Bow River downstream of Calgary were also reduced (Table 4). The level of nitrogen was largely unaltered. Reduction in phosphorus levels was not accompanied by decreased plant growth (Figure 8). In fact, although a trend towards slightly reduced macrophyte densities was apparent in 1984–85, some densities in 1983

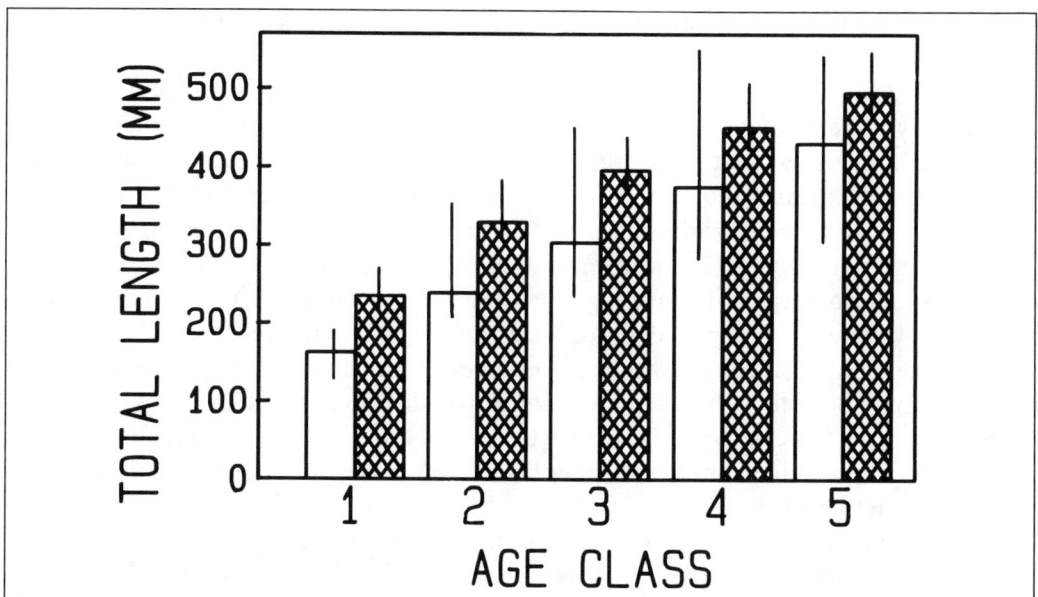

FIGURE 7.—Mean total lengths and length ranges for Bow River rainbow trout (age-classes 1–5) during the autumn of 1980–1985 (cross-hatched) compared with lengths of rainbow trout in other lotic systems by Carlander 1969 (open).

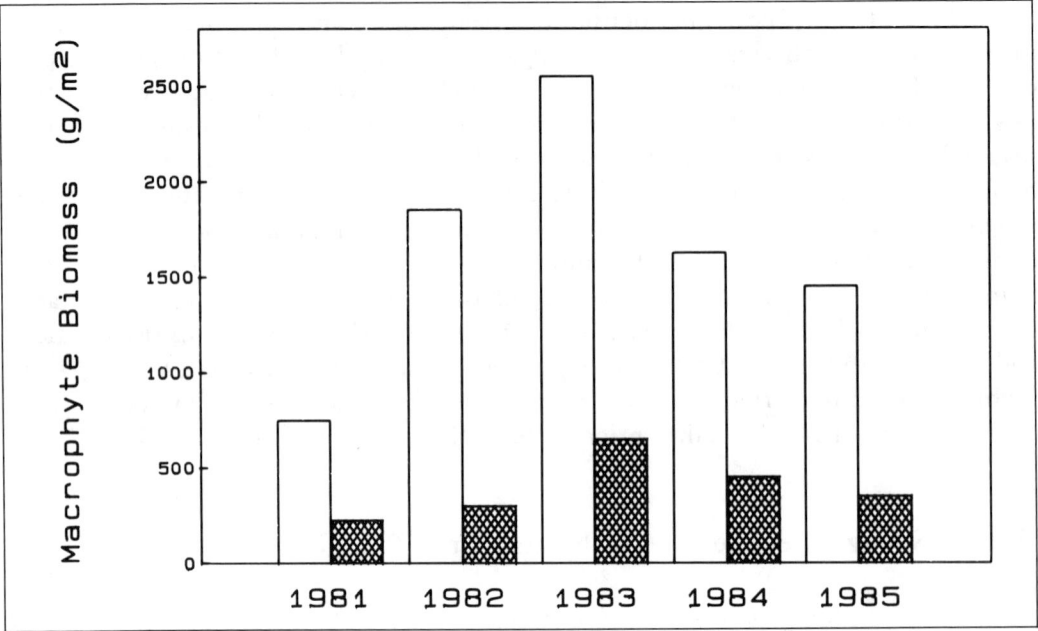

FIGURE 8.—Maximum macrophyte biomass in the Bow River downstream of the Calgary sewage effluent (open) and near Carseland (cross-hatched) before (1981–1982) and after (1983–1985) the implementation of phosphorus removal. (From Charlton and Bayne 1986.)

exceeded levels recorded prior to tertiary sewage treatment. Observations on the growth rate of rainbow trout also indicated no significant decrease in the age-specific growth rate of trout (A. J. Sosiak, Alberta Fish and Wildlife, unpublished data), suggesting that tertiary treatment has not affected secondary production, at least up to 1985.

The reason that the Bow River has not responded to phosphorus removal more dramatically appears to be related to several factors (Cross et al. 1984). First, a series of extremely low flow years since 1982 (Figure 2) may have extended the growing season, allowing greater biomass to accumulate in the years after tertiary treatment. Second, since nuisance growths can occur in the Bow River at dissolved phosphorous levels of less than 20 μg/L, post-removal levels of phosphorus may still be high enough to support a prolific macrophyte community. Third, the macrophyte community may respond to reduced phosphorus levels gradually over a number of years because macrophyte rhizomes are deeply embedded in the substrate of the river, and the plants may be able to supplement poor nutrient sources in the water with uptake from the sediments as in lentic systems (Carignan 1980, 1982). Finally, macrophytes may sustain higher than predicted growth through luxury uptake of nutrients when concentrations are elevated. It is unlikely that the macrophyte community in the zone of enrichment has reached a new equilibrium following the advent of phosphorus reduction, and scouring of the substrate by a major flood may be required to reduce biomass more rapidly. Even so, there may always be an enrichment zone below the sewage discharges that will vary in length depending upon annual variations in climatic and hydrologic conditions, as well as the nutrient input.

Conceptual Model of River Nutrient Enrichment

Results from this work on nutrients and aquatic primary producers in the Bow River can be synthesized into a conceptual model of aquatic plant dynamics for prairie river systems that are downstream from a major nutrient effluent (Figure 9). This conceptual model assumes that physical factors like current velocity and substrate, which are strongly related to the annual hydrograph regime and flow regulation, are adequate for growth. It also assumes that, below a threshold level of plant nutrients, growth rate is directly related to the concentration of the nutrient in shortest supply (i.e., generally phosphorus or nitrogen). At concentrations above threshold, nutrients exceed requirements for growth and other factors are more important.

Upstream from the point source input (ambient zone), plant growth is at equilibrium and reflects the upstream nutrient status of the river. Macrophyte growth in this zone is restricted to depositional areas where nutrients in the sediments are probably of primary significance. Effluent loading increases nutrient concentrations in the water and sediments to a maximum immediately below the outfall's mixing zone, resulting in increased plant growth.

In the zone of enrichment, plant growth is likely limited by availability of space, or self-shading, and maximum biomass reaches an equilibrium. Nutrient concentrations steadily decline with distance downstream due to settling of particulates, precipitation with calcium or iron, sorption to sediments, and biotic uptake (Brown and Bellinger 1982). The latter mechanism may often be the most significant (Gregory 1978). Based on this model, control of a point source nutrient input will

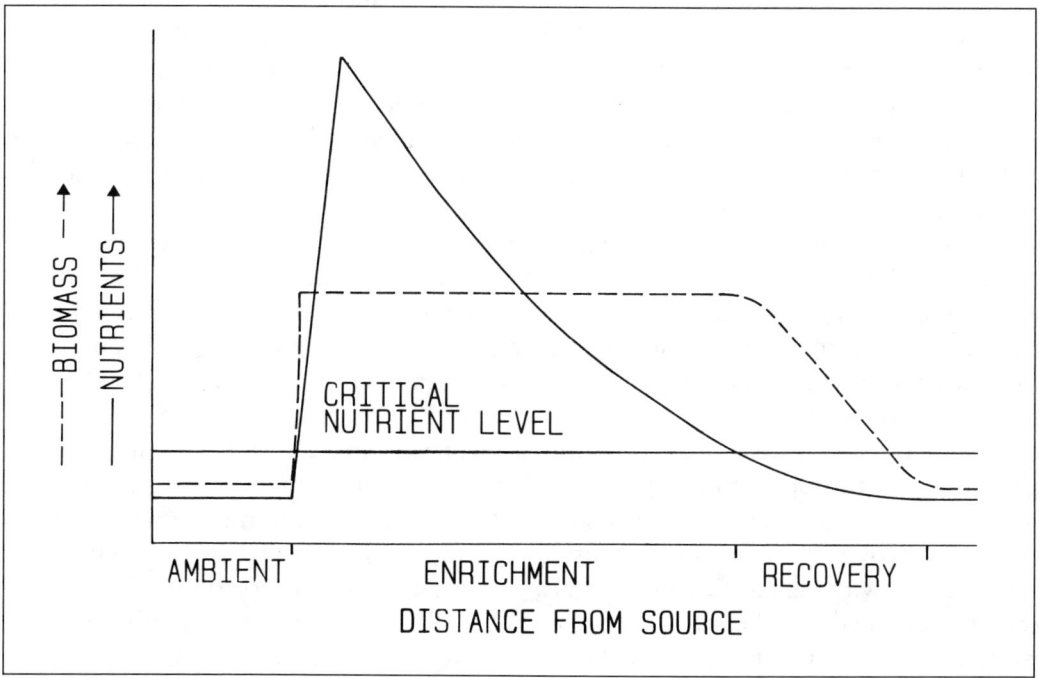

FIGURE 9.—Conceptual model of nutrient levels and plant biomass in a river downstream from a point source nutrient loading.

not eliminate a zone of enrichment except when removal efficiencies are extremely high. However, the length of the enrichment zone is also a function of ambient nutrient conditions in the river above the effluent outfall, which are determined by both nutrient loading and hydrologic conditions.

When available nutrient concentrations are depleted to less than the threshold level, aquatic growth is reduced (biomass recovery zone) and, eventually, both nutrient concentration and plant production achieve a new steady state indicative of the river system without point source nutrient loading. As there is often a natural progression in general river productivity downstream, this new steady state may be elevated relative to above the nutrient source. Because of potential luxury storage by algae and of complex seasonal and spatial patterns of nutrient storage in sediments, plant biomass may remain high for some distance past the point at which nutrient concentrations in the water column fall below critical growth threshold levels derived in the laboratory.

Water Quality Control Efforts

Jurisdiction and Legislation

Efforts to reduce negative human influences on rivers is based on regular monitoring of water quality and implementation of adequate, enforceable legislation so that the quality of the freshwater environment can be protected. In Alberta, protection of the aquatic environment is divided between provencial and federal legislation. Within provencial boundaries and outside of federal lands, management of quality and quantity of water resources is the responsibility of the Provincial Government. Furthermore, all fisheries are managed by the province under the Alberta Fishing Regulations, which are promulgated under the Federal Fisheries Act. However, the Federal Fisheries Act retains clauses for protecting fish habitat, which is generally also a provincial responsibility.

Perhaps the most important legislation in terms of monitoring and improving water quality in the South Saskatchewan basin has been the Provincial Clean Water Act introduced in 1971 and subsequently amended. The Clean Water Act is the basis of the provincial system for permitting, licensing, monitoring, and enforcing the release of contaminants into surface waters. It is administered by the Minister of the Department of Environment, and prescribes the ability of the Minister to determine contaminants and their permissible concentrations in surface waters, either generally or with respect to any part of Alberta or a water course. Thus, the Minister has the authority to set water quality standards for industrial and municipal effluents. Also, indirectly related to aquatic ecosystems, the Province of Alberta has legislative acts dealing with agricultural chemicals, environmental conservation, oil and gas conservation, hydro and electric energy, forests and forest management, public health, water resources (e.g., irrigation withdrawal), and wilderness areas.

Federal legislation, including the Federal Fisheries Act and the Canadian Water Act, perhaps have had less direct impact on improvement of water quality in the

South Saskatchewan basin. The Federal Fisheries Act established in 1970 applies to waters throughout Canada, including all inland waters, and covers "fisheries," including all commercially important secondary production (e.g., mussels, fish). Most importantly, it prohibits the harmful alteration, disruption, and destruction of "fish" habitats and the deposition of substances deleterious to waters frequented by "fish".

The intent of the Canadian Water Act is to provide for the integrated management and planning of Canada's water resources. This includes research planning, implementation of programs relating to water resources, and the conservation, development, and use of water resources to ensure optimum benefit of all Canadians. The Canadian Water Act provides for management of water resources on a watershed basis through the formulation of agreements between the federal and provincial governments.

Prospectives on River Water Quality in Southern Alberta

Sewage effluents from the city of Calgary have clearly reduced water quality in reaches of the Bow River downstream, and control efforts have now progressed to the implementation of tertiary sewage treatment. During the initial years of phosphorus removal, no immediate and corresponding decrease in aquatic plant biomass occurred. This enigma indicates that the current working model for response of plant growth to phosphorus concentration is inadequate, and a conceptual model based on empirical field data has been forwarded in this paper (Figure 9).

In order to further understand and, ultimately, manage the growth of the primary producers, several critical areas need to be addressed. These areas include examining the importance of river sediments in contributing nutrients to rooted macrophytes and the water column. Additionally, the relative requirements of aquatic plant growth for nitrogen versus phosphorus needs closer examination. Researchers are presently attempting to address these areas, as well as to determine if a new equilibrium in plant biomass will be reached after phosphorus is removed. A separate but related issue is whether current levels of trout production and the high quality fisheries in the Bow River below Calgary will be maintained if primary producers are greatly reduced.

Because periodic scouring of the river substrate by high river discharge appears to decrease macrophyte biomass by removing overwintering rhizomes, the feasibility of releasing water from hydroelectric dams annually to scour the stream bed should be investigated. This requires improved estimates of threshold current velocities that will affect the various macrophyte species, so that experiments to reduce plant growth by physical scour can be conducted. These "managed" spring floods could also remove a portion of the nutrients stored in sediments. Thus, future research should determine whether the management of water flow for hydroelectric power can be altered to enhance water quality in reaches receiving sewage effluents.

Acknowledgments

We are grateful to two anonymous referees whose constructive comments helped improve the manuscript.

References

Brown, L., and E. G. Bellinger. 1982. A case study of nutrients in the River Holme, West Yorkshire, England. Environmental Pollution (Series B) 3:81–100.

Blachford, D. P., and F. O. Ongley. 1984. Biogeochemical pathways of phosphorus, heavy metals and organochlorine residues in the Bow and Oldman rivers, Alberta, 1980–1981. Environment Canada Scientific Series 138, Ottawa.

Carignan, R. 1980. Phosphorus sources for aquatic weeds: water or sediments? Science 207:987–988.

Carignan, R. 1982. An emperical model to estimate the relative importance of roots in phosphorus uptake by aquatic macrophytes. Canadian Journal of Fisheries and Aquatic Sciences 39:243–247.

Carlander, K. D. 1969. Handbook of fishery biology, Volume 1. Iowa State University Press. Ames.

Charlton, S. E. D., and D. Bayne. 1986. Phosphorus removal: the impact upon water quality in the Bow River downstream of Calgary, Alberta. Alberta Environment, Pollution Control Division, Edmonton, Canada.

Charlton, S. E. D., H. R. Hamilton, and P. M. Cross. 1986. The limnological characteristics of the Bow, Oldman and south Saskatchewan rivers 1979–1982. Part II: the primary producers. Alberta Environment, Pollution Control Division, Edmonton, Canada.

Coupland, R. T. 1961. A reconsideration of grassland classification in the northern Great Plains of North America. Journal of Ecology 49:135–167.

Cross, P. M., H. R. Hamilton, and S. E. D. Charlton. 1984. Preliminary assessment of the effect of phosphorus reduction in Calgary's sewage treatment plants on the Bow River. Alberta Environment, Pollution Control Division, Edmonton, Canada.

Cross, P. M., H. R. Hamilton, and S. E. D. Charlton. 1986. The limnological characteristics of the Bow, Oldman and South Saskatchewan rivers. Part I: Nutrients and water chemistry. Alberta Environment, Pollution Control Division, Edmonton, Canada.

Culp, J. M., and R. W. Davies. 1982. Analysis of longitudinal zonation and the river continuum concept in the Oldman-South Saskatchewan river system. Canadian Journal of Fisheries and Aquatic Sciences 39:1258–1266.

Donald, D. B., and R. A. Mutch. 1980. The effect of hydroelectric dams and sewage on the distribution of stoneflies (Plecoptera) along the Bow River. Quaestiones Entomologicae 16:658–670.

Environment Canada. 1976. Water temperatures of selected streams in Alberta. Inland Waters Directorate, Ottawa.

Environment Canada. 1987. Historical streamflow summary of Alberta to 1986. Inland Waters Directorate, Ottawa.

Gregory, S. V. 1978. Phosphorus dynamics on organic and inorganic substrates in streams. Verhandlungen Internationale Vereinigug für Theoretische und Angewandte Limnologie 20:1340–1346.

Hamilton, H. R., and L. J. North. 1986. The Bow River water quality monitoring 1970–1980. Alberta Environment, Pollution Control Division, Edmonton, Canada.

Haslam, S. M. 1978. River plants, macrophytic vegetation of watercourses. Cambridge University Press, Cambridge, England.

Henderson, N. E., and R. E. Peter. 1969. Distribution of fishes in southern Alberta. Journal of the Fisheries Research Board of Canada 26:325–338.

Holmes, N. T. H., and B. Whitton. 1977. The macrophytic vegetation of the River Tees in 1975: observed and predicted changes. Freshwater Biology 7:43–60.

Kellerhals, R. C., C. R. Neil, and D. I. Bray. 1972. Hydraulic and geomorphic characteristics of rivers in Alberta. Alberta Environment, Cooperative Research Program in Highway and River Engineering Technical Report, Edmonton, Canada.

Longmore, L. A., and C. E. Stenton. 1981. The fish and fisheries of the South Saskatchewan River basin. Alberta Environment, Planning Division, Edmonton, Canada.

Paetz, M. J., and J. S. Nelson. 1970. The fishes of Alberta. Queen's Printer, Edmonton, Canada.

Reynoldson, T. B. 1973. Macrobenthic fauna surveys of the Oldman, Bow, Red Deer and North Saskatchewan rivers. Alberta Environment, Pollution Control Division, Edmonton, Canada.

Rowe, J. S. 1972. Forest regions of Canada. Environment Canada, Forestry Service, Ottawa.

Sosiak, A. J. 1986. A fisheries survey on the Bow River from Bearspaw Dam to Carseland, Alberta in 1985. Alberta Forestry Lands and Wildlife, Fish and Wildlife Division, Calgary, Canada.

Ward, J. V., and J. A. Stanford. 1979. Ecological factors controlling stream zoobenthos with emphasis on thermal modification of regulated streams. Pages 35–56 *in* J. V. Ward and J. A. Stanford, editors. The ecology of regulated streams. Plenum, New York.

Warner, L. A. 1973. Flood of June 1964 in the Oldman and Milk river basins in Alberta. Environment Canada Technical Bulletin 73, Ottawa.

Westlake, D. F. 1973. Aquatic macrophytes in rivers. A review. Polski Archiwum Hydrobiologii 20:31–40.

Water Quality and Biota of the Columbia River System

QUENTIN J. STOBER
U. S. Environmental Protection Agency Region IV
College Station Road, Athens, Georgia 30613, USA

ROY E. NAKATANI
Fisheries Research Institute (WH-10)
University of Washington, Seattle, Washington 98195, USA

ABSTRACT. *The Columbia River has an average annual flow of 6,657 m^3/s and drains a watershed of 670,810 km^2. Dams and water diversions have changed the river ecosystem from a cold, rapid flowing, stream to a cool, slow flowing series of impoundments. The river's production of anadromous salmonids has been greatly reduced from historical levels. Water quality and fish habitat have been degraded since the 1930s. The primary cause is hydropower development, but other causes include irrigated agriculture, logging, mining, stream channelization, and urbanization. Hydroelectric and agricultural development has changed the quantity and timing of seasonal runoff. Impoundments have modulated temperature extremes in the main stem, delaying the annual thermal maximum below Grand Coulee Dam about 30 days. Warm temperatures occur seasonally at the mouths of tributaries, which may delay the return of adult salmonids to spawning grounds. High spills at dams may supersaturate the river water with air and cause "gas bubble" disease in fish.*

Dissolved oxygen levels are adequate in the main-stem Columbia River but have been depressed in some tributaries by irrigation withdrawal and returns, and by waste loads from municipal and industrial activities. Specific conductance, nitrate-nitrite, sodium, sulfate, chloride, and temperature in the main stem generally increase downstream. The Snake and Willamette rivers account for the major input of total nitrogen and phosphorus to the Columbia River. Suspended sediment tends to increase in subbasins with logging and agriculture. Seasonal turbidity from suspended sediment has declined in the main stem since the mid-1950s because of impoundments. Toxic chemicals (pesticides, PCBs, and trace metals) have been found in fish of the Columbia River basin, resulting in at least one recent human health advisory. Aquatic production in Columbia River impoundments is related to their morphometry and retention time, as well as to nutrient availability and hydropower operations. Exotic and non-game fish have proliferated in main-stem impoundments to the detriment of native resident and anadromous fish. Water quality issues are not presently a top priority because current efforts focus on increasing the numbers of anadromous salmonids. Strong conflicts have emerged between consumptive and nonconsumptive users of Columbia River water, and possible large-scale transfers to areas deficient in water loom as a threat.

The Columbia River was discovered in 1792 by Captain Robert Gray and explored in 1805 by Lewis and Clark, setting in motion changes that profoundly altered the river and its watershed. The Columbia River basin historically produced large runs of Pacific salmon (*Oncorhynchus* spp.) and steelhead trout (*Oncorhynchus mykiss*) around which the Northwest aboriginal cultures evolved. Settlement of the basin by non-Indians began in earnest around the mid-1800s. Logging, mining, fishing, and agriculture contributed to the early economy (NPPC 1986). Water quality over the years was affected by impoundments; return flows from irrigation; soil erosion from farming, logging, and construction; discharge of domestic and industrial wastes; reactor cooling water discharges; and urbanization.

Data on Columbia River water quality were obtained as early as 1910 by Van Winkle (1914) in an initial study of Pacific Northwest streams and lakes. Little data on water quality (other than temperature) were obtained subsequently until the end of World War II. After the war, water quality received increasing attention, usually in relation to the welfare of salmonid populations.

The objectives of this paper are to describe the ecological features of the Columbia River system and to explore changes in water quality associated with past and present river management.

Morphometry

The main-stem Columbia River is 1,930 km long, of which 1,199 km are in the United States. The Columbia River basin encompasses 670,810 km^2. The Columbia River ranks fifth in drainage area and third in discharge (Table 1) among rivers in North America. It begins in the Canadian Rockies at Columbia Lake and flows northwesterly in British Columbia for about 306 km, then south 436 km across the Okanogan Highlands where it receives the Kootenai and Pend Oreille rivers near the international boundary (Figure 1). The Columbia River continues south across the semi-arid Columbia Plateau into the United States to receive the Spokane, Okanogan, Wenatchee, and Yakima rivers. The Snake River enters above the Washington-Oregon border, at which point the Columbia River turns west and flows about 483 km through the Cascade and Coast ranges. It is joined by the Umatilla, John Day, Deschutes, and Willamette rivers from Oregon and the Cowlitz and Lewis rivers from Washington before entering the Pacific Ocean near Astoria, Oregon (Neal 1972). The interior drainage extends to the continental divide in Idaho, Montana, and Wyoming.

From source to outlet the Columbia River drops 808 m and traverses several climatic zones from alpine to shrub-steppe to coastal. The interior Columbia River basin is ringed by mountains with the Bitteroot and Selkirk ranges to the east and the Cascade and Coast ranges to the west.

TABLE 1. Comparison of Columbia River with other large systems in the United States.

River system	Estimated drainage area (km^2)	Estimated average discharge (1921–1945) (m^3/s)
Mississippi	3,221,183	17,564
Missouri[a]	1,371,146	1,986
Yukon	854,700	4,249
St. Lawrence[b]	782,180	6,402
Columbia	670,810	6,657
Colorado	629,111	c
Ohio[a]	528,101	7,224
Rio Grande	444,405	283

[a]A tributary of the Mississippi River.
[b]At the Ontario-Quebec-New York border, latitude 45°N.
[c]At the Arizona-Mexico border; no dependable estimate of discharge has been made because of the large (unmeasured) consumptive uses and diversions.

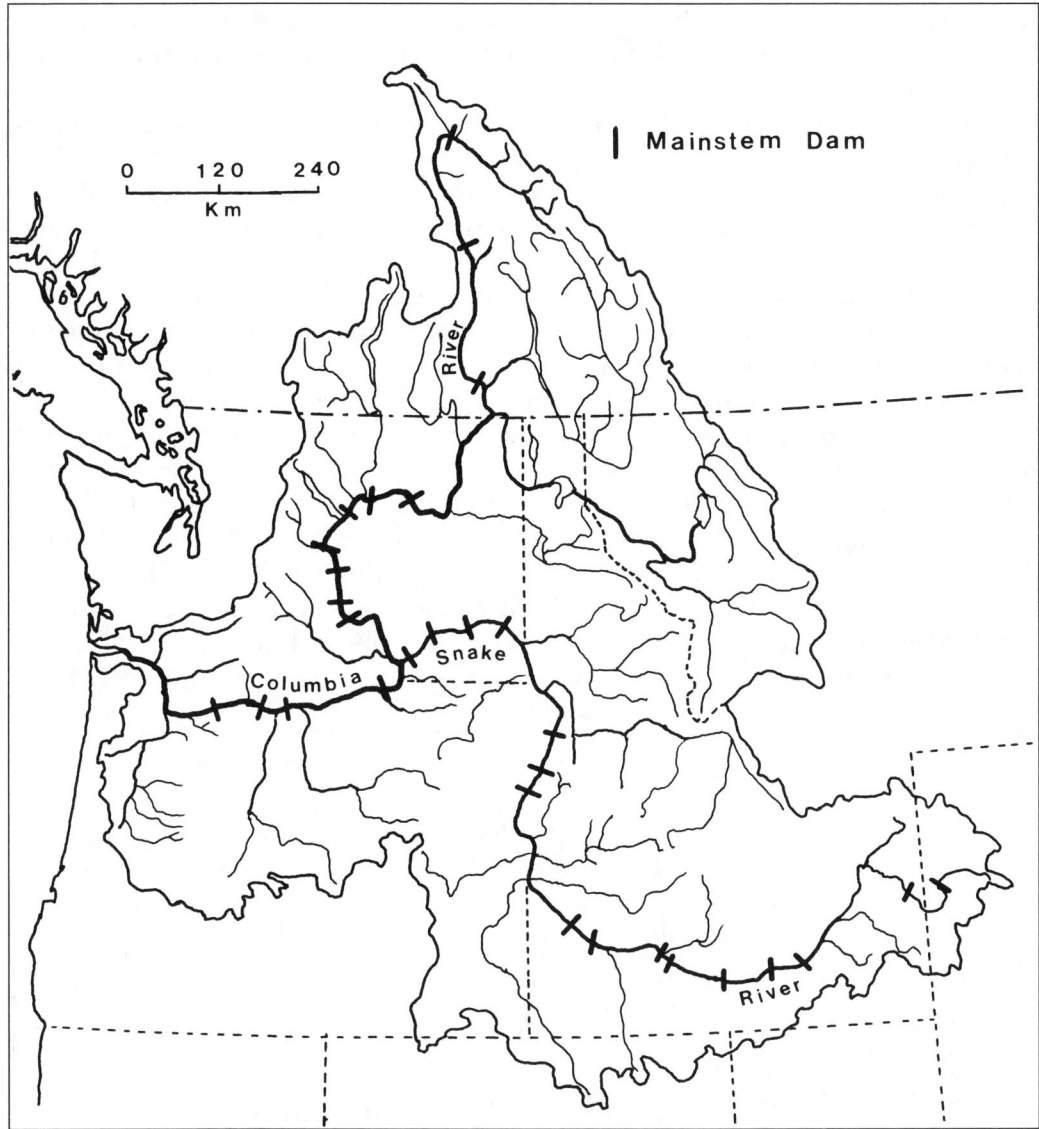

FIGURE 1. The Columbia River basin and main-stem dams. (Modified from Wydoski and Whitney 1979.)

The largest tributary, the Snake River, is 1,671 km long and drains 49% of the system's watershed in the United States. The Cascade Range forms a mountainous barrier to the passage of moisture inland, resulting in a relatively dry and open landscape to the east throughout the Columbia Plateau.

Hydrology

The average annual flow of the Columbia River at its outlet is about 6,657 m^3/s. Discharges from the Snake River average about 1,303 m^3/s annually (PNRC 1979).

Nearly one-fourth of the Columbia River's runoff originates west of the Cascade Range, an area with less than 10% of its total drainage, because rainfall is higher near the Pacific Coast.

In general, tributaries of the Columbia River originate in high, forested mountains where the climate is mesic, the gradient steep, and the stream velocity high. In the interior Columbia basin, the climate is xeric, the gradient less steep, and the stream flow reduced. The Cascade Range near the river's mouth is forested and receives heavy precipitation.

Seasonal runoff and irrigation withdrawals greatly influence flows in Columbia River tributaries. Storage and release of water from impoundments for hydropower production influence flows in the main stem. Precipitation, primarily in the form of snow, is greatest in winter, and runoff increases with snowmelt during spring and early summer. Most major floods on tributaries east of the Cascades result from rapid snowmelt. Heavy warm rains or winds often accentuate the most severe spates. Convective storms accompanied by intense rainfall may also cause local floods.

Prior to impoundment of the main stem, flows averaged 18,697 m^3/s from May through July and 1,983 m^3/s from September through March (Hickson and Rodolf 1957). After 14 dams were completed on the main-stem Columbia River and over 200 on its tributaries, flows ranged between 4,249 and 16,997 m^3/s (Lockett 1962). As a result, in most years, flows in the Columbia River have been reduced

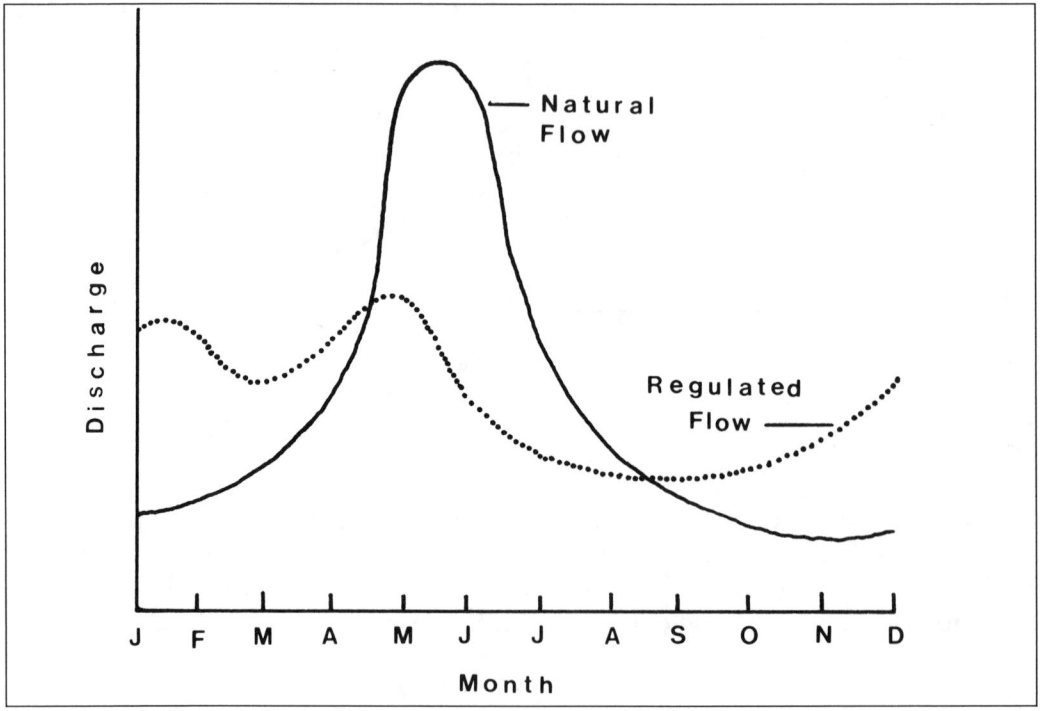

FIGURE 2. Generalized effect of reservoir operations on main-stem Columbia River flows near the Dalles, Oregon. Tributary storage of water and main-stem production of hydropower reduced the spring peak runoff that once transported juvenile salmonids downstream to the Pacific Ocean (Ebel et al. 1989).

about 50% at the time young salmonids migrate to sea in May and June (Figure 2).

By 1973, the combined storage of Mica, Keenleyside, and Grand Coulee dams on the main-stem Columbia River, and of Duncan, Albeni Falls, Libby, and Hungry Horse dams on the Kooteney, Pend Oreille, and Flathead rivers (Figure 1), provided capacity to store over $43,200 \times 10^6 m^3$ of spring runoff upstream for release later when more electricity was needed. Revelstoke Dam, completed in 1983, added $1,480 \times 10^6 m^3$ of additional storage. The system's total "active storage" is now $53,800 \times 10^6 m^3$, about a quarter of the average annual runoff (Table 2).

The upper Columbia River in British Columbia contains the Mica, Revelstoke, and Keenleyside projects, which are large cyclic storage reservoirs (Figure 3). Lake McNaughton of the Mica Project is the largest reservoir at $24,985 \times 10^6 m^3$ and, because of its size, may remain unfilled after seasonal drawdown. The Arrow Lakes (Keenleyside Dam) in Canada are also used primarily for storage.

The most important storage reservoir in the United States is Lake Roosevelt behind Grand Coulee Dam (Figure 3). It contains $6,400 \times 10^6 m^3$ of active storage but has a total volume of $11,800 \times 10^6 m^3$. The main-stem Columbia River below Lake Roosevelt is greatly influenced by storage and releases at Grand Coulee Dam. Below this point, the river-run projects most affected are Chief Joseph, Wells, Rocky Reach, Rock Island, Wanapum, and Priest Rapids, all of which have limited storage capacities.

Lake Roosevelt is operated under three seasonal patterns that include: the summer holding or storage season, the fall and winter storage or drawdown season, and the spring runoff or refill season (CRWMG 1979). The flushing rate of Lake

TABLE 2. Storage characteristics of main-stem Columbia River reservoirs (Ebel et al. 1989).

Dam	Reservoir (lake)	Location (river km)[a]	Length (km)	Total volume[b] ($10^6 m^3$)	Mean Annual discharge ($10^6 m^3$/year)	Storage ratio[c]	Flushing rate[d]
Mica	McNaughton	1,638	209	25,040	18,260	1.370	499.0
Revelstock	—	1,498	129	1,480	70,440	0.020	7.3
Keenleyside	Arrow	1,255	216	9,250	35,775	0.260	94.0
Grand Coulee	F.D. Roosevelt	960	243	11,800	96,220	0.120	45.0
Chief Joseph	Rufus Woods	877	71	616	96,470	0.007	2.6
Wells	Pateros	830	45	370	100,420	0.004	—
Rocky Beach	Entiat	761	68	493	102,390	0.005	1.8
Rock Island	Rock Island	729	34	123	105,600	0.001	—
Wanapum	Wanapum	668	61	740	105,600	0.007	2.6
Priest Rapids	Priest Rapids	639	29	247	105,720	0.002	0.7
McNary	Wallula	470	98	1,727	250,995	0.011	4.0
John Dam	Umatilla	348	122	3,084	153,960	0.020	7.3
The Dalles	Celilo	309	39	370	158,890	0.003	1.1
Bonneville	Bonneville	235	72	616	163,700	0.004	1.5

[a] Distances are measured from the river's mouth.
[b] Maximum capacity for water storage in a reservoir.
[c] Storage ratio (annual) = (total volume)/(mean annual discharge). This value is also called the exchange rate or flushing rate, and it has a value in years, convertible to days.
[d] Flushing rate = annual storage ratio × 365 (days). This is the number of days required, theoretically, to completely empty a reservoir at the mean annual discharge rate.

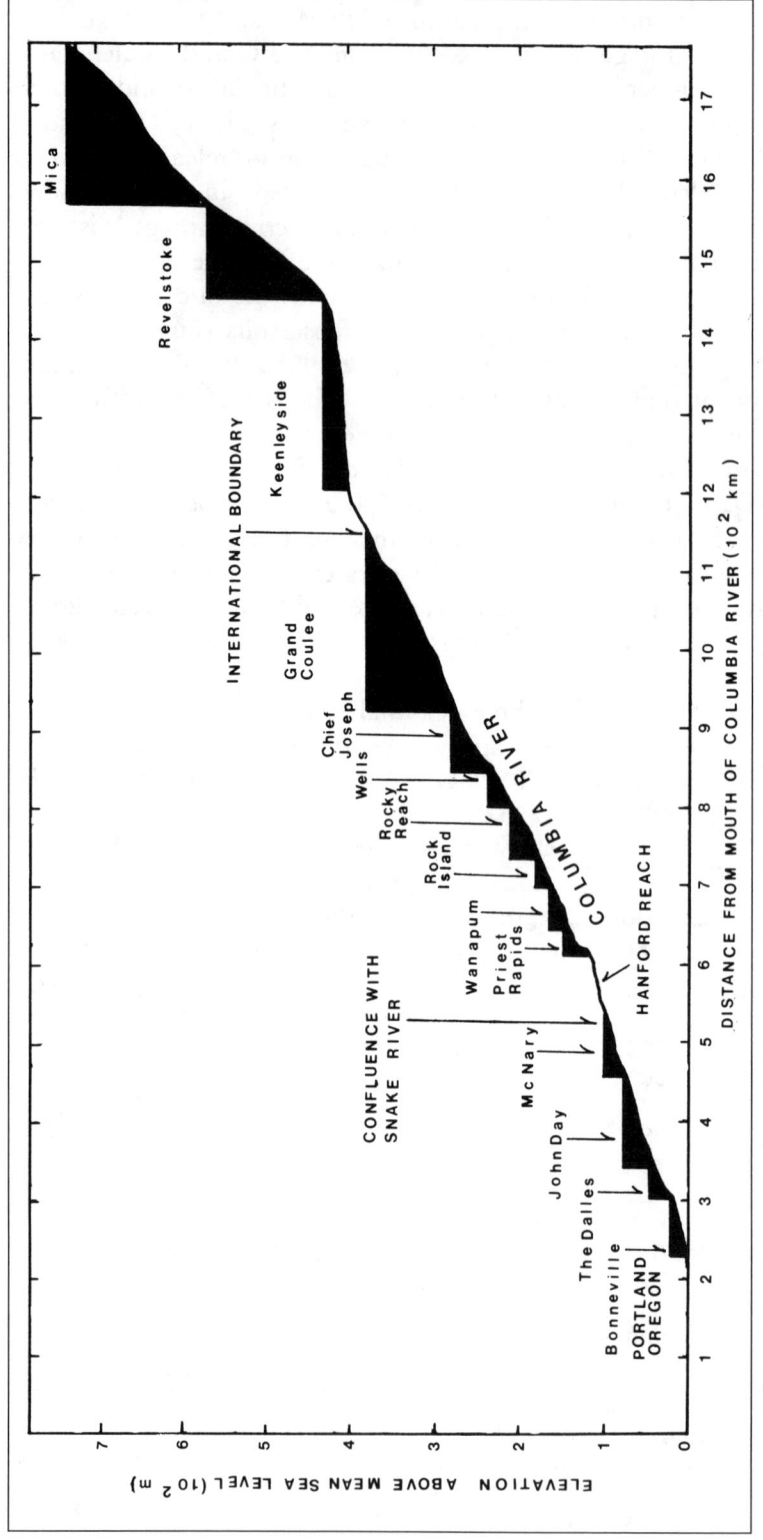

FIGURE 3. Schematic view of distance and elevation of main-stem Columbia River dams. (Modified from Becker 1985.)

Roosevelt is about 45 days. Flushing rates for river-run reservoirs below Grand Coulee Dam vary from less than 1 day (at Priest Rapids) to about 4 days (at Lake Wallula). Current velocities in these impoundments average about 0.3 m/s.

The last unimpounded section of the Columbia River in the United States is the Hanford Reach, an 80.5 km section between the head of Lake Wallula and Priest Rapids Dam (Figure 3). Flows in the Hanford Reach respond to discharges at and above Priest Rapids Dam.

Four hydroelectric reservoirs are located on the lower Columbia River (Figure 3). McNary, The Dalles, and Bonneville are river-run while John Day is a storage project (CRWMG 1979). Reservoirs on the lower Columbia River are the widest and shallowest of the system. The mean depth of Lake Bonneville, near the river's outlet, is only 9 m while that of Lake McNaughton near the river's origin in Canada is 58.5 m. Lake Umatilla behind John Day Dam, the largest of the four lower river projects, has a flushing rate of 7.3 days. Lake Umatilla is operated under four seasonal regimes: the summer holding season, the fall drawdown season, the winter flood control season, and the spring refill season (CRWMG 1979).

Discharge volumes from main-stem dams generally increase downstream because of reduced reservoir storage ratios and additional increments of water from tributaries. Annual discharges at Bonneville Dam average about $164,000 \times 10^6$ m^3. Ocean tides affect flows below Bonneville Dam.

Storage reservoirs in the Columbia River system undergo major seasonal drawdown. For example, the elevation of Lake Roosevelt decreases about 25 m each year prior to the spring spate (Stober et al. 1979). Water levels in storage reservoirs are minimum during April and May. In contrast, water levels of most river-run reservoirs may fluctuate 0.3 to 1.5 m daily in response to power generation at their outlet dams.

Water diversion from the entire system for irrigation averages about $13,276 \times 10^6$ m^3 per year (CRWMG 1986). About 56% of this is diverted from the Snake River in southern Idaho and 37% from the main-stem Columbia River at Grand Coulee Dam, the Yakima River, and other projects in Washington State. The remaining 7% is drawn from tributaries in Oregon, Montana, and Idaho. A portion of the total volume diverted returns to streams as irrigation return flow, but its quality is usually altered. Diversion seasonally creates serious instream flow problems in tributaries throughout the Columbia basin.

Water Quality and Resource Management

Temperature

Perhaps no other water quality variable has received as much attention in the Columbia River as temperature. Instream temperatures varied widely before major dams were constructed. The system of impoundments on the main-stem Columbia River has decreased the magnitude of daily, weekly and seasonal temperature fluctuations (Jaske and Goebel 1967; Jaske and Synoground 1970). Not only has the variance

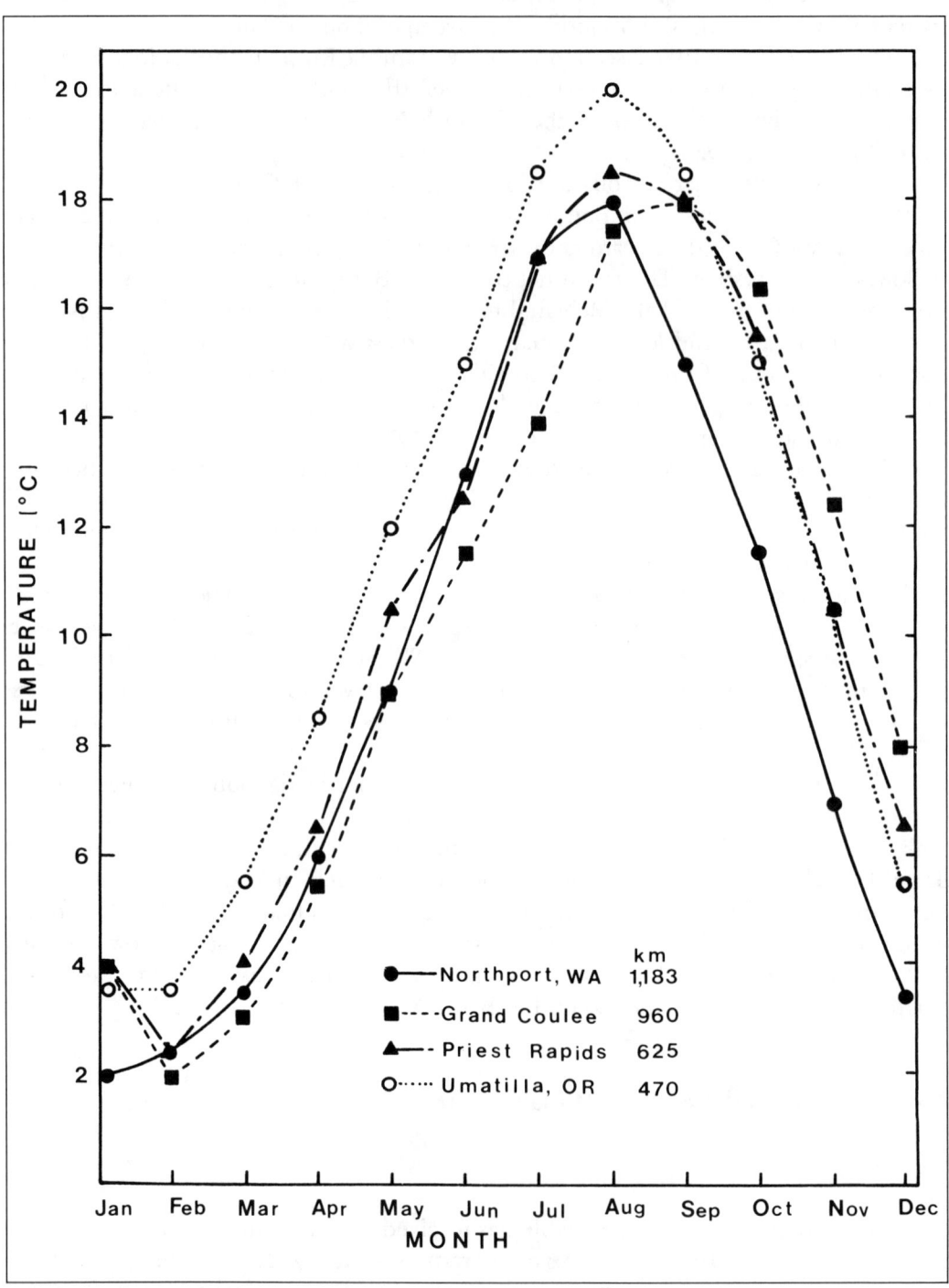

FIGURE 4. Mean monthly water temperatures at four locations on the Columbia River, October 1977 to September 1978. River kilometers are measured from the river mouth; WA is Washington; OR is Oregon. (From USGS 1978.)

been reduced, but peak summer temperatures have been delayed about 30 days in the reach below Grand Coulee Dam (Figure 4). This delay is reflected, to a lesser extent, as far downstream as Bonneville Dam (Jaske and Synoground 1970). Hence, fall chinook salmon that spawn in the main-stem Columbia River on falling temperatures (12°C or lower) are now spawning later. Such minor shifts in water temperature are unlikely to be biologically significant for sockeye salmon (*Oncorhynchus nerka*) that use the main stem only for passage, not for spawning (Mullan 1986).

From 1944 to 1971, no other large river received as much man-made heat as did the Hanford Reach from the cooling water discharges of up to eight plutonium production reactors (Foster 1972; Becker 1990). Protecting the anadromous fish of the Columbia River in the 1960's, especially fall chinook salmon that spawned in the Hanford Reach, drew much concern. Fear of detrimental effects from the added heat proved unwarranted, but caution is required whenever heated effluents are discharged near spawning areas (Nakatani 1969).

Historically, water temperatures at the mouths of tributary streams were often higher than in the main-stem Columbia River. Temperatures in the John Day, Umatilla, Walla Walla, Snake, Yakima, and Okanogan rivers were sometimes 1.1 to 3.3°C higher than in the main stem (Sylvester 1958). In 1985, the temperature of the lower Snake River was 7.2°C higher, with a maximum of 30.6°C, than in the Columbia River (CRWMG 1986). Such temperature differentials have delayed the passage of returning anadromous salmonids from the main stem to tributaries.

The states of Washington and Oregon have established water temperature standards for the Columbia River. For waters common to both states [i.e., mouth of the Columbia River upstream to river kilometer (RKm) 498], the temperature standards are the same: the river temperature shall not exceed 20°C because of human activities. When natural conditions exceed 20°C, no temperature increase will be allowed that raises the receiving water temperature more than 0.3°C, nor shall such temperature increase at any time exceed 0.3°C due to any single source or 1.1°C from all such activities combined. But the state of Washington has a different temperature standard for the Columbia River from RKm 498 to RKm 960 (to Grand Coulee Dam), which essentially uses the same language, except "... nor shall such temperature increases at any time exceed $t = 34/T + 9$," where T = highest existing temperature in degrees C outside of the mixing zone (WDOE 1982).

Gas Supersaturation

Water spilled over hydroelectric dams on the main-stem Columbia River may become supersaturated with nitrogen (78% in air), which is lethal to fish at high levels (Collins 1976). Exposure of juvenile salmonids to supersaturation during high flows from 1966 to 1978 resulted in losses ranging from 40 to 95% (Ebel et al. 1979). Supersaturation occurs when large volumes of water plunge over a spillway into a deep pool below a dam, forcing entrapped air into solution. In general, higher nitrogen saturation levels occur in reservoir tailraces than in forebay areas (Meekin and Allen 1974). When numerous reservoirs occur end-to-end, as on the main-stem Columbia and Snake rivers, supersaturation can extend through several reservoirs.

During the fall and winter when water is not spilled over main-stem dams, nitrogen levels return to normal at near 100% saturation (WDF 1974).

The energy crisis of the early 1970's resulted in a doubling of turbine capacity at lower Columbia and Snake river dams, which reduced the frequency of spill. To minimize supersaturation during high flow years when spill was necessary, the Corps of Engineers developed a spillway flow deflector (flip-lip) that created a surface flow instead of a deep plunging action below a spillway (Collins 1976). When deflectors were installed below the four lower Snake River dams and at McNary and Bonneville dams on the lower Columbia, the problem came under control (Ebel et al. 1979). Installation of additional and increased flow control at storage dams in Canada lowered gas supersaturation at middle Columbia projects that lacked deflectors. Spillway deflectors are less effective at extremely high rates of spills.

Although supersaturation is now partially controlled, it has not been eliminated. The middle and lower Columbia River and the lower Snake River remain supersaturated during high runoffs in the spring (WDF 1974; Dell et al. 1975; Bouck et al. 1976). In 1984, 1985, and 1986, concentrations of dissolved gas ranged from 113–139% in the Columbia River with some of the highest concentrations entering from the Snake River (CRWMG 1986). The overlap of the downstream migration of juvenile salmonids with high spills during spring runoff is cause for concern because juvenile fish are less tolerant of increasing water temperatures when nitrogen is present at supersaturated levels (Ebel 1969; Ebel et al. 1971). The extent of nitrogen-caused mortality among juvenile and adult salmonids depends on the degree of exposure (Beiningen and Ebel 1968; Ebel and Raymond 1976). Regulatory standards have been established at 110% saturation, but this level is frequently exceeded at peak saturation (Figure 5).

Dissolved Oxygen

Dissolved oxygen (DO) in the main-stem Columbia River rarely declines below the minimum standard of 8 mg/L. Concentrations are highest during spring runoff in May and June. Surface waters generally have higher concentrations of DO than the deeper layers in some reservoirs. However, rapid flushing and limited stratification usually prevent low DO levels in most main-stem Columbia River reservoirs.

The lower reaches of streams tributary to the main-stem Columbia River have naturally depressed DO levels during minimum stream flows in August and early fall. This effect is enhanced by irrigation withdrawal in such tributaries as the Okanogan, Yakima, and Umatilla rivers. Dissolved oxygen levels decline further in tributaries that receive large volumes of irrigation return flows (Sylvester and Seabloom 1962; Stober et al. 1979). Low flows and high solar insolation in summer combine to raise water temperatures in lower reaches of tributaries that receive irrigation returns laden with sediment, nutrients, and organic matter (Whitney and White 1984). Low DO levels that coincide with high temperatures prevent returning adult salmonids from entering such tributaries. During late summer of some years, higher water temperatures (20° to 22°C) and low dissolved oxygen (less than 6 mg/L) make living

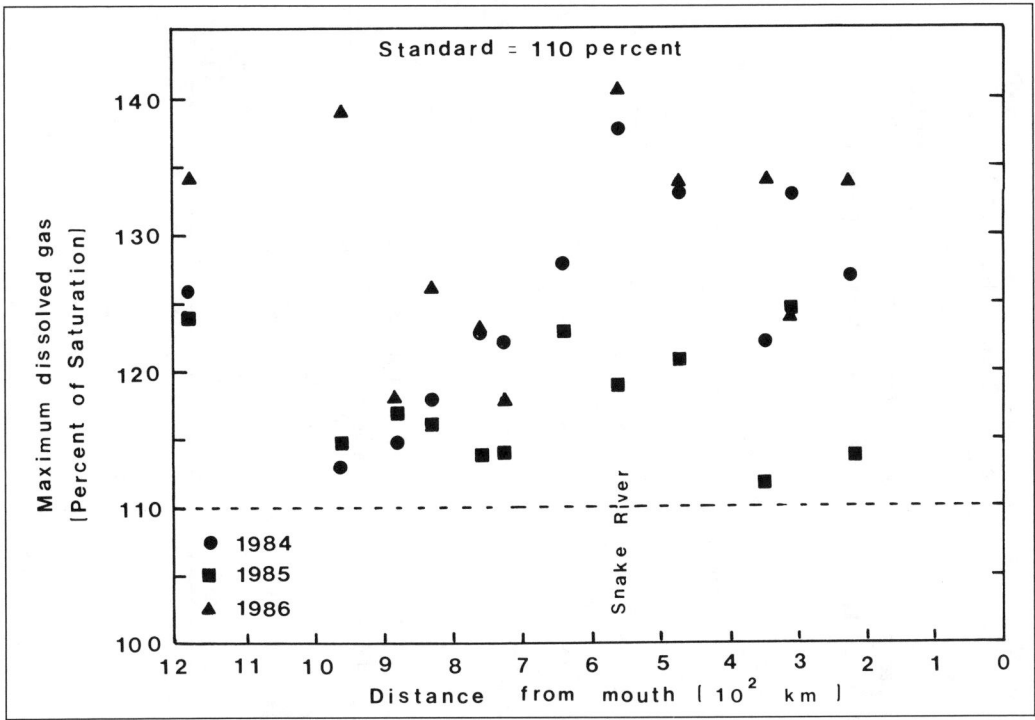

FIGURE 5. Maximum dissolved gas saturation levels in percent of saturation for 12 stations on the Columbia River during 1984, 1985, and 1986. The current federal and state standard is 110 percent. (From CRWMG 1986.)

conditions marginal for salmonids in lower Snake River reservoirs (Robeck et al. 1954; Bennett et al. 1983).

In the lower Willamette River during the 1940's, waste loads from sewage treatment plants, pulp and paper mills, and municipal and industrial point sources created extremely low DO concentrations (zero to 3 mg/L), which blocked the return of adult salmonids (Dimick and Merryfield 1945; Fish and Wagner 1950). Fish kills were common during that period and assemblages of fish species became depleted (Fish and Rucker 1950). These pollution sources were subsequently controlled. Waste loads in the Willamette River are now monitored closely to minimize the recurrence of pollution problems as the human population rapidly expands. A 1983 study in the Willamette River found more fish species present, and more fish that were intolerant of poor water quality, at all but two of eighteen sites sampled 40 years earlier (Hughes and Gammon 1987).

Salinity

Water in the Columbia River is a dilute calcium-magnesium, carbonate-bicarbonate type with a total dissolved solids content of about 90 mg/L (range 71 to 158 mg/L) from the international boundary downstream to the confluence of the Snake River (Ebel et al. 1989). Changes in water quality downstream were evident in 1954–1957 and they remain evident today (Table 3). Specific conductance (Figure 6),

TABLE 3. Mean annual values of water quality variables at stations on the main-stem Columbia River during 1954–1957, 1975–1978, and 1981–1983. Water years are denoted by their last two digits. (After Sylvester and Ruggles 1957, and USGS 1976–1979 and 1982–1984).

Variable	Northport, WA			Grand Coulee, Beebe, WA		Rock Island Dam		Priest Rapids Dam			Richland/Pasco, WA		Umatilla, OR		The Dalles, OR	Warrendale, OR	
	54-57	76-78	81-83	75-78	54-57	54-57	75-78	54-57	75-78	81-83	54-57	81-83	54-57	76-78	75-78	54-59	82-84
Specific conductance (μS/cm)	138.0	135.00	—	148.0	126.0	131.0	155.0	147.0	146.0	135.0	142.0	137.00	159.0	170.0	164.0	—	162.00
pH	7.6	8.00	—	7.7	7.7	7.7	8.0	7.9	7.9	8.0	7.8	7.70	7.9	7.9	8.0	—	8.20
Turbidity (NTU)[a]	1.3	1.40	—	3.7	16.0	12.0	4.7	11.0	4.0	1.4	12.0	1.80	20.0	—	—	—	5.50
Alkalinity (mg/L as $CaCO_3$)	59.0	60.50	—	58.0	58.0	59.0	57.0	62.0	58.7	58.0	62.0	59.20	66.0	62.0	66.0	—	67.00
Nitrite plus nitrate (μg/L as N)	90.0	118.00	—	107.0	—	—	140.0	—	117.0	182.0	—	132.00	—	190.0	168.0	—	285.00
Ammonia (μg/L as N)	50.0	97.00	—	60.0	30.0	40.0	50.0	80.0	50.0	93.0	20.0	80.00	50.0	20.0	20.0	—	100.00
Total phosphorus (μg/L as P)	40.0	35.00	—	33.0	—	—	47.0	—	37.0	32.0	—	38.00	—	40.0	45.0	—	63.00
Dissolved oxygen (mg/L)	12.0	11.60	—	11.7	12.2	12.1	12.4	12.4	12.2	11.2	11.9	11.20	11.2	11.6	11.3	—	10.70
Calcium (mg/L)	20.7	19.30	—	20.0	16.0	—	20.5	19.0	20.0	19.0	17.0	19.40	15.0	21.0	19.0	—	18.80
Magnesium (mg/L)	4.5	4.20	—	4.4	5.7	—	4.3	3.0	4.5	4.2	4.1	4.40	3.2	4.7	5.3	—	5.10
Sodium (mg/L)	1.7	1.70	—	2.1	2.1	—	2.8	4.0	2.8	2.1	2.0	2.30	7.1	6.7	6.7	—	6.70
Potassium (mg/L)	0.8	0.70	—	1.1	1.0	—	1.1	1.3	1.2	0.8	1.1	0.88	1.9	1.4	1.4	—	1.50
Sulfate (mg/L)	12.8	9.95	—	11.4	11.0	11.0	12.1	9.0	12.4	11.2	11.0	11.50	15.0	14.7	54.1	—	13.80
Chloride (mg/L)	0.9	1.0	0.97	—	1.4	—	—	—	1.6	1.39	—	—	3.4	3.6	—	3.8	3.45
Temperature (°C)	9.3	9.20	—	10.0	10.0	10.1	—	10.3	10.5	10.8	10.7	10.69	10.5	11.5	11.1	—	12.30

[a] Nephelometric turbidity units.

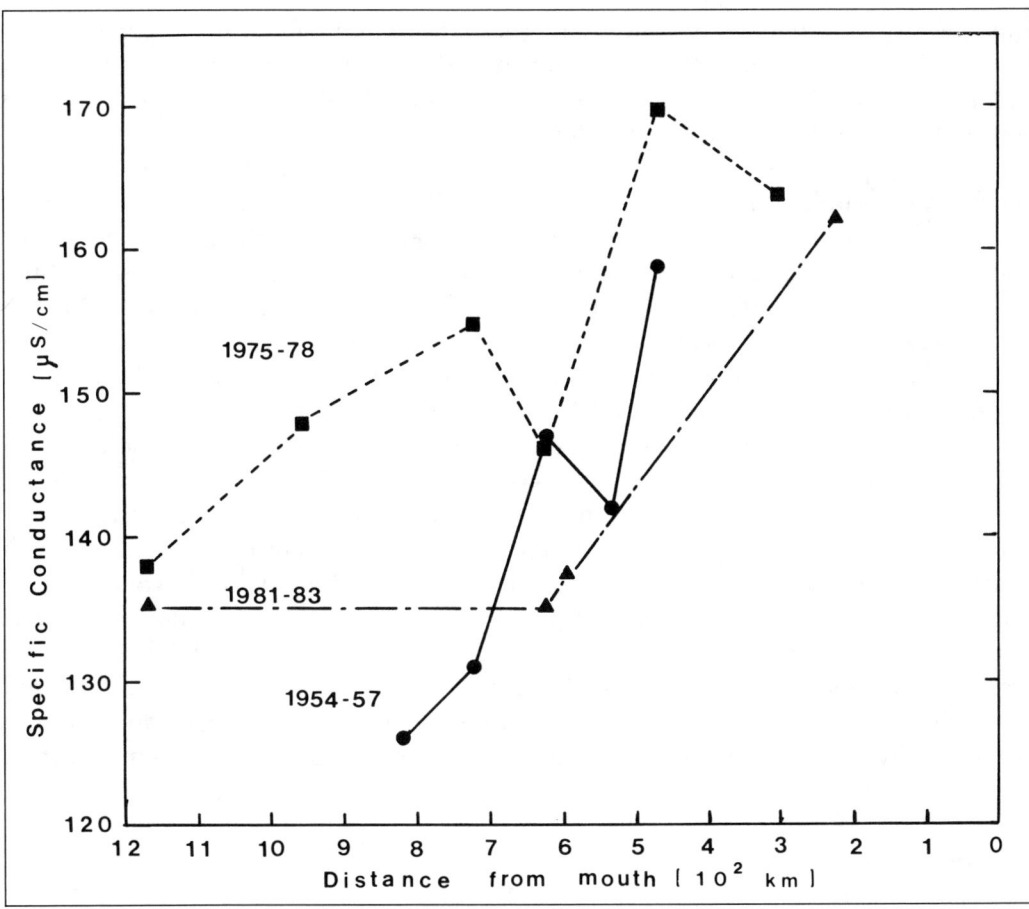

FIGURE 6. Mean annual specific conductance (μmhos) at stations on the main-stem Columbia River during 1954–1957, 1975–1978, and 1981–1983. (After Sylvester and Ruggles 1957, and USGS 1976–1979 and 1982–1984.)

sodium, chloride, and sulfate generally increase from the international boundary (RKm 1200) downstream to Warrendale, Oregon (RKm 227). The largest increases occur below the outlet of the Snake River (RKm 550). Magnesium and pH do not change appreciably downstream, and pH remains within the standard 6.5–8.5 range. Water quality monitoring above and below the Hanford Reach from 1981 to 1986 shows that operations at the government's Hanford site now have little influence on water quality in the main stem (PNL 1986; PNL 1987). A trend analysis of factors affecting the nation's surface waters from 1974 to 1981 revealed a significant increase of chloride, sulfate, and sodium in Northwest rivers (Smith et al. 1987).

Returning irrigation water often contain salts increased by evapotranspiration, leaching, and ion exchange (Sylvester and Seabloom 1962). Irrigation water commonly leaches nitrates, calcium, magnesium, sodium, potassium, chlorides, and sulfates from the soil. Specific conductance increases along with the dissolved salts. Consequently, irrigation returns can produce marked water quality changes in the receiving stream if the volume of return flow is high in relation to streamflow (Sylvester and

Ruggles 1957). Irrigation returns from the vast Columbia Basin Project enter the Columbia River between Rock Island Dam and the Snake River (USBR 1976). The Yakima and Snake rivers, which also carry dissolved salts from irrigation returns, enter the Columbia below the Hanford Reach.

Water quality has changed in the lower reaches of some extensively used tributaries. For example, water quality in the upper Yakima River near Thorp, Washington has changed only slightly since 1957 (Table 4). Pronounced changes have occurred below this point. Specific conductance, alkalinity, calcium, magnesium, sodium, and sulfate were all higher downstream at Kiona and Richland during 1975–1978 than before. While chemical constituents increased downstream, DO decreased. The entire flow of the Yakima River may be used more than once during the irrigation season (Sylvester and Seabloom 1962).

Return flows from irrigated land may flow into large rivers such as the Columbia without serious effect (Sylvester and Rambow 1967). Dissolved salts in irrigation returns may be relatively unimportant when present at non-toxic levels. Surface waters of the Columbia basin normally contain low levels of dissolved minerals, and the relatively high levels of salts in irrigation returns are diluted by the much greater flow in the Columbia River. An estimated $492,308 \times 10^6$ m^3 of water diverted for irrigation was returned by the Columbia Basin Project in 1970 (USBR 1976). This return flow amounted to only 0.46% of the mean annual flow of the Columbia River. Under average flow conditions, irrigation returns from the Project only increased conductivity by three μS/cm and nitrate by 0.01 mg/L in the main-stem Columbia River.

Nutrients

Nutrient loading generally increases in the Columbia River from the Canadian border to the mouth (Table 3). Most of the increase in total nitrogen and phosphorus is attributable to tributary inputs (Hileman et al. 1975). The Snake and Willamette rivers accounted for 57 to 80% of the total tributary nitrogen and 46 to 85% of the total tributary phosphorus in the Columbia River during four quarterly hydrologic periods. Based on yearly average concentrations, orthophosphate-phosphorus exceeded 0.01 mg/L (critical level for algae blooms) at all main-stem stations and in most tributaries except the Kettle, Methow, Chelan, Entiat, and Wenatchee rivers. Nitrate-nitrogen exceeded 0.3 mg/L (critical level for algae blooms) only at McNary Dam on the main-stem Columbia River and at seven tributary stations (Crab Creek and the Spokane, Yakima, Snake, Walla Walla, Umatilla, and Willamette rivers). Quarterly average concentrations of nitrate-nitrogen and orthophosphate-phosphorus exceeded critical levels from November to April downstream of RKm 470 on the main-stem Columbia River. Nutrients were minimum between May and October. Although nutrient levels holding potential for noxious algal blooms occur along the main stem, other factors interact to inhibit their development.

The Spokane River received intensive study throughout the 1970's because noxious algal blooms developed each year downstream from a wastewater treatment plant at the City of Spokane. Long Lake, an impoundment on the lower Spokane

TABLE 4. Mean annual water quality variables at stations on the Yakima River, Washington during 1954–57 and water years 1975–1978. (After Sylvester and Ruggles 1957 and USGS 1976–1979).

Variable	Thorp 1954-57	Thorp 1975	Parker 1975	Parker 1976	Parker 1977	Parker 1978	Kiona 1975	Kiona 1976	Kiona 1977	Kiona 1978	Richland 1975
Specific conductance (μS/cm)	71.0	73.0	130.0	113.0	171.0	132.0	223.0	206.0	345.0	223.0	246.0
pH	7.4	7.5	7.7	7.6	8.3	7.8	7.6	7.7	8.3	7.9	7.9
Turbidity (NTU)[a]	20.0	8.0	10.0	11.0	9.0	9.0	13.0	9.0	8.0	18.0	14.0
Alkalinity (mg/L as $CaCO_3$)	32.0	31.0	53.0	—	—	—	94.0	82.0	137.0	93.0	—
Nitrite plus nitrate (μg/L as N)	—	110.0	240.0	120.0	260.0	190.0	640.0	640.0	1000.0	690.0	770.0
Ammonia (μg/L as N)	30.0	60.0	120.0	120.0	150.0	110.0	—	70.0	160.0	40.0	90.0
Total phosphorus (μg/L as P)	—	20.0	100.0	90.0	160.0	90.0	150.0	110.0	180.0	150.0	140.0
Dissolved Oxygen (mg/L)	10.4	12.0	12.0	12.1	12.2	12.2	—	—	—	—	11.8
Calcium (mg/L)	18.0	7.8	12.7	—	—	—	22.0	19.0	32.0	23.0	—
Magnesium (mg/L)	1.2	2.7	4.1	—	—	—	7.6	6.6	11.8	8.2	—
Sodium (mg/L)	2.6	2.3	6.7	—	—	—	13.2	11.1	20.9	14.0	—
Potassium (mg/L)	1.6	0.4	1.4	—	—	—	2.5	2.1	3.7	2.6	—
Sulfate (mg/L)	2.0	1.5	4.3	—	—	—	12.1	9.9	23.6	13.9	—
Chloride (mg/L)	—	1.6	3.0	—	—	—	5.1	4.2	8.4	7.5	—
Temperature	6.0	7.9	10.8	9.8	9.7	10.5	13.5	10.5	12.0	12.5	11.9

[a]Nephelometric turbidity units.

River, became highly eutrophic with large standing crops of algae that decomposed, creating extensive hypolimnetic anoxia. After advanced waste treatment was implemented to remove at least 85% of the phosphorus from municipal wastewater, Long Lake reverted to more mesotrophic conditions (Soltero et al. 1979). Efforts continue to determine the maximum permissible phosphorus loading from all sources in the Spokane River basin to protect beneficial uses of Long Lake. A predictive model of phosphorus transport through the river system is used to allocate waste loads (Patmont et al. 1985).

Suspended Sediment

Suspended sediment has increased in subbasins of the Columbia River where logging and agricultural activities accelerate soil erosion (Smith et al. 1987). Loads of suspended sediment in the main-stem Columbia River, as indicated by turbidity measurements, have declined from 10–20 NTU in 1954–1957 to less than 5 NTU since 1975 (Figure 7). The numerous dams installed on the main stem and tributaries since the mid-1950's may explain the decline in turbidity.

A study of sediment dynamics in the Columbia River (Whetten et al. 1969) showed that grain size, mineralogy, and chemical composition of the sediments was highly correlated with source. Upstream sources in the Snake and Willamette rivers

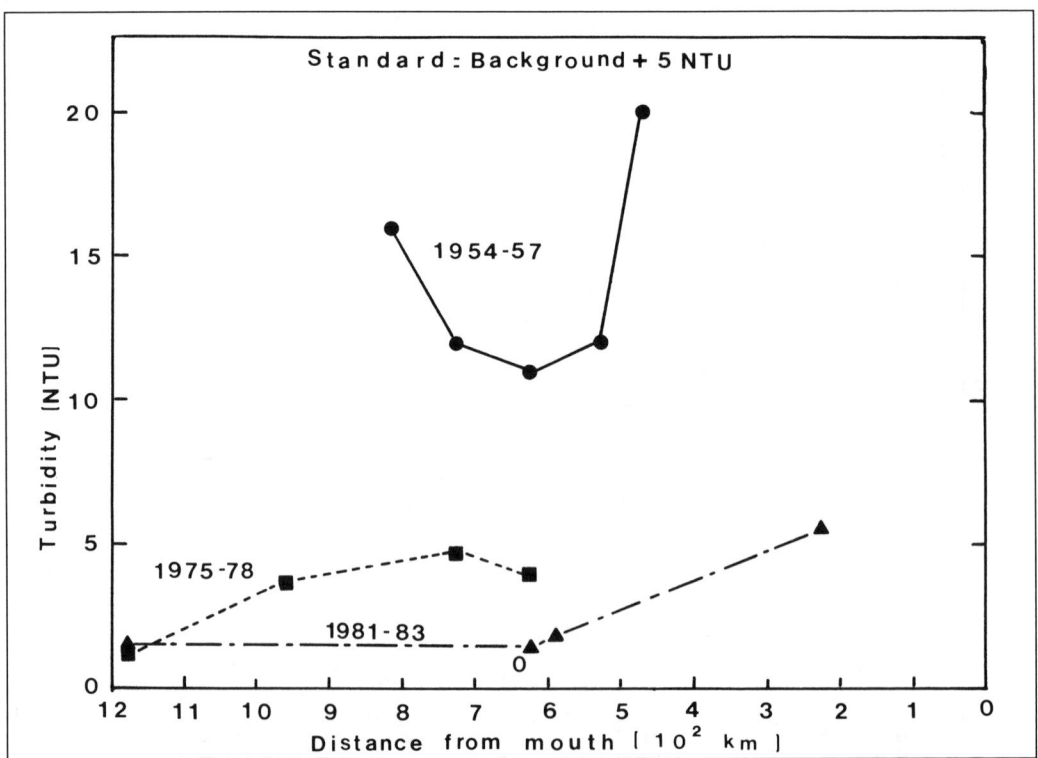

FIGURE 7. Mean annual turbidity (NTU-nephelometric turbidity units) at stations on the main-stem Columbia River during 1954–1957, 1975–1978, and 1981–1983. (After Sylvester and Ruggles 1957, and USGS 1976–1979 and 1982–1984.)

produced largely fine-grained nonvolcanic sediments carried in suspension. The upper Columbia River carried only small amounts of coarse sediment. Downstream sources produced coarse sediment of volcanic origin. Dams on the Columbia River temporarily slow the downstream flow of sediment, but do not act as permanent barriers to movement of fine particles, except at Grand Coulee. Fine sediment deposited in low-head reservoirs is resuspended during high flows and passed downstream.

After Mount St. Helens erupted in May 1980, many streams in the lower Columbia basin carried greatly increased loads of suspended sediment from the fall of volcanic ash and its subsequent erosion (Collins et al. 1982; Turton 1982), and from mudflows in streams directly draining the volcano (Cummans 1981). The mudflows entered the upper estuary of the Columbia River via the Toutle and Cowlitz rivers at RKm 109, below all main-stem dams. Tolerance of salmonids to high loads of suspended sediment and recovery of tributary habitats (Stober et al. 1981b; Martin et al. 1984; Jones and Salo 1986) allowed portions of the devastated stream to support wild salmonid stocks one year after the eruption (Martin et al. 1984). Because sediment can have a high impact on salmonid reproduction, the state of Washington established a 5 NTU increment as the standard that cannot be exceeded when background turbidity is 50 NTU or less, and a 10% increment as the standard that cannot be exceeded when background turbidity is more than 50 NTU. The standards are designed to protect salmonid habitats in the Pacific Northwest (Lloyd 1987).

Toxic Chemicals

Many types of pesticides are used on agricultural lands in the Columbia River basin. A pesticide-monitoring program initiated in 1973 detected DDT, DDE, Ronnel, and Thiodon (endosulfan) at minute concentrations in some samples from open irrigation channels and buried pipe drains. The residues measured were of minor significance (USBR 1976).

Knowledge of pesticide levels in the main-stem Columbia River remains incomplete. Files at various agencies contain data on pesticides in the main stem, but this information has not been assessed for possible impacts on biota.

Pesticides are monitored in the Yakima River at Kiona by the U. S. Geological Survey. None of the 24 major "problem" pesticides were detected during water year 1978 (USGS 1976–1979). Routine monitoring between 1979 and 1984 showed higher levels of DDT and its metabolites, DDE and DDD, in Yakima River fish than elsewhere in Washington State. Analyses for organochlorine compounds revealed dieldrin, PCB-1260, endosulfan, and endrin in fish tissue. Fish in the lower Yakima River were more contaminated than those upstream and returning anadromous salmonids had lower body burdens than resident fishes (Johnson et al. 1986). Concentrations of t-DDT, dieldrin, PCB-1260, and mercury in edible fish tissue were well below "action levels" (USFDA 1987).

Pesticide levels are probably lower in the Columbia River, which receives proportionately less return flow than does the Yakima River, yet the propensity for many synthetic chemicals to bioconcentrate creates special hazards. Minor releases of polychlorinated biphenyls (PCBs) have occurred in the Columbia River basin. Some salmonids and resident fishes now carry small body burdens of PCBs and other organochlorine compounds, presumably because they have been widely used in agriculture. Recent studies of fish assemblages on the Willamette River showed that disease and morphological anomalies increased markedly in the lower river. An associated decrease in biomass downstream suggested increased levels of sublethal stress, possibly from toxic chemicals (Hughes and Gammon 1987).

The world's largest lead-zinc smelter and refinery is located on the upper Columbia River at Trail, British Columbia, about 16 km above the international boundary. The facility releases iron, manganese, zinc, copper, lead, arsenic, cadmium, and mercury via liquid effluents and leachate from slag. This plant is the primary source of metal inputs to the Columbia River in British Columbia (Ministry of Environment 1979; Smith 1987; Scheehan and Lamb 1987). Elevated concentrations of zinc, copper, lead, and cadmium in bridgelip sucker (*Catostomus columbianus*) and sediment have been found in the upper reaches of Lake Roosevelt (Hopkins et al. 1985).

In 1978 and 1980, cadmium and lead concentrations in fish at Grand Coulee (analyzed whole) were among the highest in the nation (Lowe et al. 1985). Sporadic water quality alerts (when USEPA Quality Criteria for Water 1986 are exceeded) have been issued by the U. S. Geological Survey because of zinc, copper, lead, cadmium, and mercury in water samples from the Columbia River near the international boundary. The threat to aquatic life and human health warrant long-term monitoring (Johnson et al. 1988). However, metals concentrations in edible tissue of walleye (*Stizostedion vitreum*), rainbow trout (*Oncorhynchus mykiss*), and other sport fish from Lake Roosevelt were within limits set by the U.S. Food and Drug Administration (USFDA) for fish marketed commercially. Canada has adopted an action level of 0.5 ppm for mercury, half the USFDA level, in sport-caught fish. The B. C. Ministry of Environment and Parks found levels between 0.5 and 1 ppm in a few walleye from the Columbia River above the border. As a result, they released an "information bulletin" in September 1987 suggesting that meals of Columbia River walleye weighing over two pounds should be eaten no more than once a week (Johnson et al. 1988).

Sublethal concentrations of toxicants may negatively affect fish behavior in the main-stem Columbia River. Fluoride at low levels (0.3–0.5 mg/L) in effluents from an aluminum plant once delayed upstream migrating salmonids at John Day Dam (Damkaer and Dey 1985). After fluoride discharges were reduced in 1983, delays no longer occurred.

Documentation of toxic chemical discharges in the Columbia River ecosystem is not comprehensive. Although safe limits do not appear to be exceeded, and probably have not been frequently exceeded in the past, comprehensive long-term monitoring is needed. Because fish bioaccumulate chemicals that pose significant risks to human health and terrestrial wildlife, the introduction of trace amounts of toxicants must be controlled.

Biota and Ecosystem Properties

Biological communities in impoundments on the Columbia River are influenced by the specific morphometry and limnology of each reservoir. The main stem has changed from lotic to primarily lentic with the recent linkage of storage and run-of-river reservoirs. Storage reservoirs provide some habitat stability because water is retained longer. However, large annual changes in water level of reservoirs are common and they disrupt production of biota in the littoral zone.

The flushing rate provides a measure of stability in each reservoir (Table 2). Near-field effects associated with inflow and outflow generally dominate circulation in run-of-river reservoirs. Far-field effects, such as thermal circulation and wind-generated currents, are more common in storage reservoirs and natural lakes (Baxter 1977). Run-of-river reservoirs do not often stratify. But thermal stratification has been documented in Roosevelt Lake, the large storage reservoir behind Grand Coulee Dam (Stober et al. 1977b; Stober et al. 1981a).

When water is not spilled from the surface of Columbia River reservoirs at dams, it is drawn from below the surface. If the reservoir is stratified, water withdrawn at the outlet level will induce internal currents with little vertical movement. Density currents may develop in storage reservoirs when the inflow water has a different temperature or sediment load, and when water is withdrawn deep at the dam. Density currents occur in Lake Roosevelt (Jaske 1966; Jaske and Snyder 1967) on the Columbia River and in Brownlee Reservoir (Raleigh and Ebel 1967) on the Snake River. Density currents of any magnitude probably do not develop in run-of-river reservoirs on the main-stem Columbia or Snake rivers. When present, stratification and density currents can stabilize epilimnetic zones, allowing increased biological production.

Aquatic Productivity

Development of phytoplankton populations in reservoirs is related to water retention time and nutrient availability. A maximum retention time of four days is not conducive to building up phytoplankton populations in Rufus Woods Reservoir, below Grand Coulee Dam (Erickson et al. 1977). Reduced primary production may be a feature of all Columbia River reservoirs that are run-of-river.

The Upper Arrow, Lower Arrow, and McNaughton reservoirs on the upper Columbia River in Canada are oligotrophic with dissolved oxygen near saturation at all depths. Nutrient levels are low, as is typical of oligotrophic lakes, and diatoms are the predominant phytoplankton. Thermal stratification in these lakes is limited and may not occur in most years (B. C. Research 1977).

Phytoplankton abundance in main-stem Columbia River reservoirs peaks in the spring or early summer (April–June), and is usually followed by a second bloom in the fall (September–October). Levels of chlorophyll *a* are minimum in winter.

Diatoms dominate phytoplankton communities in the main-stem Columbia River (Stober et al. 1979). Blue-green algae comprise over 50% of the phytoplankton in the lower Yakima River during July and August.

Zooplankton have been sampled at numerous locations along the main-stem Columbia River (Stober et al. 1979). The cladocerans *Bosmina longirostris* and *Daphnia* sp. and the copepods *Cyclops bicuspidatus* and *Diaptomus ashlandi* are the most abundant species. Cladocerans that live less than 20 days are more adapted to survival in reservoirs or rivers than copepods, which have lifespans over 150 days (Pederson 1974). Zooplankton densities along the main-stem Columbia River are generally low, except during the seasonal peak, which varies at different locations from June to September. Maximum densities of zooplankton range from 60,000/m^3 in Roosevelt Lake (Stober et al. 1979) to 25,195/m^3 in Rufus Woods Reservoir (Erickson et al. 1977), 12,000/m^3 in the Hanford Reach (Page and Neitzel 1976), and 12,500/m^3 in the lower Columbia River (Clark and Snyder 1970).

Zooplankton abundance in the Columbia River appears to be correlated with retention of water in reservoirs. Higher densities occur in Roosevelt Lake, which has an average retention time of 45 d, than in Rufus Woods Reservoir (just below), which has a maximum retention time of only 4 d (Erickson et al. 1977). Zooplankton densities are lowest in the riverine Hanford Reach and the lower Columbia River.

The benthic fauna in Columbia River reservoirs has received only limited attention. Oligochaeta and Chironomidae predominate in Lake Roosevelt (Earnest et al. 1966) and nearby Banks Lake (Stober et al. 1977a), and they are probably the most common benthos in other main-stem Columbia River reservoirs. Chironomids are most abundant from August to September, while oligochaetes maintain low but relatively constant numbers throughout the year. The density of bottom organisms in Roosevelt Lake is low because water levels alternate widely with season (Earnest et al. 1966).

Fish Populations

Sixty-one species of fish have been identified in the main-stem Columbia River from Bonneville Dam to Lake Roosevelt (Stober et al. 1979). The species fall into four categories: anadromous salmonids, cold-water game fish, warm-water game fish, and non-game fish.

Four species make up the bulk of the anadromous salmonid runs: chinook salmon (*Oncorhynchus tshawytscha*), sockeye salmon, coho salmon (*O. kisutch*), and steelhead trout (*O. mykiss*). The river's capacity to produce salmonids was greatest prior to 1930. Predevelopment estimates of salmon and steelhead runs in the Columbia River basin range from about 7.5 to 8.9 million fish annually (Chapman 1986). Current runs average about 2.5 million fish (NPPC 1986), indicating a basin-wide reduction of about 5 to 6.4 million fish each year. Chief Joseph Dam on the Columbia River and Hells Canyon Dam on the Snake River blocked return runs and eliminated all production of anadromous fish further upstream. Declines in salmonids runs have been greatest in the upper Columbia and Snake rivers because of habitat loss and mortalities of upstream and downstream migrants at the numerous dams.

Sixteen species of cold-water game fish occur in the main-stem Columbia River. Kokanee (freshwater race of sockeye salmon), cutthroat trout (*Oncorhynchus clarki*), and brown trout (*Salmo trutta*) generally reside only in Rufus Woods and Roosevelt reservoirs. Bull trout (*Salvelinus confluentus*) are somewhat more widespread, especially in tributaries entering the main stem from Roosevelt Lake downstream to Priest Rapids Dam. The American shad (*Alosa sapidissima*), an introduced anadromous species, now migrates upstream to spawn in reaches of the mid-Columbia and lower

Snake rivers. Rainbow trout, mountain whitefish (*Prosopium williamsoni*), and white sturgeon (*Acipenser transmontanus*) are the only cold-water gamefish species common throughout the main stem.

Dams have improved the survival, and most likely the habitat, for the introduced American shad. In 1956, 8,056 American shad were counted over Bonneville Dam but only four reached upstream as far as McNary Dam. The counts rapidly increased. In 1961, not only did 265,697 American shad pass Bonneville Dam but they penetrated further upstream. In 1984, the count was 1,274,737 American shad at Bonneville Dam, 269,733 at McNary Dam, and 25,188 at Priest Rapids Dam, located at RKm 625 (USACE 1984).

The twelve species of warm-water fish in the Columbia River are introduced. Species that occur throughout the main stem include walleye, yellow perch (*Perca flavescens*), and bass (*Micropterus* spp.). The crappies (*Pomoxis* spp.), bluegill (*Lepomis macrochirus*), pumpkinseed (*L. gibbosis*) and ictalurids (i.e., catfish) are common in the mid- and lower reaches of the Columbia River. A few warm-water exotics inhabit the flowing Hanford Reach and some predominate among reservoir fish. Some exotic species are predatory and, as they increased in numbers and spread, may have subjected native fish fauna to increased predation (Li et al. 1987).

Non-game species constitute about half of the fish in the Columbia River. The squawfish (*Ptyocheilus oregonensis*), peamouth (*Mylocheilus caurinus*), chiselmouth (*Acrocheilus alutaceus*), largescale sucker (*Catostomus macrocheilus*) and bridgelip sucker, redside shiner (*Richardsonius balteatus*), and the introduced carp (*Cyprinus carpio*) occur throughout the main stem. The exotic tench (*Tinca tinca*), and the native Pacific lamprey (*Entosphenus tridentatus*) and sandroller (*Percopsis transmontana*), occur in the mid- and lower portions of the main stem.

Warm-water game fish often hosted parasites when introduced (largely from the mideastern U.S.) but, because of host-specific requirements, they have had little effect on native fish. At least nineteen diseases are endemic to fish in the Columbia River, and they may be considered an ecosystem characteristic. Harmful diseases include ceratomyxosis, caused by the myxosporidian *Ceratomyxa shasta*; bacterial kidney disease, caused by *Renibacterium salmoninarum*; and infectious hematopoietic necrosis, caused by a rhabdovirus. Each disease is highly destructive among salmonids in hatcheries and difficult or impossible to treat with anti-microbial agents (Fryer 1986). Modern hatchery practices require strict controls to minimize outbreaks of fish disease.

The bacterial pathogen *Flexibacter columnaris* sometimes becomes epizootic among Columbia River fish (e.g., in 1970 and 1973), usually in association with unusually low flows and warm water temperatures (Becker and Fujihara 1978). Columnaris is readily transmitted to returning adult salmon in fish ladders at Columbia River dams (Fujihara and Hungate 1971). However, it is unlikely that columnaris has greatly influenced the decline of anadromous salmonid runs to the Columbia River (Gould and Wedemeyer 1981).

Reservoirs and Fish Production

Stability of water levels in Columbia River reservoirs influences aquatic productivity, which is generally highest in shoreline littoral zones. In storage reservoirs, the timing of annual drawdown in relation to the stage of fish living in the littoral zone can select what species will flourish. In the Columbia River system, kokanee that spawn among talus shoals in the fall have become less abundant in Banks Lake and Lake Roosevelt when the annual drawdown increased (Stober and Tyler 1982), but the number of walleye that spawn in the spring in Lake Roosevelt have increased (Stober et al. 1977b). Squawfish, which also spawn in reservoirs in the spring, remain abundant.

Lake Umatilla above John Day Dam is the only storage impoundment on the lower Columbia River. The pool elevation is reduced for flood control during April, but the drawdown is limited to 3.4 m. Populations of cold-water game fish in Lake Umatilla are affected less by drawdown than by reduced access to streams suitable for spawning. Warm-water gamefish and non-game fish generally spawn in the spring and summer, often in shoreline sloughs where water level fluctuations influence spawning success. The influence of dams on white sturgeon that spawn in the main-stem Columbia River is less certain.

Productivity of resident fishes in run-of-river reservoirs is influenced by short water retention times and frequent fluctuations in water level. These characteristics, combined with narrow, steep-sided basins and sloughing shorelines, can reduce fish productivity. Most run-of-river reservoirs have a flowing area at the upper end, a central area characteristic of both river and lake, and a lagoon area of quiet water at the dam forebay where the depth is greatest (Baxter 1977). Passage of juvenile and adult anadromous salmonids through these reservoirs and over dams exposes them to numerous hazards such as turbine mortalities, delayed migration, gas supersaturation, effects from regulated flows and temperatures, susceptibility to predators, and effects of power peaking operations.

Hanford Reach

The Hanford Reach, between the headwaters of Lake Wallula and Priest Rapids Dam, remains unimpounded and flowing. Benthic periphyton in this reach are predominantly diatoms. Dry weight of periphyton and total organic matter peak in September and bottom out in March. Chlorophyll *a* concentrations ranged from 68 mg/m^2 in March 1977 to 129 mg/m^2 in June 1978. Diversity values for microflora in the Hanford Reach are low (Page and Neitzel 1978; Page et al. 1979).

Benthic macroinvertebrates in the Hanford Reach consist largely of trichopterans (77%). Chironomids rarely form more than 25% of the community (exclusive of a freshwater sponge). Larvae of aquatic insects are abundant in the fall and winter and low in the summer. Diversity of macroinvertebrates is low, reflecting single dominance and lack of complexity (Wolf 1976; Page and Neitzel 1976, 1977, and 1978; Page et al. 1979).

The Hanford Reach is shallower and it contains a greater variety of habitats than do impounded sections of the main stem. Habitats include a cool and flowing channel, numerous gravel bars and islands, and sloughs and coves along the shoreline where water is warmed by insolation. Current velocities in one deep section of the Hanford Reach average 91 to 152 cm/s (PNL 1976). High water velocities favor salmonids and white sturgeon over suckers and centrarchids. The flow in the Hanford Reach allows spawning of fall chinook salmon, steelhead, and white sturgeon and, thus, is crucial for maintaining midriver stocks of these fish (Gray and Dauble 1977). Numbers of fall chinook salmon spawning in the Hanford Reach (Figure 8) have increased since 1947 (Becker 1985; Dauble and Watson 1990), a span when eight dams were constructed along the main-stem Columbia River.

Fish with varied diets appear to be successful in the Hanford Reach, especially those feeding on the abundant caddisflies. Although predam data are limited, it appears that more fish are produced in the Hanford Reach per unit area than in main-stem reservoirs. The challenge is to maximize the potential of reservoirs for producing sport and commercial fish.

Water Quality Control and Fishery Mitigation

Efforts to control water pollution in the Columbia River basin arise from the federal Clean Water Act of 1972. This legislation provided the impetus and backing for passage of water pollution control legislation in states. The Act authorized control of point source pollution under National Pollutant Discharge Elimination System (NPDES) permits, which eventually resulted in the rehabilitation of many miles of streams in the basin. In addition, the Act authorized state water pollution control agencies and the federal Environmental Protection Agency to enforce regulations pertaining to all existing and new point source discharges. Non-point source pollution in the Columbia River system, however, is still an unresolved issue.

Improving water quality is important for all fish in the Columbia River ecosystem, but most recent efforts have been to maintain or recover runs of anadromous

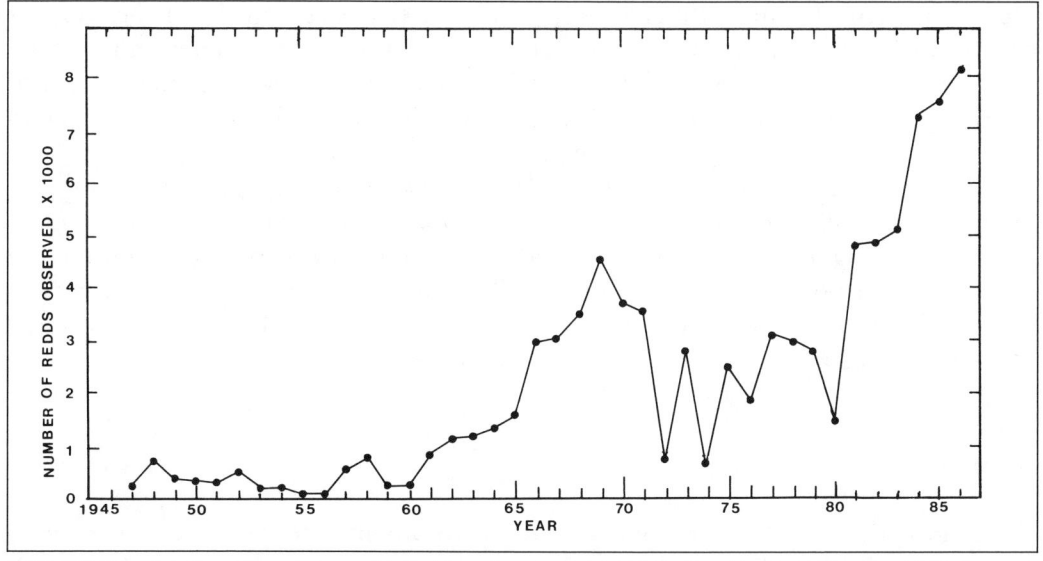

FIGURE 8. Relative abundance of fall chinook salmon redds in the Hanford Reach, 1947 through 1986. (Modified from Becker 1985 and Dauble and Watson 1990.)

salmon through mitigation of losses caused by hydroelectric projects. Three major types of mitigation have occurred: federal multipurpose mitigation programs, fish and wildlife program measures of the Northwest Power Planning Council (NPPC), and mitigation funded under Federal Energy Regulatory Commission (FERC) licenses for nonfederal hydropower projects.

In 1938, Congress passed the Mitchell Act, authorizing artificial propagation of anadromous salmonids to compensate for destruction of habitat by hydroelectric development and other environmental changes. The Act was amended in 1946 to allow the states to participate in stream improvement efforts (i.e., passage of migratory fish), screening of irrigation diversions, cleaning up stream debris, and research to improve the quality of hatchery fish). Since the early 1960's, treaty tribes have participated in decisions affecting management of the fishery through the Northwest Indian Fisheries Commission, the Columbia River Intertribal Fish Commission, and other liaison groups.

The federal government assumed fisheries management authority in 1977 with passage of the Fishery Conservation and Management Act. The Northwest Power Planning Act (NPPA) of 1980 and its fish and wildlife program represented another milestone, particularly in improving degraded habitat in the Columbia River. The program includes a "water budget" that sets aside $5,748 \times 10^3$ m^3 of water each year to increase flows and spills at main-stem dams from April 15 to June 15 in order to reduce turbine mortality of outmigrating smolts. Spill plans are drafted and implemented by fishery and tribal personnel working with public power utilities. These utilities and the Corps of Engineers have made substantial progress in mitigating or compensating for salmon and steelhead losses at their dams on the Columbia River.

This large effort does not place water quality high on the list of priorities for salmonid mitigation. Yet water quality is of basic importance. As long as the federal Clean Water Act is diligently enforced, we expect that water quality issues (except for gas supersaturation) will remain at relatively low priority for the foreseeable future in the Columbia River. Knowledge of toxic contaminant levels (aromatic hydrocarbons, organochlorine pesticides/PCB's, dioxin/furans, and trace metals) should be increased through long-term, systematic monitoring programs.

Strong conflicts have recently emerged between consumptive and nonconsumptive users of Columbia River water, particularly in regard to the water budget and maintaining instream flows in headwater tributaries during the summer to enhance production of young salmonids. The possibility of large-scale transfers to water deficient areas in Arizona and California as their population base continues to grow remains a threat.

Acknowledgments

We express our appreciation to John Yearsley, U. S. Environmental Protection Agency, Region X; Wesley J. Ebel, National Marine Fisheries Service, Seattle; Donald G. Watson and C. Dale Becker, Pacific Northwest Laboratories, Richland, Washington;

Lynwood S. Smith, School of Fisheries, University of Washington; James W. Mullan, U. S. Fish and Wildlife Service, Leavenworth, Washington; and Art F. Johnson and John Bernhardt, Washington Department of Ecology, for providing documents on important aspects of Columbia River water quality. We also thank D. W. Chapman, Don Chapman Associates, Boise, Idaho; and Hiram W. Li, Oregon Cooperative Fishery Research Unit, Oregon State University for their critical review of this manuscript.

Partial support was provided the senior author by USEPA Region IV. The junior author was supported by the Northwest College and University Association for Science (University of Washington under contract DE-AM06–76-RLO-RLO02225 with the U. S. Department of Energy).

References

Baxter, R. M. 1977. Environmental effects of dams and impoundments. Annual Review of Ecological Systematics 8:225–283.

B. C. Research. 1977. Limnology of Arrow, McNaughton, Upper Campbell, and Williston Lakes. Report for British Columbia Hydro and Power Authority, Vancouver, Canada.

Becker, C. D. 1985. Anadromous salmonids of the Hanford Reach, Columbia River: 1984 status. PNL-5371, Pacific Northwest Laboratories, Richland, Washington.

Becker, C. D. 1990. Aquatic bioenvironmental studies: the Hanford experience 1944–1984. Studies in Environmental Science 39. Elsevier Press, New York.

Becker, C. D., and M. P. Fujihara. 1978. The bacterial pathogen *Flexibacter columnaris* and its epizootiology among Columbia River fish. Monograph 2. American Fisheries Society, Washington, D. C.

Beiniingen, K. T., and W. J. Ebel. 1968. Effect of John Day Dam on dissolved nitrogen concentrations and salmon in the Columbia River. Transactions of the American Fisheries Society 99:664–671.

Bennett, D. H., P. M. Bratovick, W. Knox, D. Palmer, and H. Ansel. 1983. Status of the warmwater fishery and the potential of improving warmwater fishery habitat in lower Snake River reservoirs. Report to U. S. Army Corps of Engineers, Walla Walla District, Walla Walla, Washington.

Bouck, G. R., A. V. Nebeker, and D. G. Stevens. 1976. Mortality, salt water adaptation and reproduction of fish exposed to gas supersaturated water. U. S. Environmental Protection Agency, EPA-600/3-76-054, Corvallis, Oregon.

Chapman, D. W. 1986. Salmon and steelhead abundance in the Columbia River in the nineteenth century. Transactions of the American Fisheries Society 115:662–670.

Clark, S. M., and G. R. Snyder. 1970. Limnological study of lower Columbia River, 1967–68. U. S. Fish and Wildlife Service Special Scientific Report—Fisheries 610.

Collins, G. B. 1976. Effects of dams on Pacific salmon and steelhead. U. S. National Marine Fisheries Service, Marine Fisheries Review 38:39–46.

Collins, B., T. Dunne, and A. K. Lehre. 1982. Sediment influx to the Toutle River from erosion of tephra, May 1980-May 1981. Pages 82-91 *in* W. H. Funk, editor. Proceedings from Conference, Mt. St. Helens: Effects on water resources. Report No. 41 State of Washington Water Research Center, WSU-UW, Pullman, Washington.

CRWMG (Columbia River Water Management Group). 1979. Columbia River water management report for water year 1978. CRWMG, Portland, Oregon.

CRWMG (Columbia River Water Management Group). 1986. Columbia River water management report for water year 1985. CRWMG, Portland, Oregon.

Cummans, J. 1981. Mudflows resulting from the May 18, 1980, eruption of Mount St. Helens, Washington. U. S. Geological Survey Circular 850-B:16. USGS, Alexandria, Virginia.

Damkaer, D. M., and D. B. Dey. 1985. Effects of water-borne pollutants on salmon-passage at John Day Dam, Columbia River (1982–1984). Final Report to U. S. Corps of Engineers and U. S. National Marine Fisheries Service, Seattle, Washington.

Dauble, D. D., and D. G. Watson. 1990. Spawning and abundance of fall chinook salmon (*Oncorhynchus tshawytscha*) in the Hanford Reach of the Columbia River, 1948–1988. PNL-7289, Pacific Northwest Laboratory, Richland, Washington.

Dell, M. B., M. W. Erho, and B. D. Leman. 1975. Occurrence of gas bubble disease symptoms on fish in Mid-Columbia River reservoirs. Public Utility District of Grant, Douglas and Chelan Counties, Washington.

Dimick, R. E., and F. Merryfield. 1945. The fishes of the Willamette River system in relation to pollution. Oregon State College, Engineering Experiment Station, Corvallis, Oregon.

Earnest, D. E., M. H. Spence, R. W. Kiser, and W. D. Brunson. 1966. A survey of the fish populations, zooplankton, bottom fauna, and some physical characteristics of Roosevelt Lake. Report to Washington Department of Game, Olympia, Washington.

Ebel, W. J. 1969. Supersaturation of nitrogen in the Columbia River and its effects on salmon and steelhead trout. U. S. Fish Wildlife Service Fisheries Bulletin 68:1–9.

Ebel, W. J., C. D. Becker, J. W. Mullan, and H. L. Raymond. 1989. The Columbia River—toward a holistic understanding. Pages 205–219, *in* D. P. Dodge, editor. Proceedings of the International Large Rivers Symposium, Canadian Special Publication of Fisheries and Aquatic Sciences 106. Department of Fish and Oceans, Ottawa.

Ebel, W. J., E. M. Dawley, and B. H. Monk. 1971. Thermal tolerance of juvenile Pacific salmon and steelhead trout in relation to supersaturation of nitrogen gas. U. S. Fish Wildlife Service Fisheries Bulletin 69:833–843.

Ebel, W. J., and H. L. Raymond. 1976. Effect of atmospheric gas supersaturation on salmon and steelhead trout of the Snake and Columbia rivers. U. S. National Marine Fisheries Service, Marine Fisheries Review 38:1–14.

Ebel, W. J., G. K. Tanonaka, G. E. Monan, H. L. Raymond, and D. L. Park. 1979. Status report—1978; the Snake River salmon and steelhead crisis: its relation to dams and the national energy shortage. U. S. National Marine Fisheries Service, Northwest and Alaska Fisheries Center, Seattle, Washington.

Erickson, A. W., Q. J. Stober, J. J. Brueggeman, and R. L. Knight. 1977. An assessment of the impact on the wildlife and fisheries resources of Rufus Woods Reservoir expected from the raising of Chief Joseph Dam from 946 to 956 feet mean sea level. University of Washington, College of Fisheries, Seattle.

Fish, F. F., and R. R. Rucker. 1950. Pollution in the lower Columbia Basin in 1949 with particular reference to the Willamette Basin. U. S. Fish and Wildlife Service, Special Scientific Report—Fisheries 30.

Fish, F. F., and R. A. Wagner. 1950. Oxygen block in the main-stem Willamette River. U. S. Fish and Wildlife Service, Special Scientific Report—Fisheries 41.

Foster, R. F. 1972. The history of Hanford and its contribution of radionuclides to the Columbia River. Pages 3-18 *in* A. T. Pruter and D. L. Alverson, editors. The Columbia River estuary and adjacent ocean waters, bioenvironmental studies. University of Washington Press, Seattle.

Fryer, J. L. 1986. Epideminology and control of infectious diseases of salmonids in the Columbia River basin. Annual Report to U. S. Department of Energy, Bonneville Power Administration, Portland, Oregon.

Fujihara, M. P., and F. P. Hungate. 1971. *Chondrococcus columnaris* diseases of fishes: influence of Columbia River fish ladders. Journal of the Fisheries Research Board of Canada, 28:533–536.

Gould, R. W., and G. A. Wedemeyer. 1981. The role of fish disease in the decline of Columbia River anadromous salmonid populations. USFWS/NMFS Columbia River Endangered Species Project, National Fisheries Research Center, Seattle, Washington.

Gray, R. H., and D. D. Dauble. 1977. Checklist and relative abundance of fish species from the Hanford Reach of the Columbia River. Northwest Science 51:208–215.

Hickson, R. E., and F. W. Rodolf. 1957. History of the Columbia River jetties. Pages 283–298 *in* Proceedings of the first conference on coastal engineering. The Engineering Foundation, Council on Wave Research.

Hileman, J., R. Cunningham, and V. Kollias. 1975. Columbia River Nutrient Study—1972. Working Paper No. EPA-910-9-75-011, in cooperation with Washington State Department of Ecology and the Oregon State Department of Environmental Quality. EPA, Region X, Seattle, Washington.

Hopkins, B. S., D. K. Clark, M. Schlender, and M. Stinson. 1985. Basic water monitoring program, fish tissue and sediment sampling for 1984. Washington Department of Ecology 85-7, Olympia, Washington.

Hughes, R. M., and J. R. Gammon. 1987. Longitudinal changes in fish assemblages and water quality in the Willamette River, Oregon. Transactions of the American Fisheries Society 116:196–209.

Jaske, R. T., 1966. An evaluation of the use of selective discharges from Lake Roosevelt to cool the Columbia River. BNWL-208, Pacific Northwest Laboratories, Richland, Washington.

Jaske, R. T., and J. B. Goebel. 1967. Effects of dam construction on temperatures of the Columbia River. Journal of the American Water Works Association 59:935–942.

Jaske, R. T., and G. R. Snyder. 1967. Density flow regime of Franklin D. Roosevelt Lake. American Society of Civil Engineers, Sanitary Engineering Division, Journal 93:15–28.

Jaske, R. T., and M. O. Synoground. 1970. Effect of Hanford plant operation on the temperatures of the Columbia River 1964 to present. BNWL-1345, Pacific Northwest Laboratories, Richland, Washington.

Johnson, A., D. Norton, and D. Yake. 1986. Occurrence and significance of DDT compounds and other contaminants in fish, water and sediment from the Yakima River. Washington Department of Ecology, Water Investigations Section, 86-5, Olympia.

Johnson, A., D. Norton, and D. Yake. 1988. An assessment of metals contamination in Lake Roosevelt. Washington State Department of Ecology, Water Quality Investigations Section, Olympia. Segment No.: 26-00-04.

Jones, R. P., Jr., and E. O. Salo. 1986. The status of anadromous fish habitat in the north and south fork Toutle River watersheds, Mount St. Helens, Washington, 1984. Fisheries Research Institute, University of Washington, Project Completion Report, FRI-UW-8601, Seattle.

Li, H. W., C. B. Schreck, C. E. Bond, and E. Rexstad. 1987. Factors influencing changes in fish assemblages of Pacific Northwest streams. Pages 193–202 *in* W. J. Matthews and D. C. Heins, editors. Community and evolutionary ecology of North American stream fishes. University of Oklahoma Press, Norman.

Lloyd, D. S. 1987. Turbidity as a water quality standard for salmonid habitats in Alaska. North American Journal of Fisheries Management 7:34–45.

Lockett, J. B. 1962. Phenomena affecting improvement of the lower Columbia estuary and entrance. Pages 695–755 *in* Proceedings 8th conference on coastal engineering. The Engineering Foundation, Council on Wave Research.

Lowe, T. P., T. W. May, W. G. Brumbaugh, and D. A. Kane. 1985. National Contaminant Biomonitoring Program: concentrations of several elements in freshwater fish, 1978–1981. Archives of Environmental Contamination and Toxicology 14: 363–388.

Martin, D. J., L. J. Wasserman, R. P. Jones, and E. O. Salo. 1984. Effects of Mount St. Helens eruption on salmon populations and habitat in the Toutle River. Fisheries Research Institute, University of Washington, Technical Completion Report, FRI-UW-8412, Seattle.

Meekin, T. K., and T. L. Allen. 1974. Nitrogen saturation levels in the mid-Columbia River, 1965–1971. Washington Department of Fisheries, Technical Report 12:32–77, Olympia.
Ministry of Environment. 1979. Kootenay air and water quality study, Phase II, Victoria, British Columbia.

Mullan, J. W. 1986. Determinants of sockeye salmon abundance in the Columbia River, 1880's–1982: a review and synthesis. U. S. Fish and Wildlife Service, Biological Report 86(12).

Nakatani, R. E. 1969. Effects of heated discharges on anadromous fishes. Pages 294–317 *in* Peter A. Krenkel and Frank L. Parker, editors. Biological Aspects of Thermal Pollution, Vanderbilt University Press, Nashville, Tennessee.

Neal, V. T. 1972. Physical aspects of the Columbia River and its estuary. Pages 19-40 *in* A. T. Pruter and D. L. Alverson, editors. The Columbia River estuary and adjacent ocean waters, bioenvironmental studies. University of Washington Press, Seattle.

NPPC (Northwest Power Planning Council). 1986. Council staff compilation of information on salmon and steelhead losses in the Columbia River basin. Portland, Oregon.

Page, T. L. and D. A. Neitzel. 1976. Zooplankton: seasonal distribution, relative abundance and entrainment effects. Pages 3-1 to 3-20 *in* Final report on aquatic ecological studies conducted at the Hanford Generating Project, 1973–74. WPPSS Columbia River Ecology Studies, Volume 1. Pacific Northwest Laboratories, Richland, Washington.

Page, T. L., and D. A. Neitzel. 1977. Relative abundance of Columbia River macrobenthos near WNP-1, 2 and 4 in October and December 1975. Pages 4.1-4.9 *in* Aquatic ecological studies near WNP-1, 2, and 4 October 1975 through February 1976. WPPSS Columbia River Ecology Studies, Volume 3. Pacific Northwest Laboratories, Richland, Washington.

Page, T. L., and D. A. Neitzel. 1978. Columbia River benthic macrofauna and microflora near WNP-1, 2 and 4: January through Decemnber 977. Pages 4.1–4.36 *in* Aquatic ecological studies near WNP-1, 2 and 4, January through December 1977. WPPSS Columbia River Ecology Studies, Volume 5. Pacific Northwest Laboratories, Richland, Washington.

Page, T. L., D. A. Neitzel, and R. W. Hanf. 1979. Columbia River benthic macrofauna and microflora near WNP-1, 2 and 4: January through August 1978. Pages 4.1–4.27 *in* Aquatic ecological studies near WNP-1, 2 and 4 January through August 1978, WPPSS Columbia River Ecology Studies, Volume 6. Pacific Northwest Laboratories, Richland, Washington.

Patmont, C. R., G. J. Pelletier, M. E. Harper. 1985. Phosphorus attenuation in the Spokane River. Completion Report to State of Washington, Department of Ecology, Olympia.

Pederson, G. L. 1974. Plankton, secondary production and biomass; seasonality and relation to trophic state in three lakes. Doctoral dissertation. University of Washington, Seattle.

PNL (Pacific Northwest Laboratory). 1976. WPPSS Columbia River ecology studies, Volume 1. Final report to United Engineers and Constructors for Washington Public Power Supply System, Richland, Washington.

PNL (Pacific Northwest Laboratory). 1986. Environmental monitoring at Hanford for 1985. Report to U. S. Department of Energy, PNL-5817/UC-47 and UC-11, Richland, Washington.

PNL (Pacific Northwest Laboratory). 1987. Environmental monitoring at Hanford for 1986. Report to U. S. Department of Energy, PNL-6120/UC-41, UC-11, Richland, Washington.

PNRC (Pacific Northwest Regional Commission). 1979. Water today and tomorrow, volume 2. The Region. Pacific Northwest River Basins Commission, Vancouver, Washington.

Raleigh, R. F., and W. J. Ebel. 1967. Effect of Brownlee reservoir on migrations of anadromous salmonids. Pages 415–443 *in* Reservoir fishery resources symposium. American Fisheries Society, Southern Division, Reservoir Committee, Bethesda, Maryland.

Robeck, G. G., C. Henderson, and R. C. Palange. 1954. Water quality studies on the Columbia River. U. S. Department Health, Education and Welfare. Robert A. Taft Sanitary Engineering Center, Cincinnati, Ohio.

Scheehan, S. W., and M. Lamb. 1987. Water chemistry of the Columbia and Pend d'Oreille Rivers near the international boundary. Environment Canada, Vancouver, British Columbia.

Smith, A. L. 1987. Levels of metals and metallothionein in fish of the Columbia near the international boundary. British Columbia Ministry of Environment and Parks, and Environment Canada.

Smith, R. A., R. B. Alexander, and M. G. Wolman. 1987. Water quality trends in the nation's rivers. Science (Washington, D. C.) 235:1607–1615.

Soltero, R. A., D. G. Nichols, G. P. Burr, and L. R. Singleton. 1979. The effect of continuous advanced wastewater treatment by the City of Spokane on the trophic status of Long Lake, Washington. Eastern Washington University, Cheney, and State of Washington, Department of Ecology, Olympia.

Stober, Q. J., M. E. Kopache, and T. J. Jagielo. 1981a. The limnology of Lake Roosevelt. Final report FRI-UW-8106 to U. S. Fish and Wildlife Service. Fisheries Research Institute, University of Washington, Seattle.

Stober, Q. J., B. D. Ross, C. L. Melby, P. A. Dinnel, T. H. Jagielo, and E. O. Salo. 1981b. Effects of suspended volcanic sediment on coho and chinook salmon in the Toutle and Cowlitz rivers. Technical Completion Report FRI-UW-8124. Fisheries Research Institute, University of Washington, Seattle.

Stober, Q. J., et al. (seven coauthors). 1979. Columbia River irrigation withdrawal environmental review: Columbia River fishery study. Report FRI-UW-7919. Fisheries Research Institute, University of Washington, Seattle.

Stober, Q. J., and R. W. Tyler, 1982. Rule curves for irrigation drawdown and kokanee salmon (*Oncorhynchus nerka*) egg to fry survival. Fisheries Research (Amsterdam) 1:195–218.

Stober, Q. J., R. W. Tyler, J. A. Knutzen, D. Gaudet, C. E. Petrosky, and R. E. Nakatani. 1977a. Operational effects of irrigation and pumped storage on the ecology of Banks Lake, Washington. Final report to U. S. Bureau of Reclamation, FRI-UW-7732. Fisheries Research Institute, University of Washington, Seattle.

Stober, Q. J., R. W. Tyler, C. E. Petrosky, T. J. Carlson, D. Gaudet, and R. E. Nakatani. 1977b. Survey of fishery resources in the forebay of Franklin D. Roosevelt Reservoir, 1976–77. Report to U. S. Bureau of Reclamation, FRI-UW-7724. Fisheries Research Institute, University of Washington, Seattle.

Sylvester, R. O. 1958. Water quality studies in the Columbia Basin. U. S. Fish and Wildlife Service, Special Scientific Report — Fisheries 239.

Sylvester, R. O., and C. A. Rambow. 1967. Water quality of the State of Washington, Volume 4. Washington Water Research Center, Pullman, Washington.

Sylvester, R. O., and C. P. Ruggles. 1957. A water quality and biological study of the Wenatchee and middle Columbia rivers prior to dam construction. Department of Civil Engineering, University of Washington, Seattle.

Sylvester, R. O., and R. W. Seabloom. 1962. A study on the character and significance of irrigation return flows in the Yakima River basin. Department of Civil Engineering, University of Washington, Seattle.

Turton, D. J. 1982. Sediment transport in a small mountain stream draining a volcanic ash covered watershed in the Cascade foothills of southwestern Washington. Pages 378–394 *in* Proceedings from conference on Mt. St. Helens: effects on water resources. Report No. 41, Washington Water Research Center, WSU-UW, Pullman.

USACE (U. S. Army Corps of Engineers). 1984. Annual fish passage report. U. S. Army Engineer Districts, Portland and Walla Walla.

USBR (U. S. Bureau of Reclamation). 1976. Final environmental statement, proposed Columbia basin project, Washington. U. S. Bureau of Reclamation, Pacific Northwest Region, Boise, Idaho.

USEPA (U. S. Environmental Protection Agency). 1986. Quality Criteria for Water 1986. Office of Water Regulations and Standards, EPA 440/5-86-001. Washington, D.C.

USFDA (U. S. Food and Drug Administration). 1987. Action levels for poisonous or deleterious substances in human food and animal feed. Center for Food Safety and Applied Nutrition, Industry Programs Branch, Washington, D. C.

USGS (U. S. Geological Survey). 1976–1979. Water resources data for Washington, Volume 2, eastern Washington, water years 1975–1978, 4 reports. USGS, Tacoma, Washington.

USGS (U. S. Geological Survey). 1982–1984. Water resources data for Washington, Volume 2, eastern Washington, water years 1981–1983, 3 reports. USGS, Tacoma, Washington.

Van Winkle, W. 1914. Water Supply Paper, U. S. Geological Survey. National Technical Information Service, Springfield, Virginia.

WDOE (Washington Department of Ecology). 1982. Water quality standards for waters of the state of Washington. Washington Administrative Code, Chapter 173-303, Olympia.

WDF (Washington Department of Fisheries). 1974. Nitrogen supersaturation investigations in the mid-Columbia River. Technical Report 12. Washington Department of Fisheries, Olympia.

Whetten, J. T., J. C. Kelley, and L. G. Hanson. 1969. Characteristics of Columbia River sediment and sediment transport. Journal of Sedimentary Petrology 39:1149–1166.

Whitney, R. R., and S. T. White. 1984. Estimating losses caused by hydroelectric development and operation, and setting goals for the fish and wildlife program of the Northwest Power Planning Council. Completion report, Appendix A, University of Washington, School of Fisheries, Seattle.

Wolf, E. G. 1976. Characterization of the benthos community. Pages 4.1–4.23 *in* Final report on aquatic ecological studies conducted at the Hanford Generating Project, 1973–74. WPPSS Columbia River ecology studies, Volume 1. Pacific Northwest Laboratories, Richland, Washington.

Wydoski, R. S., and R. R. Whitney. 1979. Inland fishes of Washington. University of Washington Press, Seattle.

Water Quality and Water Management Sacramento-San Joaquin River System

T. R. MONGAN
Consulting Engineer, 84 Marin Avenue
Sausalito, California 94965, USA

B. J. MILLER
Consulting Engineer, P. O. Box 5995
Berkeley, California 94705, USA

ABSTRACT. *The Sacramento-San Joaquin watershed provides more than half of the surface water used in California. Morphology and hydrology of the watershed have been greatly modified over the last two hundred years. Major ecological shifts caused by introduction of foreign aquatic species have continued to the present, making it pointless to attempt to maintain a static ecosystem. There is controversy over water quality in the system, and over the amount of water that can be diverted for urban and agricultural use without compromising instream flow for fisheries, recreation, and other environmental uses. Key issues include drinking water quality, environmental effects of agricultural return flows, and the detailed relationship between stream flows and fish abundance. Voluntary transfers of water (water marketing) have been proposed to obtain the water necessary for California's growing urban needs from irrigated agriculture, and to circumvent the need for additional water storage projects. Increasing urbanization in the Central Valley will shift some water from agricultural to urban use within the watershed. However, more extensive transfers of water from agriculture to urban users must be carefully managed to avoid adverse social and economic consequences.*

The Sacramento and San Joaquin rivers drain the Central Valley of California and the surrounding mountains, joining in the Sacramento-San Joaquin Delta and emptying into the Pacific Ocean through San Francisco Bay. Water quality in these rivers is a key issue in a major controversy in California resource management. That controversy involves the appropriate balance between water provided for urban and agricultural use and water provided for instream environmental uses, such as fisheries and recreation.

From a water supply standpoint, the Sacramento-San Joaquin system is the most important river system in California. Further water development is precluded on the Klamath and Eel rivers on the north coast, because they are designated as Wild and Scenic Rivers. However, the Colorado River along the southeastern border of the state supplies an important fraction of the water used in southern California.

Runoff from the Sacramento-San Joaquin watershed averages about 26 million acre-feet/year (32×10^9 m^3/year). The Sacramento River alone carries more than 30% of the runoff of all the rivers in California. The Sacramento-San Joaquin watershed supplied 55% of the water used in the state in 1985, including 60% of the irrigation water and 40% of the drinking water. Water from the Sacramento-San Joaquin system is moved throughout the state in some of the world's largest man-made water systems (Figures 1 and 2), with 27 million acre-feet (33×10 m^3) of reservoir storage and about 2100 kilometers of aqueducts (CDWR 1983).

The Sacramento-San Joaquin watershed (including the Tulare Lake basin, which only rarely drains to the delta) was home to 4.4 million people (17% of California's population) in 1985. By 2010, the population is expected to reach 7.2 million, comprising 20% of the expected state population at that time (CDWR 1987).

Irrigated agriculture is the main industry in the Sacramento-San Joaquin watershed. The California Department of Water Resources (CDWR) estimates net agricultural water use in the watershed at about 20 million acre-feet/year (25×10^9 m^3/year). No large increases in agricultural water use are projected because substantial increases in irrigated acreage are not anticipated (CDWR 1987). The importance of

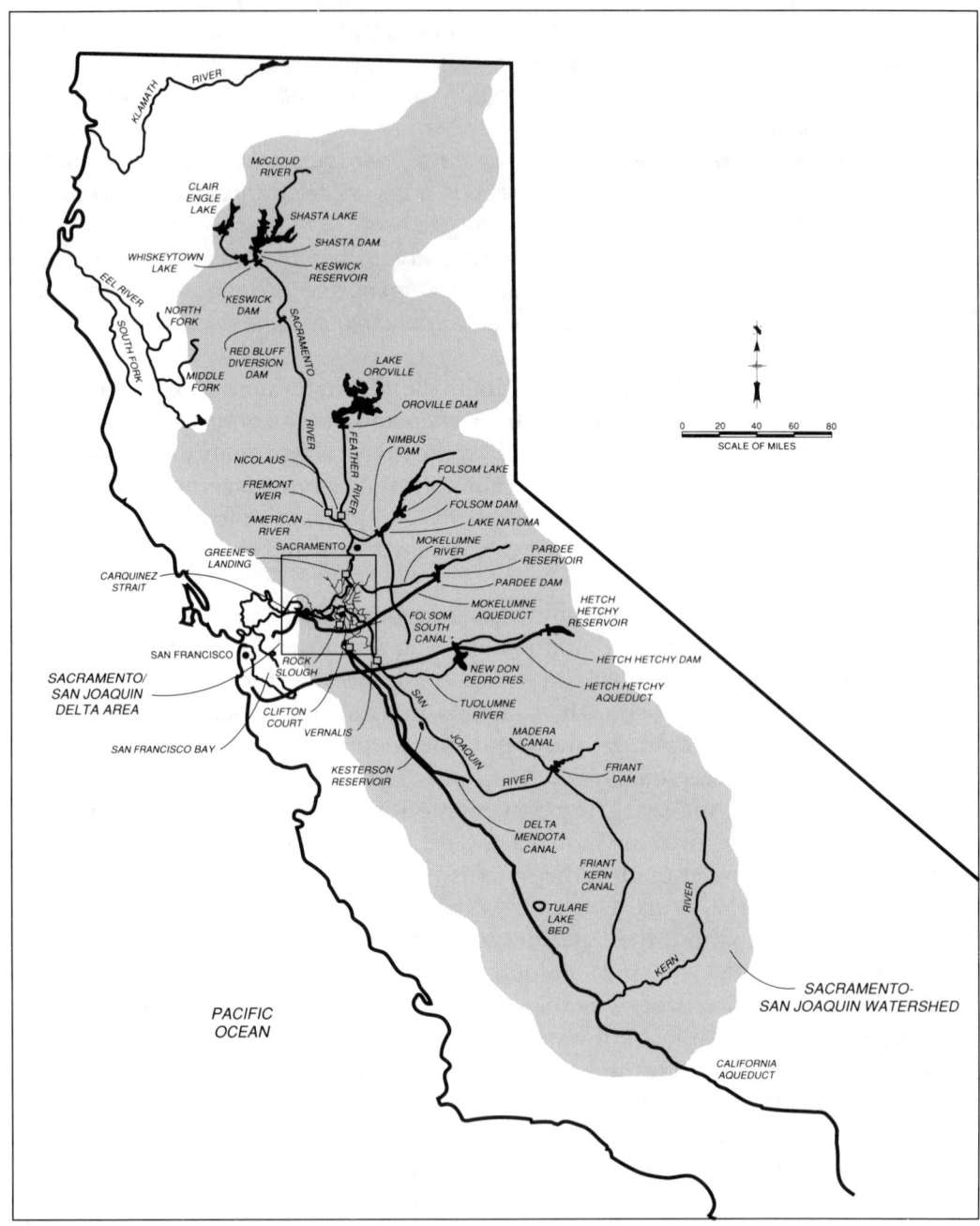

FIGURE 1. Key locations and facilities relevant to California water supply and use. The shaded area indicates the Sacramento-San Joaquin watershed and the internal drainage of the Tulare Lake basin, which only occasionally drains to the San Joaquin River.

FIGURE 2. Key locations and facilities in the Sacramento-San Joaquin Delta that are relevant to California water supply and use.

agriculture in California is indicated by the $16 billion annual value of California's agricultural products. Agriculture-related activities contribute about $64 billion annually to California's economy, and irrigation is likely to remain a large, but declining, fraction of California water use.

Important Sacramento-San Joaquin fishery resources include native chinook salmon *Oncorhynchus tshawytscha*; striped bass *Morone saxatilis*, introduced in 1879; and American shad *Alosa sapidissima*, also introduced in the late 1800's. Other important native fish include steelhead *Oncorhynchus mykiss*, white sturgeon *Acipenser transmontanus*, and green sturgeon *Acipenser medirostris*.

One of the four races of chinook salmon in the Sacramento-San Joaquin system, the winter run, has been declared threatened by the federal government and endangered by the state of California. The state is also considering the delta smelt *Hypomesus transpacificus* as a candidate for listing as an endangered species.

Morphology and Hydrology

Morphology and hydrology of the Sacramento-San Joaquin system have been greatly modified since European settlers arrived in California in the late 1700's. In fact, one of the few things that has not changed is the great variability of yearly and seasonal stream flows caused by variations in California's climate.

Morphological changes along the rivers began with construction of an extensive system of levees, converting the delta and the Central Valley lowlands from a vast marshland into some of the richest agricultural land in the world. The delta became a network of channels surrounding numerous diked islands. Dams, pumping plants, and canals were constructed to meet the urban and agricultural water needs of the state, creating many large artificial lakes in the process.

River flows under natural, undisturbed conditions prior to the arrival of Europeans cannot be estimated with certainty. It is unlikely that annual average discharge from the Sacramento-San Joaquin Delta to San Francisco Bay (delta outflow) was much greater than it is today. This is because evapotranspiration by the lush native riparian vegetation that lined the rivers and the extensive tule swamps in the delta and valley lowlands was probably roughly equivalent to the sum of pumped export and present in-basin urban and agricultural consumptive use. However, seasonal distribution of flows undoubtedly differed from today, because dual-purpose flood control and water supply projects now reduce winter and spring flows while releases from storage now increase late summer-early fall flows.

Water resource discussions in California have often been complicated by the assumption that CDWR's calculated "unimpaired" flows were a good estimate of flows under natural undisturbed conditions. In fact, "unimpaired" flows are computer simulations of the present highly developed and channelized system that assume no dams or export pumps are operating. These simulations provide an estimate of the total water available in the watershed for use in a given year. However, because losses that would occur to riparian and swamp vegetation under natural conditions are not included, unimpaired flow estimates are probably considerably higher than natural flow. In particular, unimpaired flow data cannot be used to support the conclusion that the average annual delta outflow under undisturbed natural conditions was substantially higher than it is today.

Reliable data on flows in the Sacramento-San Joaquin Delta have been available since 1921. Since that time, depletions have increased substantially (Figure 3). Depletions are the total amount of water removed from the system for export and in-basin consumptive use. Despite increasing depletions, detailed statistical analysis of the delta outflow data in Figure 3 demonstrates that the actual delta outflow has trended

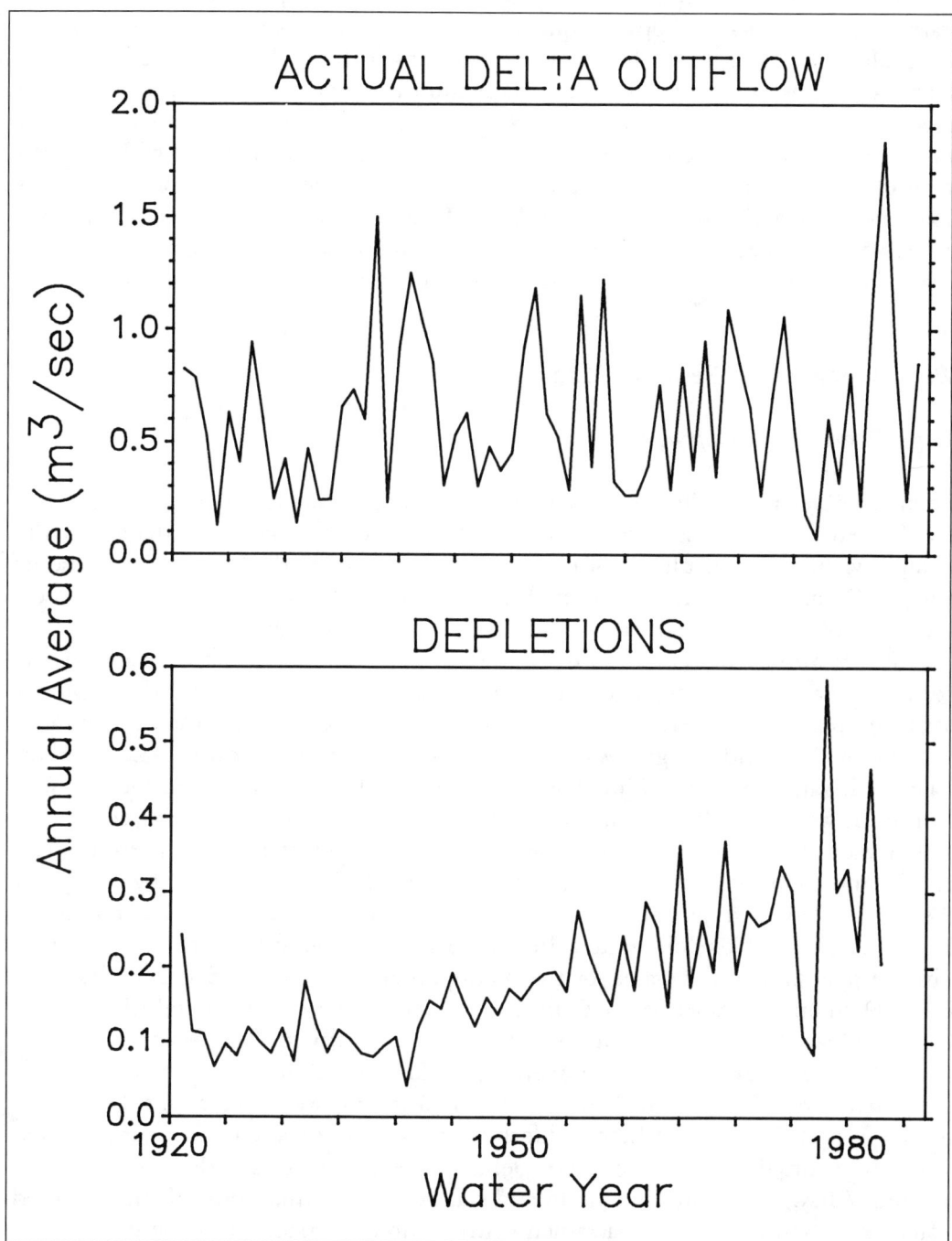

FIGURE 3. Annual average outflow from the Sacramento-San Joaquin Delta and depletions in Sacramento-San Joaquin River flows (Fox et al. 1990). Depletions are the total amount of water removed upstream and within the delta for export and consumptive use within the watershed.

slightly upward on an annual average basis, in part because of an increasing trend in precipitation in the watershed (Fox et al. 1989).

Water exports from the southern Sacramento-San Joaquin Delta through the State Water Project's California Aqueduct and the federal Central Valley Project's Delta-Mendota Canal have altered flow conditions in the delta. The export pumps draw water across the delta from the Sacramento River, and sometimes lead to reverse flows in the lower San Joaquin River and other delta channels. These cross-delta and reverse flows can interfere with fish migrations and exacerbate problems with entrainment of fish eggs and larvae through intake pumps. Figure 4 shows monthly exports from the delta and monthly flows at Jersey Point on the lower San Joaquin River since 1959. The negative Jersey Point flows on Figure 4 are the upstream "reverse flows," which are most likely to occur in June or July, especially in dry years.

Water Quality and Resource Management

Water Resource Development

Extensive development in the Sacramento-San Joaquin watershed began in the late 1800's. Levees were constructed to control floods and to reclaim the delta and the valley lowlands for agriculture. Stream beds were choked with debris from hydraulic mining. Dams were constructed for mining, flood control, and water supply purposes, and some of them blocked the migration routes of anadromous fish.

Water Supply Systems. Most of the precipitation in California falls in the northern part of the state, and much of the urban and agricultural development has taken place in southern California. Therefore, this century saw the development of some of the world's largest water supply systems in the Sacramento-San Joaquin watershed. Large dams, including Shasta, Oroville, Folsom and others in the northern part of the watershed (Figure 1), store water from winter storms and spring snowmelt. Stored water is later released for power generation, downstream use, prevention of salinity intrusion into the Sacramento-San Joaquin Delta, and export by pumps of the Central Valley Project and State Water Project in the southern delta (Figure 2). Big dams on the Mokelumne and Tuolumne rivers, tributaries of the San Joaquin River, supply part of the water for the metropolitan area around San Francisco Bay. Friant Dam on the upper San Joaquin River diverts water north through the Madera Canal and south down the Friant-Kern Canal to the Tulare Lake basin.

The Delta Link. The Sacramento-San Joaquin Delta is a key element in California's water system. It is the source of drinking water for about 17 million people in northern and southern California, and much of the state's irrigation water. Water is exported from the southern delta to the San Francisco Bay area, the San Joaquin Valley, the Tulare Lake basin and southern California. Most of this exported water passes south from the Sacramento River through channels in the delta.

Many levees in the Sacramento-San Joaquin system are constructed of poorly consolidated dredged materials. As the peat soils of the delta islands oxidize and erode, the delta islands subside and hydraulic head increases against the levees. The

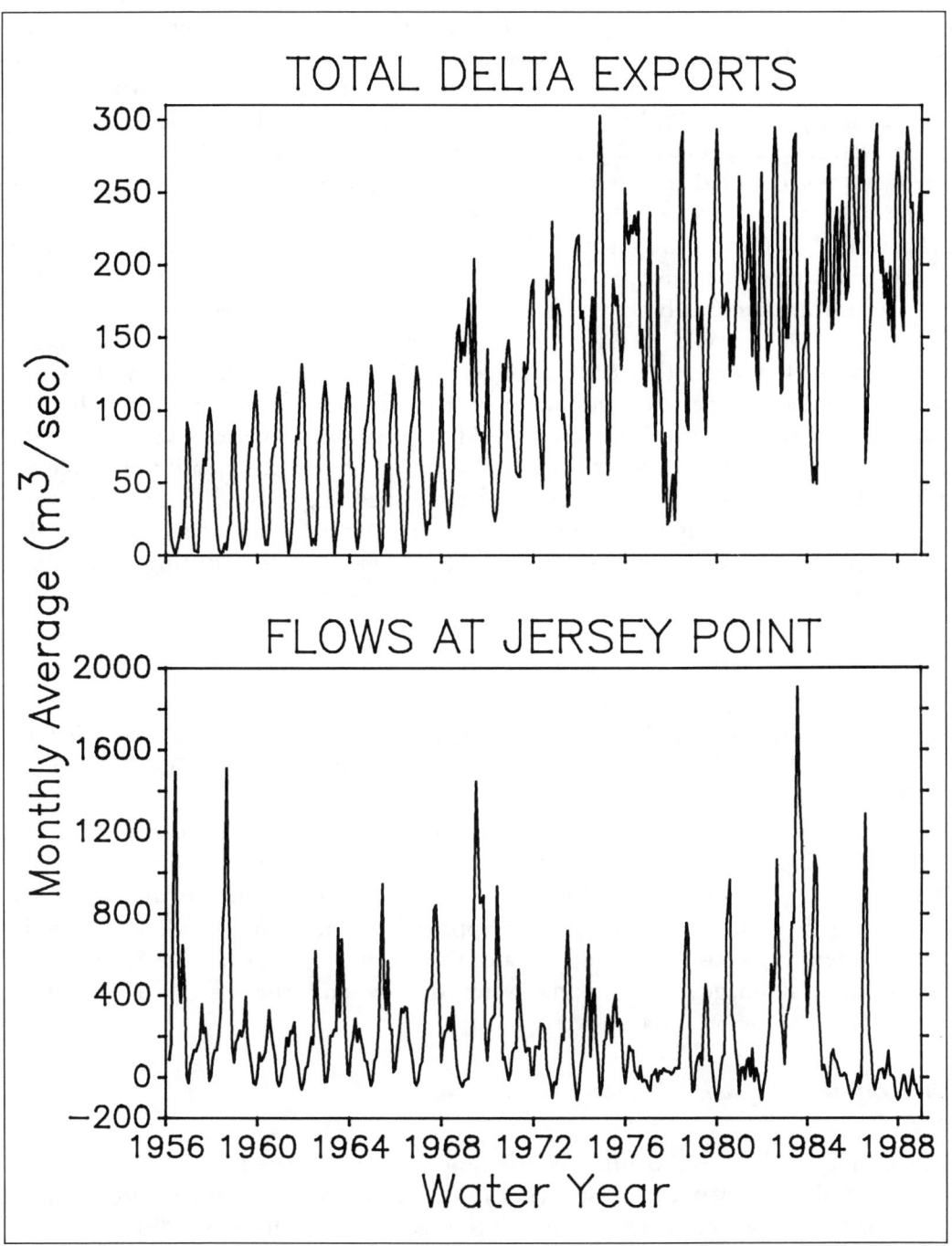

FIGURE 4. Total monthly exports from the Sacramento-San Joaquin Delta, and monthly flows at Jersey Point. Negative flows are reverse flows in the lower San Joaquin River (J. P. Fox, unpublished).

long-term trend of rising sea level also slowly increases hydraulic head against the levees.

In many places, levees in the Sacramento-San Joaquin Delta overlie sand lenses that could liquify in a major earthquake, leading to collapse of the levees. Work done for East Bay Municipal Utility District (EBMUD) indicates that an earthquake with a lateral acceleration of 0.15g (0.15 times the acceleration of gravity) would collapse many of the delta levees, and that such an earthquake should be expected on an average of once every 30 years. Although earthquake induced collapse of delta levees has not occurred in this century, the 1900's have been seismically quiescent in the delta as compared to the 1800's. Three large earthquakes occurred near the delta in the 1800's that might have caused extensive soil liquefaction and collapse of the levees present today.

An earthquake that broke levees surrounding islands in the western Sacramento-San Joaquin Delta would increase the tidal prism, which is the amount of brackish San Francisco Bay water that enters and leaves the delta in each tidal cycle. There is a delicate balance between freshwater outflow from the rivers and brackish tidal flows in the delta. So, until the levees could be repaired, increased river flow would be needed to prevent salt water from intruding to protect the quality of water for drinking and agriculture. Quick repair of the levees might be difficult if a major earthquake created competition for construction equipment and materials. So, the delta is physically a weak link in the long chain of facilities constituting California's water supply system.

Water Quality Overview

Present water quality in the Sacramento-San Joaquin drainage is exemplified by the data in Table 1, taken from the "Delta Drinking Water Quality Study" (CUWA 1989).

Historically, as upstream water use increased, salinity intrusion into the Sacramento-San Joaquin Delta became a growing problem in the 1920's and 1930's (Figure 5). Prevention of salinity intrusion during summers and low flow years was one of the main reasons for approval of the Central Valley project in 1933. Currently, the two major (and related) water quality issues in the Sacramento-San Joaquin drainage are drinking water quality and the effects of irrigation return flows on water quality.

Drinking Water Quality

Public concern about the quality of drinking water is reflected in a steady increase in the number of chemical constituents in drinking water that are subject to government regulation. For example, under the federal Safe Drinking Water Act, the U. S. Environmental Protection Agency (EPA) must regulate 83 drinking water contaminants by June 19, 1989, and an additional 25 contaminants every three years thereafter. Along with salinity, trihalomethanes and pesticides are focuses of concern.

TABLE 1. Water quality data from selected points in the Sacramento-San Joaquin system. The data are from CUWA (1989) and were collected at various times between 1975 and 1988. The following abbreviations are used: NTU = Nephelometric Turbidity Units, mF/L = Million Fibers per liter, MPN = Most Probable Number, and THMFP = Trihalomethane Formation Potential. THMFP-CDWR, determined by the CDWR method, indicates the *maximum* amounts of trihalomethanes that could be produced. THMGP-EBMUD, determined by the EBMUD method, reflects trihalomethane levels likely to be present in chlorinated water.

Constituents (units)	Number of samples	Range	Mean	Standard deviation
Feather River (at Nicolaus): annual data				
Total dissolved solids (mg/L)	135	40-92	63.00	10.00
Chloride (mg/L)	72	0-7	2.00	1.00
Sodium (mg/L)	72	2.7-7	4.00	0.90
Turbidity (NTU)	87	1-26	6.00	5.00
Lead (mg/L)	11	0-10	0.90	3.00
Selenium (mg/L)	11	0-10	4.00	5.00
Total organic carbon (mg/L) carbon (mg/L)	20	1-3.3	2.20	0.60
Total phosphorus (mg/L)	126	0-0.39	0.04	0.04
Total Kjeldahl nitrogen (mg/L)	133	0-0.54	0.19	0.09
Feather River (at Nicolaus): dry season (April-September)				
Total dissolved solids (mg/L)	70	45-91	61.00	9.00
Chloride (mg/L)	34	0-4	2.00	0.70
Sodium (mg/L)	34	2.7-6	3.70	0.80
Turbidity (NTU)	45	1-19	5.00	4.00
Lead (mg/L)	7	0-10	1.00	4.00
Selenium (mg/L)	6	0-10	4.00	5.00
Total organic carbon (mg/L)	17	1-3.3	1.90	0.70
Total phosphorus (mg/L)	62	0-0.39	0.04	0.05
Total Kjeldahl nitrogen (mg/L)	66	0-0.4	0.20	0.09
Feather River (at Nicolaus): wet season (October-March)				
Total dissolved solids (mg/L)	65	40-92	66.00	10.00
Chloride (mg/L)	38	0-7	2.00	2.00
Sodium (mg/L)	38	3-7	4.00	1.00
Turbidity (NTU)	42	1-26	8.00	7.00
Lead (mg/L)	4	0	0.00	0.00
Selenium (mg/L)	5	0-10	3.00	6.00
Total organic carbon (mg/L)	3	1.7-2.5	2.10	0.40
Total phosphorus (mg/L)	64	0-0.3	0.04	0.04
Total Kjeldahl nitrogen (mg/L)	67	0-0.54	0.19	0.10
Total dissolved solids (mg/L)	138	66-190	119.00	23.00
Chloride (mg/L)	124	2-24	7.00	3.00
Sodium (mg/L)	123	4.6-26	12.00	5.00
Turbidity (NTU)	67	0-230	24.00	34.00
Total organic carbon (mg/L)	14	1.7-4.7	2.80	0.80
Total phosphorus (mg/L)	131	0.03-0.54	0.10	0.07
Total Kjedahl nitrogen (mg/L)	128	0.1-1	0.30	0.10
Sacramento River (at Fremont Weir): annual data				
Total dissolved solids (mg/L)	138	66-190	119.00	23.00
Chloride (mg/L)	124	2-24	7.00	3.00
Sodium (mg/L)	123	4.6-26	12.00	5.00
Turbidity (NTU)	67	0-230	24.00	34.00
Total organic carbon (mg/L)	14	1.7-4.7	2.80	0.80
Total phosphorus (mg/L)	131	0.03-0.54	0.10	0.07
Total Kjeldahl nitrogen (mg/L)	128	0.1-1	0.30	0.10

TABLE 1. Continued.

Constituents (units)	Number of samples	Range	Mean	Standard deviation
Sacramento River (at Fremont Weir): dry season (April-September)				
Total dissolved solids (mg/L)	70	76-190	124.00	22.00
Chloride (mg/L)	62	2-16	7.00	2.00
Sodium (mg/L)	62	4.6-26	14.00	4.00
Turbidity (NTU)	35	2.5-48.5	17.00	12.00
Total organic carbon (mg/L)	14	1.7-4.7	2.80	0.80
Total phosphorus (mg/L)	66	0.04-0.24	0.09	0.04
Total Kjeldahl nitrogen (mg/L)	66	0.1-0.7	0.30	0.10
Sacramento River (at Fremont Weir): wet season (October-March)				
Total dissolved solids (mg/L)	68	66-184	113.00	23.00
Chloride (mg/L)	62	2-24	6.00	3.00
Sodium (mg/L)	61	5-26	11.00	4.00
Turbidity (NTU)	32	2-230	33.00	46.00
Total phosphorus (mg/L)	65	0.03-0.54	0.11	0.09
Total Kjeldahl nitrogen (mg/L)	62	0-0.1	0.30	0.10
American River (at Nimbus): annual data				
THMFP-CDWR (μg/L)	44	120-387	237.00	62.00
THMFP-EBMUD (μg/L)	42	41-100	58.00	15.00
Total dissolved solids (mg/L)	89	27-44	36.00	5.00
Bromide (mg/L)	41	0.004-0.06	0.03	0.02
Chloride (mg/L)	42	1-4	2.00	0.70
Sodium (mg/L)	29	1.4-2.7	2.20	0.31
Turbidity (NTU)	42	0.5-130	6.00	21.00
Asbestos (mF/L)	14	12-2200	343.00	608.00
Lead (mg/L)	16	0.001-0.005	<0.01	<0.01
Selenium (mg/L)	16	<0.002-<0.009	<0.01	<0.01
Total organic halides (mg/L)	24	4-500	71.00	107.00
Total organic carbon (mg/L)	24	1-8.3	3.40	1.90
Chlorophyll a (μg/L)	42	0.05-6	1.60	1.40
Total coliforms (MPN/100ml)	42	2-3500	484.00	819.00
Total phosphorus (mg/L)	31	0-0.06	0.02	0.01
Total Kjeldahl nitrogen (mg/L)	34	0.04-1.31	0.32	0.30
American River (at Nimbus): dry season (April-September)				
THMFP-CDWR (μg/L)	24	164-387	244.00	60.00
THMFP-EBMUD (μg/L)	20	41-80	55.00	10.00
Total dissolved solids (mg/L)	44	27-39	34.00	4.00
Bromide (mg/L)	19	0.01-0.05	0.03	0.02
Chloride (mg/L)	20	1-3	2.00	0.60
Sodium (mg/L)	14	1.4-2.7	2.10	0.40
Turbidity (NTU)	20	0.6-5	1.30	1.00
Asbestos (mF/L)	5	12-190	53.00	77.00
Lead (mg/L)	9	0.001-0.005	<0.01	<0.01
Selenium (mg/L)	9	<0.002-<0.009	<0.01	<0.01
Total organic halides (mg/L)	11	6-500	95.00	155.00
Total organic carbon (mg/L)	11	1.6-8.3	4.00	2.00
Chlorophyll a (μg/L)	20	0.05-6	1.40	1.30
Total coliforms (MPN/100ml)	20	2-1300	231.00	326.00
Total phosphorus (mg/L)	19	0.01-0.03	0.02	0.01
Total Kjeldahl nitrogen (mg/L)	19	0.04-1.31	0.40	0.40

TABLE 1. Continued.

Constituents (units)	Number of samples	Range	Mean	Standard deviation
American River (at Nimbus): wet season (October-March)				
THMFP-CDWR (μg/L)	20	120-375	229.00	65.00
THMFP-EBMUD (μg/L)	22	41-100	60.00	18.00
Total dissolved solids (mg/L)	45	27-44	38.00	5.00
Bromide (mg/L)	22	0.004-0.06	0.03	0.02
Chloride (mg/L)	22	1-4	2.00	0.80
Sodium (mg/L)	15	1.9-2.5	2.20	0.20
Turbidity (NTU)	22	0.5-130	10.00	29.00
Asbestos (mF/L)	9	70-2200	504.00	718.00
Lead (mg/L)	7	0.002-0.004	<0.01	<0.01
Selenium (mg/L)	7	<0.002-<0.005	<0.01	<0.01
Total organic halides (mg/L)	13	4-110	50.00	33.00
Total organic carbon (mg/L)	13	1-5.9	3.00	2.00
Chlorophyll a (μg/L)	22	0-6	2.00	1.60
Total coliforms (MPN/100ml)	22	49-3500	714.00	1048.00
Total phosphorus (mg/L)	12	0-0.06	0.02	0.01
Total Kjeldahl nitrogen (mg/L)	15	0.1-0.85	0.24	0.19
Sacramento River (at Greene's Landing): annual data				
THMFP-CDWR (μg/L)	57	131-1005	307.00	158.00
THMFP-EBMUD (μg/L)	42	55-230	84.00	27.00
Total dissolved solids (mg/L)	209	54-151	100.00	20.00
Bromide (mg/L)	4	0-0.05	0.02	0.03
Chloride (mg/L)	257	1.5-18	7.00	3.00
Sodium (mg/L)	76	3-18	11.00	3.00
Turbidity (NTU)	137	3-53	12.00	10.00
Asbestos (mF/L)	28	0-3200	512.00	684.00
Selenium (mg/L)	6	0-0.005	<0.01	<0.01
Total organic halides (mg/L)	6	0-310	76.00	121.00
Total organic carbon (mg/L)	13	1-14	6.40	5.00
Chlorophyll a (μg/L)	241	0-28	3.00	4.00
Total coliforms (MPN/100ml)	42	41-17000	2129.00	3631.00
Total phosphorus (mg/L)	170	0.05-0.26	0.12	0.04
Sacramento River (at Greene's Landing): dry season (April-September)				
THMFP-CDWR (μg/L)	32	131-712	305.00	142.00
THMFP-EBMUD (μg/L)	20	55-110	80.00	14.00
Total dissolved solids (mg/L)	111	60-143	101.00	20.00
Bromide (mg/L)	1	0	0.00	0.00
Chloride (mg/L)	136	2-15	7.00	3.00
Sodium (mg/L)	38	3-18	11.00	3.00
Turbidity (NTU)	75	5-36	10.00	7.00
Asbestos (mF/L)	13	0-680	272.00	166.00
Selenium (mg/L)	4	0-0.001	<0.01	<0.01
Total organic halides (mg/L)	1	0	0.00	0.00
Total organic carbon (mg/L)	6	1.8-14	9.50	5.00
Chlorophy a (μg/L)	141	0-28	4.00	4.00
Total coliforms (MPN/100ml)	20	41-7900	1301.00	1804.00
Total phosphorus (mg/L)	93	0.05-0.24	0.11	0.03

TABLE 1. Continued.

Constituents (units)	Number of samples	Range	Mean	Standard deviation
Sacramento River (at Greene's Landing): wet season (October-March)				
THMFP-CDWR (µg/L)	25	184-1005	311.00	178.00
THMFP-EBMUD (µg/L)	22	57-230	89.00	35.00
Total dissolved solids (mg/L)	98	54-151	100.00	20.00
Bromide (mg/L)	3	0-0.05	0.03	0.03
Chloride (mg/L)	121	2-18	7.00	3.00
Sodium (mg/L)	38	5-15	10.00	3.00
Turbidity (NTU)	62	3-54	13.00	12.00
Asbestos (mF/L)	15	0-3200	721.00	882.00
Selenium (mg/L)	2	0-0.001	<0.01	<0.01
Total organic halides (mg/L)	5	0-310	91.00	128.00
Total organic carbon (mg/L)	7	1-8.3	4.00	3.00
Chlorophyll a (µg/L)	100	0-23	3.00	3.00
Total coliforms (MPN/100ml)	22	70-17000	2883.00	4642.00
Total phosphorus (mg/L)	77	0.06-0.26	0.13	0.04
San Joaquin River (at Vernalis): annual data				
THMFP-CDWR (µg/L)	60	207-1476	502.00	211.00
Total dissolved solids (mg/L)	166	88-1014	390.00	245.00
Chloride (mg/L)	211	10-354	75.00	55.00
Sodium (mg/L)	46	11-100	54.00	24.00
Turbidity, (NTU)	104	3-75	20.00	8.00
Asbestos (mF/L)	14	270-3300	1153.00	760.00
Lead (mg/L)	1	0.01	0.01	0.00
Selenium (mg/L)	32	0-<0.003	<0.01	<0.01
Chlorophyll a (µg/L)	208	2-337	38.00	54.00
Total phosphorus (mg/L)	170	0.1-0.6	0.30	0.10
San Joaquin River (at Vernalis): dry season (April-September)				
THMFP-CDWR (µg/L)	29	267-773	504.00	123.00
Total dissolved solids (mg/L)	90	94-1014	418.00	237.00
Chloride (mg/L)	108	10-354	81.00	55.00
Sodium (mg/L)	23	15-89	53.00	23.00
Turbidity (NTU)	57	9-75	27.00	11.00
Asbestos (mF/L)	5	900-3300	1700.00	938.00
Selenium (mg/L)	14	0-<0.002	<0.01	<0.01
Chlorophyll a (µg/L)	126	5-337	51.00	63.00
Total phosphorus (mg/L)	93	0.1-0.5	0.30	0.09
San Joaquin River (at Vernalis): wet season (October-March)				
THMFP-CDWR (µg/L)	31	207-1476	501.00	271.00
Total dissolved solids (mg/L)	76	88-958	379.00	253.00
Chloride (mg/L)	203	11-286	73.00	54.00
Sodium (mg/L)	23	11-100	54.00	26.00
Turbidity (NTU)	47	3-34	16.00	4.00
Asbestos (mF/L)	9	270-1800	849.00	456.00
Lead (mg/L)	1	0.01	0.01	0.00
Selenium (mg/L)	18	0-<0.003	<0.01	<0.01
Chlorophyll a (µg/L)	82	2-220	26.00	56.00
Total phosphorus (mg/L)	77	0.1-0.6	0.30	0.10

TABLE 1. Continued.

Constituents (units)	Number of samples	Range	Mean	Standard deviation
Clifton Court Forebay: annual data				
THMFP-CDWR (μg/L)	46	174-910	497.00	142.00
THMFP-EBMUD (μg/L)	42	100-250	163.00	40.00
Total dissolved solids (mg/L)	201	80-713	237.00	140.00
Bromide (mg/L)	40	0.02-0.48	0.14	0.11
Chloride (mg/L)	247	11-307	64.00	64.00
Sodium (mg/L)	74	17-108	41.00	23.00
Turbidity (NTU)	86	4.7-35	12.70	6.40
Asbestos (mF/L)	14	230-960	591.00	209.00
Lead (mg/L)	16	0.001-0.018	<0.01	<0.01
Selenium (mg/L)	37	<0.001-<0.009	<0.01	<0.01
Total organic halides (mg/L)	19	10-750	94.00	148.00
Total organic carbon (mg/L)	24	2.8-16	8.00	4.00
Chlorophyll a (μg/L)	240	0-63	8.00	8.00
Total coliforms (MPN/100ml)	42	23-5000	652.00	985.00
Total phosphorus (mg/L)	163	0.03-0.29	0.13	0.04
Clifton Court Forebay: dry season (April-September)				
THMFP-CDWR (μg/L)	22	174-686	476.00	121.00
THMFP-EBMUD (μg/L)	20	100-230	158.00	31.00
Total dissolved solids (mg/L)	106	80-678	221.00	130.00
Bromide (mg/L)	20	0.02-0.4	0.20	0.10
Chloride (mg/L)	129	11-307	59.00	64.00
Sodium (mg/L)	36	18-79	35.00	18.00
Turbidity, (NTU)	42	4.9-35	15.60	7.20
Asbestos (mF/L)	5	500-960	720.00	196.00
Lead (mg/L)	9	0.001-0.018	<0.01	<0.01
Selenium (mg/L)	20	<0.001-<0.009	<0.01	<0.01
Total organic halides (mg/L)	10	12-180	62.00	51.00
Total organic carbon (mg/L)	11	2.8-16	9.00	4.00
Chlorophyll a (μg/L)	140	3-63	11.00	10.00
Total coliforms (MPN/100ml)	20	33-2800	518.00	826.00
Total phosphorus (mg/L)	90	0.03-0.21	0.12	0.03
Clifton Court Forebay: wet season (October-March)				
THMFP-CDWR (μg/L)	24	339-910	517.00	157.00
THMFP-EBMUD (μg/L)	22	110-250	168.00	47.00
Total dissolved solids (mg/L)	95	98-713	253.00	148.00
Bromide (mg/L)	20	0.04-0.48	0.13	0.10
Chloride (mg/L)	118	16-300	67.00	65.00
Sodium (mg/L)	38	17-108	46.00	26.00
Turbidity (NTU)	44	4.7-23.5	10.20	4.50
Asbestos (mF/L)	9	230-910	519.00	188.00
Lead (mg/L)	7	0.001-0.005	<0.01	<0.01
Selenium (mg/L)	17	<0.001-<0.005	<0.01	<0.01
Total organic halides (mg/L)	9	10-750	121.00	195.00
Total organic carbon (mg/L)	13	3-16	7.00	4.00
Chlorophyll a (μg/L)	100	0-22	4.00	4.00
Total coliforms (MPN/100ml)	22	23-5000	780.00	1224.00
Total phosphorus (mg/L)	73	0.08-0.29	0.13	0.04

TABLE 1. *Continued.*

Constituents (units)	Number of samples	Range	Mean	Standard deviation
Rock Slough: annual data				
THMFP-CDWR (μg/L)	59	225-775	479.00	126.00
Total dissolved solids (mg/L)	56	111-706	244.00	150.00
Chloride (mg/L)	53	13-277	61.00	69.00
Sodium (mg/L)	56	13-154	43.00	37.00
Turbidity (NTU)	56	3-22	11.00	5.00
Asbestos (mF/L)	16	140-1100	568.00	227.00
Selenium (mg/L)	16	0-0.001	0.00	<0.01
Rock Slough: dry season (April-September)				
THMFP-CDWR (μg/L)	29	225-638	458.00	102.00
Total dissolved solids (mg/L)	26	111-574	212.00	117.00
Chloride (mg/L)	23	14-210	45.00	52.00
Sodium (mg/L)	26	15-125	36.00	29.00
Turbidity (NTU)	26	5-22	13.00	5.00
Asbestos (mF/L)	5	140-1100	602.00	357.00
Selenium (mg/L)	7	0	0.00	0.00
Rock Slough: wet season (October-March)				
THMFP-CDWR (μg/L)	30	256-775	498.00	145.00
Total dissolved solids (mg/L)	30	122-706	272.00	170.00
Chloride (mg/L)	30	13-277	73.00	78.00
Sodium (mg/L)	30	13-154	49.00	42.00
Turbidity, (NTU)	30	3-18	10.00	4.00
Asbestos (mF/L)	11	260-950	553.00	160.00
Selenium (mg/L)	9	0-0.001	<0.01	<0.01
Mokelumne River: annual data				
THMFP-CDWR (μg/L)	18	115-425	251.00	69.00
Total dissolved solids (mg/L)	18	34-44	40.00	4.00
Chloride (mg/L)	18	1-2	1.00	0.20
Sodium (mg/L)	18	1-2	2.00	0.20
Turbidity, (NTU)	18	1-9	3.00	2.00
Asbestos (mF/L)	11	10-200	56.00	66.00
Mokelumne River: dry season (April-September)				
THMFP-CDWR (μg/L)	9	204-425	280.00	73.00
Total dissolved solids (mg/L)	9	35-44	41.00	4.00
Chloride (mg/L)	9	1	1.00	0.00
Sodium (mg/L)	9	2	2.00	0.00
Turbidity, (NTU)	9	1-3	2.00	0.70
Asbestos (mF/L)	4	10-53	30.00	22.00
Mokelumne River: wet season (October-March)				
THMFP-CDWR (μg/L)	9	115-295	222.00	53.00
Total dissolved solids (mg/L)	9	34-43	40.00	3.50
Chloride (mg/L)	9	1-2	1.00	0.30
Sodium (mg/L)	9	1-2	2.00	0.30
Turbidity (NTU)	9	2-9	5.00	2.00
Asbestos (mF/L)	7	17-200	71.00	79.00

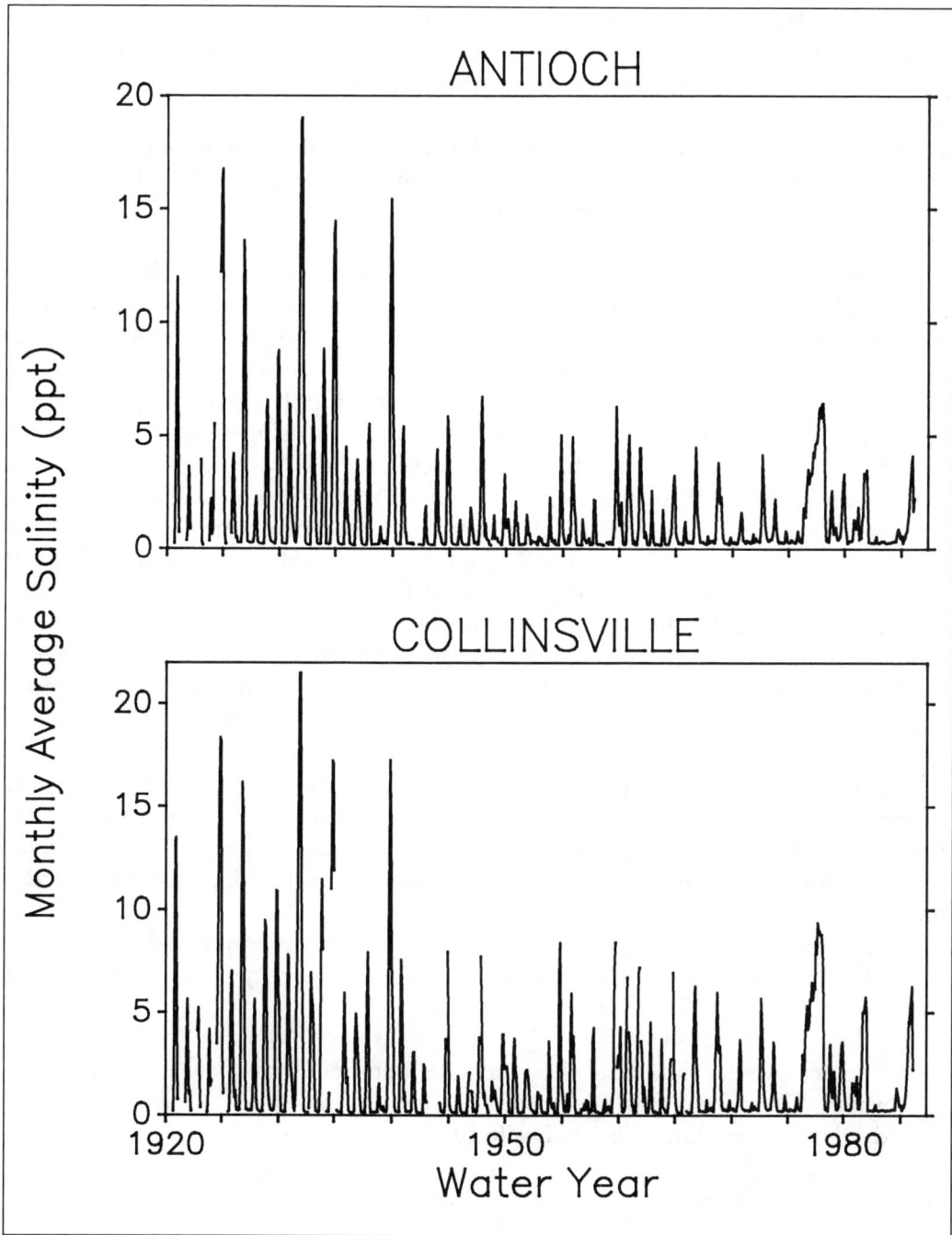

FIGURE 5. Monthly average surface salinity at Antioch and Collinsville (Fox et al. 1990).

Trihalomethanes and MX. Trihalomethanes are carcinogenic compounds formed when water containing common naturally-occurring organic chemicals, such as fulvic and humic acids, is disinfected with chlorine. Four species of trihalomethanes occur in drinking water: chloroform, bromodichloromethane, dibromochloromethane, and bromoform. There is some evidence that brominated forms are more carcinogenic than chloroform forms. Brominated trihalomethanes occur in treated water of the Sacramento-San Joaquin Delta because salinity intrusion carries bromide ions from sea water. Agricultural return flows from delta islands with peat soils significantly increase the potential for trihalomethane formation in water that passes through the channels to export pumps in the southern delta. Water from the Clifton Court forebay supplies drinking water to southern California through the California Aqueduct.

The potential for trihalomethane formation at Clifton Court, as determined by an EBMUD method reflecting trihalomethane levels expected in water disinfected with chlorine alone, averages about 160 $\mu g/L$ (CUWA 1989). The EPA Maximum Contaminant Level for total trihalomethanes in treated drinking water is 100 $\mu g/L$ in finished drinking water supplies. Municipal water supply agencies drawing water from the delta are currently meeting the THM standard by using disinfection processes that do not rely exclusively on chlorine. However, EPA is considering lowering the THM levels to 60 $\mu g/L$ or less, and it is not clear that this new standard can be met using delta water and any feasible combination of treatments.

Recently a compound called MX has been identified in chlorinated drinking water taken from sources containing trihalomethane precursors. Based on Ames mutagenicity tests, MX [3-chloro-4-(dichloromethyl)-5-hydroxy-2(5H)-furanone] is the strongest mutagen commonly existing in chlorinated surface water supplies. MX also has a geometric isomer called E-MX [E-2-chloro-3(dichloromethyl)-4-oxo-butenoic acid], which is about one-tenth as mutagenic. The discovery of MX and E-MX highlights the concern over trihalomethane precursors in drinking water supplies.

The high cost of shifting to other forms of drinking water disinfection (such as ozonation), and the possible adverse public health consequences of giving up the benefits of residual chlorine protection at the point of use, suggest that ways should be investigated to minimize trihalomethane precursors in drinking water drawn from the delta.

Pesticides. Responding to concerns about pesticide residues in drinking water drawn from the Sacramento-San Joaquin Delta, the Interagency Delta Health Aspects Monitoring Program (IDHAMP) reported that:

> "Pesticides and industrial chemicals are detected infrequently in delta water. When detected, concentrations are very low and do not exceed drinking water standards" (CDWR 1989).

Agricultural Return Flows

In the past, concerns over the effects of agricultural return flows on water quality centered on increases in total dissolved solids. Now, in addition to trihalomethanes and pesticide residues, the issue also involves selenium concentrations and possible adverse effects of warm irrigation return flows on salmon.

Selenium. Selenium is necessary to the health of most animals and plants in small amounts, but toxic in larger amounts. Soils in some areas in the western San

Joaquin Valley are rich in selenium. Organisms in and around Kesterson Reservoir concentrate the selenium carried in drainage water from the western San Joaquin lands, leading to a severe adverse effect on waterfowl that has become a national issue. Although delivery of irrigation return flows to Kesterson Reservoir has been stopped and extensive mitigation measures have been instituted, drainage from selenium-rich areas in parts of the western San Joaquin Valley eventually reaches the San Joaquin River. This created fears of selenium contamination in the river and downstream in the Sacramento-San Joaquin Delta and San Francisco Bay. However, IDHAMP says that:

> "Selenium is barely measurable in delta export water. The San Joaquin River does contain measurable amounts of selenium, but it does not exceed drinking water standards (10 $\mu g/L$)" (CDWR 1989).

Adverse environmental effects of selenium have not been found in the delta or in the San Joaquin River. Elevated levels of selenium in biota (but no adverse effects) have been found in parts of San Francisco Bay. Elevated levels of selenium in biota of the bay may result from local sources around the south bay and oil refinery discharges near Carquinez Strait, rather than from sources in the San Joaquin Valley.

Water Temperature. A warming trend has occurred in the Sacramento River. Juvenile salmon reared in the cooler upper Sacramento River and its tributaries may encounter warmer water temperatures when they emigrate downstream to the ocean, which may adversely affect their survival. It is not clear how much warming is caused by irrigation return flows and how much is caused by other human activities.

Water Allocation Issues

The key issue faced by water resource management in the Sacramento-San Joaquin watershed, and in California as a whole, is the allocation of water between urban use, agricultural use, and instream use for fisheries, recreation, and other environmental purposes. Most people wanting more water for instream use see obtaining the additional water by constructing more water development facilities as a contradiction. Water conservation, reclamation, and reuse may delay the immediate need for additional water development projects. However, without additional water development, more water for instream use must eventually be obtained by redirecting it from urban or agricultural use.

The demand for water to supply about 9 million acres of irrigated cropland is so large (CDWR 1987) that California agriculture "consumes" about five and a half times as much water as do the state's urban areas. Three facts have generated political pressure to shift water from agriculture to urban and instream uses: agriculture uses a lot of water, the political power of urban areas make the growing need for more waterl difficult to deny, and environmental issues are of increasing concern.

As the Central Valley becomes more urbanized, some permanent shifts of water from agriculture to urban use will take place. Mutually acceptable shifts of water from agriculture to urban use, particularly in dry years, are also likely to continue. Proponents of more extensive transfers of agricultural water see the transfers as a way to meet growing urban water needs while leaving more water in the streams for aquatic biota without constructing additional dams and water development facilities.

Pressures to decrease agricultural water use focus on problems created by irrigation return flows, the pricing of irrigation water, and the costs of crop support programs. People proposing extensive redirection of water from agricultural use to urban and environmental uses claim the redirection can be accomplished with little social disruption by expanded water marketing. Water marketing involves selling agricultural water to urban or instream users willing to pay higher prices for the water. Factors affecting the feasibility of more extensive sales of agricultural water to urban users include the strategic and economic importance of California agriculture to the nation and the state, the likelihood of adverse socio-economic side effects on agricultural communities, the availability of facilities to carry out water transfers, and the legal and institutional constraints inhibiting free marketing of water in California. In any case, water marketing promises to be a major political issue in California in years to come.

Biota and Ecosystem Properties

Overview

The most striking thing about the biota and ecosystem of the Sacramento-San Joaquin watershed is the extensive modifications that have taken place in the last 200 years. This is particularly evident in the Sacramento-San Joaquin Delta, where diking of the islands created an ecosystem totally unlike the original swampy marshland encountered by the early Spanish. Flow regimes are different, seasonal salinity regimes are different, and many introduced species (including striped bass) now inhabit delta waters. The terrestrial ecology of the Sacramento-San Joaquin watershed has been modified by removal of much of the native tule, wetland, and riparian vegetation on the valley floor, and by displacement of native plants and animals through introducing exotic plants and animals characteristic of farming, stock raising, and urban living. These extensive modifications make it impossible to return the present Sacramento-San Joaquin watershed to the undisturbed ecosystem that existed before the early Spanish arrived. Instead, Californians must work to keep the existing, highly-modified ecosystem in good condition.

Public concerns over biological resources in the Sacramento-San Joaquin watershed center on species of recreational and economic importance, such as salmon, striped bass, and waterfowl. These key species must be considered in terms of the changed ecosystem they now inhabit.

Effects of Recent Species Introductions

Within the last twenty years, major changes have taken place in the zooplankton, phytoplankton, and fish populations of the Sacramento-San Joaquin Delta. The consequences of these extensive changes on the delta food web are not clear, but they may be partly responsible for recent declines of fish populations, including the striped bass. Abundance of the copepod *Eurytemora affinis*, the preferred food of striped bass larvae at the critical time when they begin to feed, has declined since the early 1970's. An Asian copepod *Sinocalanus doerii* (introduced in 1968) has largely displaced *Eurytemora* in delta waters. *Sinocalanus* is not a preferred food for juvenile bass, perhaps because its escape behavior is more efficient than that of *Eurytemora*, making it more difficult for young bass to catch. A more recently introduced copepod of the genus *Pseudodiaptomous* is becoming abundant, but the effects of this change are yet uncertain. Algal blooms in the delta have been dominated by *Melosira granulata* since the 1976-1977 drought, and *Melosira* does not seem to be easily utilized as food by delta zooplankton. The yellowfin goby, *Acanthogobius flavimanus*, a voracious feeder on invertebrates and small fish, was accidentally introduced in the early 1960s and is now abundant in the Sacramento-San Joaquin estuary.

White bass (*Morone chrysops*) were found in Lake Kaweah in the Tulare Lake basin in 1977. Adult white bass are fish predators that could impact salmon and striped bass populations if they became established in the Sacramento-San Joaquin Delta. Pumping floodwaters from Tulare Lake to the San Joaquin River in 1983 threatened to introduce white bass to the Sacramento-San Joaquin system. A white bass eradication program, begun in 1987, is aimed at defusing this threat.

A major ecological change is also taking place in San Francisco Bay, the lower part of the Sacramento-San Joaquin estuary. This change involves a tremendous increase in the bay population of an Asian clam *Potamocorbula amurensis*, which was introduced in 1986. Since then, this clam has become the dominant benthic organism in large parts of northern San Francisco Bay, with population densities as high as 25,000 per square meter (Schemel 1989), and it is spreading fast. The consequences of this ecological change for fisheries, and for the delta ecosystem immediately upstream from San Francisco Bay, are uncertain. *Potamocorbula* is an efficient filter feeder and tolerates a very wide range of salinity. The new clam may be depleting the zooplankton that support larval striped bass in San Francisco Bay, and this could eventualy be detrimental to the fishery for adult bass.

Effects of Physical Modifications

From a basin-wide perspective, the extensive physical modifications of the Sacramento-San Joaquin watershed have had far-reaching biological consequences. Modified flow regimes have caused problems for estuarine resident and anadromous fish. Salmon runs in the San Joaquin River have been seriously depleted. Filling of tidelands in the estuary, disposal of dredge spoils, removal of riparian vegetation, erection of diversion dams, and projects to protect stream banks have had major effects on biota. However, dams and reservoirs have created extensively-used recreation and warm-water fishing opportunities.

Lawsuits have been started to preserve the environmental conditions resulting from long-standing physical modifications of the ecosystem. For example, during the last thirty years, releases from Folsom and Nimbus dams (Figure 1) resulted in average summer flows in the lower American River that are six times larger than during pre-dam days. This enabled development of a recreational boating industry and a linear park, the American River Parkway, along the 37 km stretch of the lower American River that runs through the heart of the Sacramento area, between Nimbus Dam and the Sacramento River. EBMUD had contracted with the Bureau of Reclamation, in advance of their need for the water, to divert flow from the American River, via Nimbus Dam and Folsom South Canal, for municipal use in counties just east of San Francisco Bay. The Environmental Defense Fund, Sacramento County, and others unsuccessfully attempted to block EBMUD's diversion, claiming that the instream uses developed in the interim in the American River Parkway would be harmed. EDF et al. based their case on an evolving legal doctrine that requires protection of instream uses as part of the public trust. Water quality issues played a big role in the case, because EBMUD stressed that prudent public policy required them to obtain drinking water from the best available source, and that water quality in the American River is better than water quality at alternative diversion points below the confluence of the American and Sacramento rivers.

Salmon Problems

The Salmon Resource. Since 1957, when estuarine gill-net fishing was outlawed, the ocean commercial troll harvest of Sacramento-San Joaquin chinook salmon has averaged about 324,000 fish annually, constituting about 57% of California's total catch. The ocean recreational catch of chinook salmon from the Sacramento-San Joaquin system averages about 60,000 fish each year. Estimates for the average inland sport harvest range between 6,000 and 35,000 fish. The Sacramento River and its tributaries contribute about 80% of the salmon produced in the watershed (CSWRCB 1988).

The annual salmon escapement to the Sacramento-San Joaquin basin and the commercial and ocean sport catch have been relatively stable over the last 20 years (Figure 6), partly because of an extensive hatchery program. CDWR estimates that, between 1978 and 1984, hatcheries on the Feather and American rivers produced about 48% of the total spawning escapement to the Sacramento-San Joaquin watershed and about 44% of the ocean harvest of chinook salmon from the Central Valley. Trucking immature salmon from the hatcheries on the Feather and American rivers downstream to release points in San Francisco Bay has helped. Survival rates for trucked fish are six to eight times as high as for hatchery fish that migrate down the Sacramento River and through the delta because trucked fish avoid many hazards of the out-migration.

Decline of River-Produced Salmon. The abundance of river-produced salmon seems to have decreased about 50% in the last 20 years (CSWRCB 1988). Possible factors in the decline of river-produced salmon include dams and upstream development that eliminated much salmon spawning and rearing habitat, and reduced recruitment from spawning gravels; regulation of rivers to store both fall storm flows that trigger spawning runs and spring freshets that facilitate emigration of smolts;

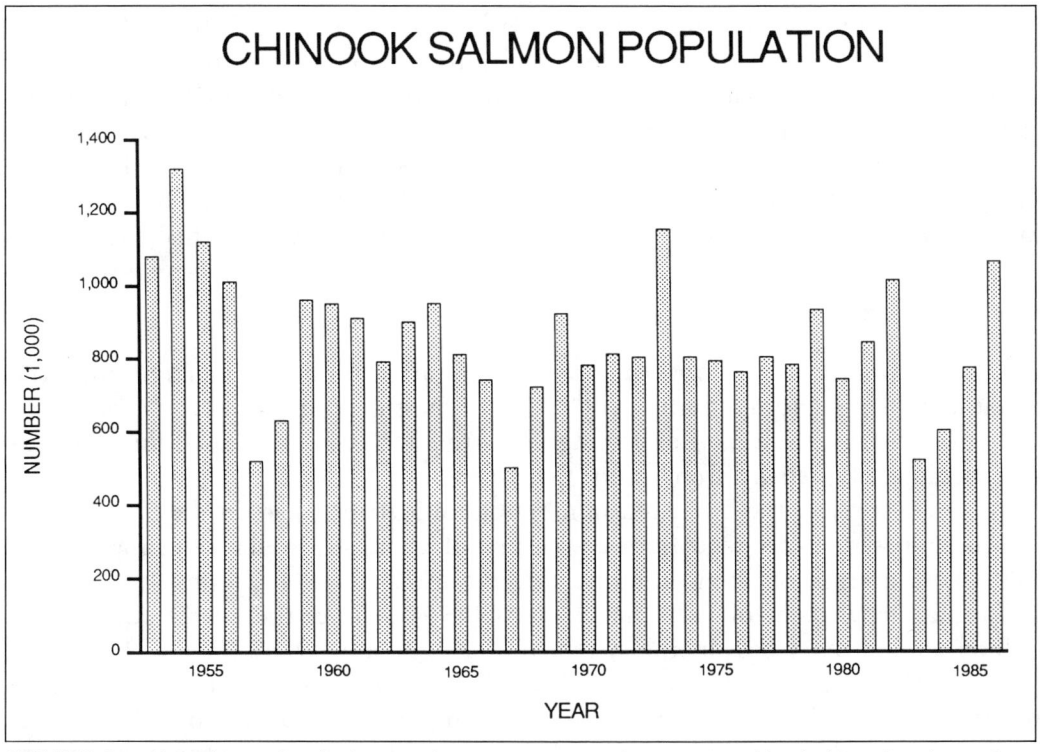

FIGURE 6. Abundance (total of annual escapement and ocean catch) of chinook salmon from 1953 to 1986 (Dettman et al. 1987).

warmer water temperatures that disrupt salmon migration and adversely affect their survival; diversion structures that create diversion losses and provide a haven for predatory fish; reductions in riparian vegetation; and stream bank protection programs that eliminate near-shore habitat.

Winter-Run Chinook. A small winter run of chinook salmon once spawned in the headwaters of the McCloud River, a northern tributary of the Sacramento River. Their spawning migration was blocked by construction of Shasta and Keswick dams in the early 1940's. However, the release of cool water from the dams created more spawning habitat for winter-run chinook in the main-stem Sacramento River than had existed previously in the upper McCloud River. As a result, the winter run grew from a small return of several hundred fish to over 80,000 fish by the mid 1960's. Several factors, including construction of the Red Bluff Diversion Dam about 48 km below Keswick Dam, subsequently contributed to a decline in the winter run to about 550 fish in 1989. The decline led to designation of the winter run as a threatened (federal) and endangered (state) species. State and federal agencies are attempting to restore the winter run under the "Ten-Point Winter Run Restoration Plan" developed in 1988.

Salmon Management. An important salmon management issue in California is whether to manage stocks to protect river spawners and risk under-utilization of hatchery fish, or to fully utilize hatchery fish and risk depleting river spawners (with attendant failure to saturate spawning and rearing habitat). Policy of the California

Resources Agency (CRA) is that the upper Sacramento River system should be "managed to optimize... natural and wild fish populations, even though this policy may result in 'surplus' populations of returning hatchery fish" (CRA 1989). California's Salmon, Steelhead Trout, and Anadromous Fisheries Program Act (1988) aims to double the current natural production of salmon and steelhead trout by the end of this century, and CRA has instituted the Upper Sacramento River Fisheries and Riparian Habitat Management Plan as part of a program to meet this goal.

Striped Bass Problems

Declining Abundance of Striped Bass. Soon after their introduction into the Sacramento-San Joaquin estuary in 1879, striped bass begin to support a commercial and sport fishery. Commercial fishing for striped bass was banned in 1935, in part because of pressure from sport fishermen.

In recent years, the estimated number of adult striped bass in the Sacramento-San Joaquin system declined from an average of about 1.7 million during 1969 to 1976 to an average of about 1 million during 1977 to 1984 (Figure 7). However, California Department of Fish and Game (CDFG) believes there were about three million adult striped bass in the system during the early 1960's (CDFG 1989).

CDFG calculates a 38 mm Striped Bass Index, based on trawl data, to estimate the abundance of juvenile striped bass each summer. The number of juveniles has declined, particularly since the 1976-1977 drought (Table 2), in both the lower estuary (Suisun Bay) and in the Sacramento-San Joaquin Delta.

Theories About the Bass Decline. The reason for the decline of striped bass in the Sacramento-San Joaquin system is not known, although several theories have been advanced. The alternative theories involve presence of toxic chemicals; a reduced egg supply resulting from lower abundance of older female bass; reverse flows that transport juveniles to inhospitable habitat in the central delta and increase entrainment in agricultural diversions, Central Valley Project pumps, and State Water Project pumps; and changes in the aquatic food web.

There is no conclusive evidence regarding the effects of toxic substances on striped bass abundance in the Sacramento-San Joaquin system. CDFG estimates that twice as many eggs were shed in 1985, when the Striped Bass Index was 6.3, as in 1986 when the Striped Bass Index rose to 65.0. However, proponents of the egg theory argue that production of more eggs increases the chance that patches of larval bass will occur where food resources are plentiful, leading to a stronger juvenile year class. CDFG data (CDFG 1987) indicate that the survival of striped bass larvae has decreased since the 1976–1977 drought. Lower survival might result from changes in the aquatic food web, from flow modifications leading to increased entrainment and transport of larvae to inhospitable areas in the interior delta, or even from the toxic effects of chemicals.

Striped Bass Management. Hatchery rearing of striped bass only recently became feasible. Large releases of hatchery fish began in 1984, and over a million fish are now released each year. Hatchery-reared striped bass are just beginning to contribute to the fishery, and initial results indicate that the releases may help increase

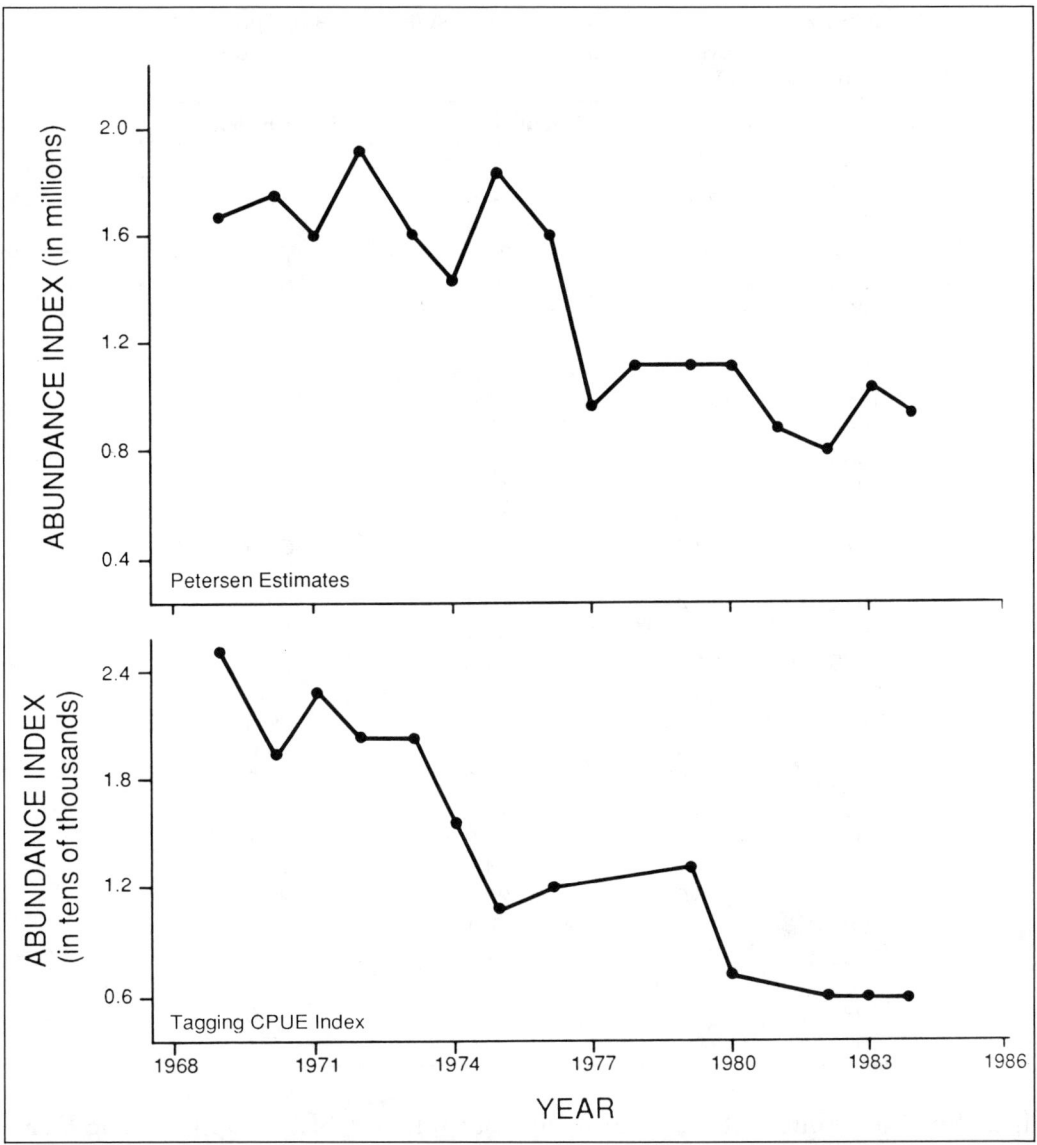

FIGURE 7. Estimated trends in adult striped bass abundance in San Francisco Bay and the Sacramento-San Joaquin Delta from 1969 to 1984 (Stevens et al. 1985).

the abundance of adult striped bass. The State Water Project negotiated an agreement to replace bass lost since 1986 because of its operations with hatchery-reared bass and, in 1989, CDFG released a "Striped Bass Restoration and Management Plan for the Sacramento-San Joaquin Estuary" (CDFG 1989).

Decline of Other Delta Fish

The abundance of other fish in the Sacramento-San Joaquin Delta seems to have declined along with abundance of striped bass. The annual catch of striped bass,

TABLE 2. Striped bass index by year (CSWRCB 1988). The 1989 data were provided by James Sutton of CSWRCB (personal communication).

Year	Delta index	Suisun Bay index	Total Index
1959	30.7	3.0	33.7
1960	32.0	13.6	45.6
1961	25.2	6.4	31.6
1962	46.8	32.1	78.9
1963	38.2	43.5	81.7
1964	54.7	20.7	75.4
1965	49.4	67.8	117.2
1966		No data collected	
1967	35.1	73.6	108.7
1968	39.6	17.7	57.3
1969	33.6	40.2	73.8
1970	36.6	41.9	78.5
1971	24.6	45.0	69.6
1972	13.4	21.1	34.5
1973	15.6	47.1	62.7
1974	17.4	63.4	80.8
1975	23.4	42.1	65.5
1976	21.1	14.8	35.9
1977	8.3	0.7	9.0
1978	16.5	13.1	29.6
1979	5.4	11.5	16.9
1980	2.8	11.2	14.0
1981	15.4	13.7	29.1
1982	9.5	39.2	48.7
1983	1.2	14.2	15.4
1984	6.3	20.0	26.3
1985	2.2	4.1	6.3
1986	23.8	41.1	64.9
1987	7.3	5.3	12.6
1988	3.9	0.7	4.6
1989	3.1	2.0	5.1

threadfin shad, white catfish, delta smelt, bluegill, and black crappie at the Delta Fish Protection Facility averaged 97% lower in 1980 than in 1970. The fact that other species appear to have declined in parallel with striped bass suggests that declining striped bass abundance results from changes in the environment, rather than from factors specific to striped bass. Introduced species, flow changes, and trace amounts of toxic substances are potential causes.

Wetlands

Problems caused by selenium at Kesterson Reservoir drew attention to the availability and suitability of wetland habitat for wildfowl in the Sacramento-San Joaquin watershed. This wetland habitat has been greatly reduced and altered from its undisturbed natural condition by urban and agricultural development.

Changes in the Estuary

The ecosystem of the Sacramento-San Joaquin estuary, including San Francisco Bay, has also been greatly modified since European settlers arrived. Modifications include elimination of large marsh and wetland areas by filling and dredging to accommodate the bay area population, pollutant discharges from urban and industrial activities around the bay, and tremendous biological changes from accidental and deliberate introduction of hundreds of terrestrial and aquatic species. As witnessed by the changes now resulting from introduction of *Potamocorbula amurensis* as recently as 1985, far-reaching ecological changes are continuing.

Concerns have been expressed about the effects of upstream water developments on outflow in the Sacramento-San Joaquin Delta, based on claims that more outflow is needed to reduce environmental degradation in San Francisco Bay. Evidence of a general decline in the bay ecosystem is lacking, but localized pollution does occur near wastewater discharge points. However, in the 1988 draft Water Quality Control Plan for the San Francisco Bay-Sacramento-San Joaquin Delta Estuary, the California State Water Resources Control Board (CSWRCB) declined to establish delta outflow standards for San Francisco Bay "because the evidence presented (in hearings) was judged insufficient as a basis for water quality objectives." CSWRCB also noted that using delta outflow to flush pollutants (other than ocean derived salts) out of the estuary is not a reasonable use of water.

Water Quality Control Efforts

Legal Jurisdiction

The United States Congress and the California Legislature pass the laws that set the stage for human activity in the Sacramento-San Joaquin watershed. The California Department of Water Resources is responsible for operation of the State Water Project, while the Bureau of Reclamation runs the Central Valley Project. However, the three government bodies with the most direct responsibility for water quality and water resource management in the Sacramento-San Joaquin system are the CSWRCB, and the Regional Water Quality Control Boards of the Central Valley and San Francisco Bay. Extensive environmental and water quality monitoring programs are conducted in the watershed by state and federal agencies. As part of California's continuing water quality control planning, CSWRCB is conducting a multi-year hearing, planning, and review process to prepare a Water Quality Control Plan and a Pollutant Policy Document for San Francisco Bay and the Sacramento-San Joaquin Delta. Draft versions of these plans were released in November 1988. After a storm of controversy, CSWRCB decided to redraft both documents.

Key Legislation

Legislation directly affecting water quality in the Sacramento-San Joaquin system includes the federal Clean Water Act, the state Porter-Cologne Act, the federal and state Safe Drinking Water acts, and the state Safe Drinking Water and Toxic Enforcement Act of 1986 (Proposition 65). Proposition 65 prohibits businesses and industry from discharging chemicals known to cause cancer or reproductive problems into drinking water sources. Proposition 65 will be important in future controversies over the safety of drinking water because 242 chemicals were listed as "known to the state to cause cancer or reproductive toxicity" as of November 1, 1988.

The federal Reclamation Act and its amendments play a key role in controversies over agricultural water use. Any listing of additional aquatic species of the Sacramento-San Joaquin system under the state and federal Endangered Species Acts could also significantly affect water resource planning for the watershed.

Trends in Water Quality Control

Regulations arising from federal and state Safe Drinking Water Acts and Proposition 65 reflect intense public concern with drinking water quality. This concern will probably draw more attention to water quality in the Sacramento-San Joaquin Delta, the stability of delta levees, and the effect of delta agricultural drains on drinking water. The Clean Water Act is encouraging more attention to non-point discharges, and increased efforts to regulate and control such pollutant sources can be expected. CSWRCB's position that "use of delta outflow solely to flush pollutants, other than ocean derived salts, out of the estuary is not reasonable" puts renewed emphasis on pollution control at the source.

Future Prospects

Overview

Urban growth in the Central Valley of California will lead to more concerns about urban and industrial discharges. Increased urbanization means that a certain amount of water will be transferred from agriculture to urban use. Controversy over renewal of Bureau of Reclamation irrigation contracts, and political pressures to review the pricing of irrigation water by crop support programs, may combine with increased urbanization to force a change in the present system of agricultural water withdrawal and irrigation returns. Work will continue on agricultural drainage problems that affect water quality and agricultural productivity. The controversy over the potential environmental impacts of a master canal to take agricultural drain water from the San Joaquin valley, exacerbated by the events at Kesterson Reservoir, is likely to prevent development of a master drainage canal in the near future.

Controversies over environmental impacts of water development projects will continue to make it difficult to build new dams and conveyances. We can expect continued emphasis on maintaining drinking water quality and growing appreciation of the problems involved in using the sensitive and unstable Sacramento-San Joaquin delta as a key link in California's water transport system. This may eventually lead to reconsideration of an isolated cross-delta conveyance facility, though this option has been in disfavor since a peripheral canal around the delta was rejected by California voters several years ago.

Controversy will continue over the appropriate balance between urban and agricultural water use and instream water use for fisheries, recreation, and the stream environment. More detailed information on the quality and quantity of water needed by fish, and more reliable information on the possibility of saving water from urban and agricultural use, are both needed. In the absence of better information, important resource allocation decisions may be unduly influenced by public emotion and political pressure.

Consequences of Changing Weather Patterns

There is a trend for more precipitation in the Sacramento-San Joaquin watershed to appear as rain instead of snow (Roos 1988). This trend could have serious consequences for water resource management in California. Reduced storage of water in the winter snowpack would require additional facilities to store runoff in order to sustain California's present level of water use. Any climate change that maintains the trend from snow to rain or to increase the frequency of floods and droughts, such as that associated with global warming, will further accentuate the need for additional water storage.

Role of Fisheries Issues

Fishery issues will play a central role in balancing instream water needs with urban and agricultural needs. Environmental arguments for increased (or differently timed) instream flows will continue to stress benefits to fisheries because increased fish abundance is an issue with great public appeal.

Fishery biologists will need to provide clear, reliable guidelines for water resource allocations that critically affect future development in California. To serve as a reliable basis for resource allocation, estimates from biological sampling programs should always be accompanied by error estimates that indicate their precision. Estimates from calculations or computer models should be accompanied by sensitivity analyses showing how the results will change if key assumptions are varied. If fisheries biologists can provide well-supported recommendations, the success of managing water and fishery resources will increase. However, if their recommendations do not reflect consensus within their profession and are subject to extensive criticism by knowledgeable individuals, the recommendations may be discounted. This will increase the danger that public policy may be based on preconceived notions and political pressures.

Conclusion

From a public policy standpoint, Californians must limit discharges of pollutants to their rivers wherever possible, and they must continue to increase the efficiency of urban and agricultural water use. To support management decisions to balance instream water use with urban and agricultural water requirements of California's growing population, we need more complete information on the amounts of water required to sustain fish populations and the potential benefits for fish if water allocations are changed. Finally, to maintain and enhance fishery resources in the Sacramento-San Joaquin system, California must continue to improve instream habitats and continue an aggressive hatchery program for both salmon and striped bass.

References

CDFG (California Department of Fish and Game). 1987. Factors affecting striped bass abundance in the Sacramento-San Joaquin River system. Exhibit 25 for the CSWRCB 1987 Water Quality/Water Rights Proceeding. CDFG, Sacramento.

CDFG (California Department of Fish and Game). 1989. Striped bass restoration and management plan for the Sacramento-San Joaquin estuary-Phase I. CDFG, Sacramento.

CDWR (California Department of Water Resources). 1989. The delta as a source of drinking water. CDWR, Sacramento.

CDWR (California Department of Water Resources). 1987. California water: looking to the future. Bulletin 160-87, CDWR, Sacramento.

CDWR (California Department of Water Resources). 1983. The California water plan-projected use and available water supplies to 2010. Bulletin 160-83, CDWR, Sacramento.

CRA (California Resources Agency). 1989. Upper Sacramento River fisheries and riparian habitat management plan. Document, CRA, Sacramento.

CSWRCB (California State Water Resources Control Board). 1988. Water quality control plan for salinity and pollutant policy document (draft), San Francisco Bay/Sacramento-San Joaquin Delta Estuary. CSWRCB, Sacramento.

CSWRCB (California State Water Resources Control Board). 1987. Regulation of agricultural drainage to the San Joaquin River - SWRCB Order No. W. Q. 85-1. Technical Committee Report, CSWRCB, Sacramento.

CUWA (California Urban Water Agencies). 1989. Delta drinking water quality study. Report by Brown and Caldwell, CUWA, Sacramento.

Dettman, D., D. Kelley, and W. Mitchell. 1987. The influence of flow on Central Valley salmon. Exhibit 561 for the CSWRCB 1987 Water Quality/Water Rights Proceeding. California State Water Resources Control Board, Sacramento.

EBMUD (East Bay Municipal Utility District). 1989. Water supply management program, Volumes I and II. EBMUD, Oakland.

Fox, J. P., T. R. Mongan, and W. J. Miller. 1990. Trends in freshwater inflow to San Francisco Bay from the Sacramento-San Joaquin Delta. Water Resources Bulletin 26:101–116.

Roos, M. 1988. Climate change and changes in California snowmelt-runoff patterns. Pages 507–516 *in* M. Waterston and R. Burt, editors. Proceedings of the Symposium on Water-Use Data for Water Resources Management. Water Resources Association, Bethesda, Maryland.

Schemel, L. 1989. *Potamocorbula amurensis* discovered in San Francisco Bay. Newsletter, California Interagency Ecological Study Program, Sacramento.

Stevens, D. E., D. W. Kohlhorst, L. W. Miller, and D. W. Kelley. 1985. The decline of striped bass in the Sacramento-San Joaquin estuary, California. Transactions of the American Fisheries Society 114:12–30.

Some Geohydrological Features of the Santo Domingo Basin, Sierra San Pedro Mártir, Baja California Norte, Mexico

CARLOS YRURENTAGOYENA UGALDE
Patronato Para la Proteccion Y Reforestacion Iturbide 172-B Ensenada Baja California, Mexico

ABSTRACT. *Some physical and ecological features of the Santo Domingo Basin, in the San Pedro Mártir National Park of northern Baja California, Mexico, are described. The basin is a major, self-sustaining water district in this hot and arid region and receives its precipitation at high elevations in the San Pedro Mártir Mountains. The endemic San Pedro Mártir trout,* Oncorhynchus mykiss nelsoni, *occurs in the upper reaches of the basin where relative isolation provides its greatest protection. Construction of proposed water storage dams on streams of the Santo Domingo Basin to meet the region's urgent need for water may adversely affect the San Pedro Mártir trout and other natural resources.*[1]

The state of Baja California Norte, Mexico, occupies the northern part of the Baja California (BC) Peninsula, and is oriented north and south roughly parallel to the continental shelf (Figure 1). The climate of the BC Peninsula is generally characterized as "hot and dry". The geographical resources of the BC Peninsula include approximately 70,113 km^2 of arid land supplied with 3,250 million m^3 of water annually (Paredes 1984). An estimated 88% of this water comes from the Colorado River and the aquifer under the 3,000 km^2 Mexicali Valley (SARH 1983).

The origin and interrelations of land and sea life on the BC Peninsula are complex and difficult to characterize. Both land and sea communities developed in association with the extended geological history of the region. Contrasts are captured and exemplified by the five distinct bioregions on the BC Peninsula (Figure 2). These bioregions account for the presence of an interesting and extensive number of endemic species in relatively small geographical areas (Lindsay 1970; Wiggins

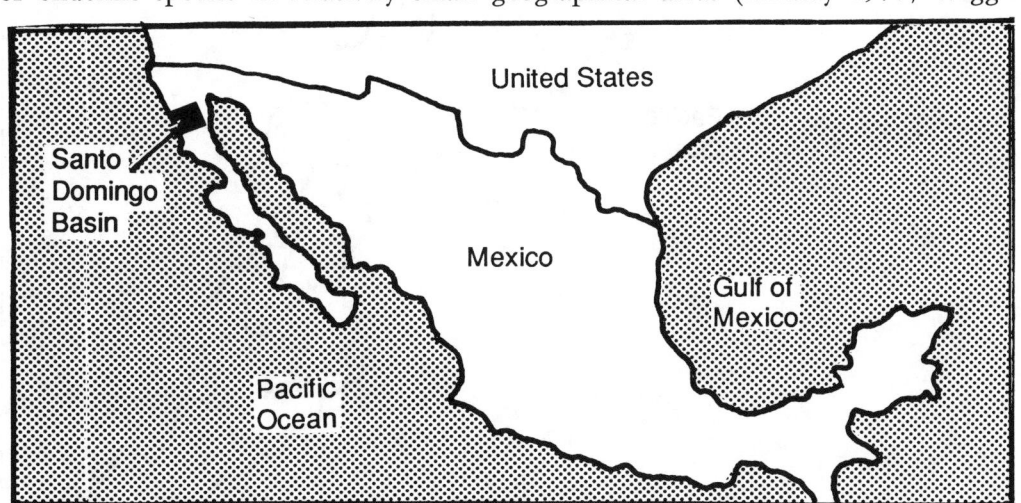

FIGURE 1. Location of the Santo Domingo Basin on the northern Baja California Peninsula (Baja California Norte), Mexico.

---

1. Unrelenting population growth, water shortages, and water pollution create critical situations in many parts of Mexico. Sound management policies, conservation measures, and enforceable regulations are, for the most part, lacking. Funds and resources to evaluate water problems and to seek solutions on a watershed or regional basis are very limited. —The Editors—

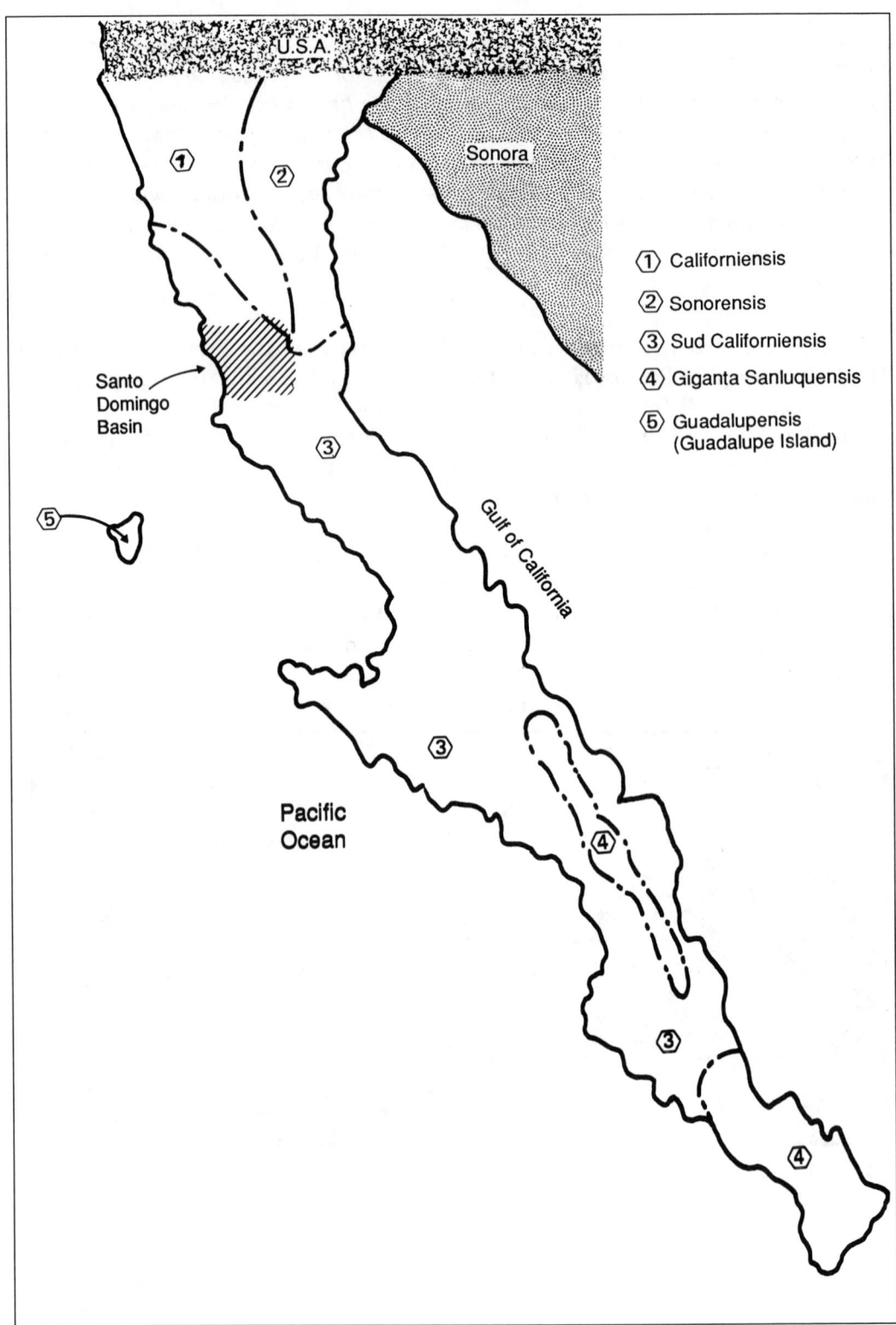

FIGURE 2. Bioregions in Baja California. (After Pineda 1983.)

1980; Murphy 1983; Pinera 1983). Mountain ranges, geological barriers, and atmospheric currents provide the BC Peninsula as a whole, and the Santo Domingo Basin as a specific entity, with many miniature climates.

This article focuses on the geology, hydrology, ecology, and future prospects of the Santo Domingo Basin, San Pedro Mártir Park. Special reference is given to a proposed water dam project that was initially scheduled to begin in early 1990, and to possible environmental impacts of the project on the natural resources of the basin.

Physiography

The Santo Domingo Basin (Figure 3) is located on the northwest side of BC Peninsula, about 160 km southwest of the city of Ensenada (between 30°22'00", 31°00'15"

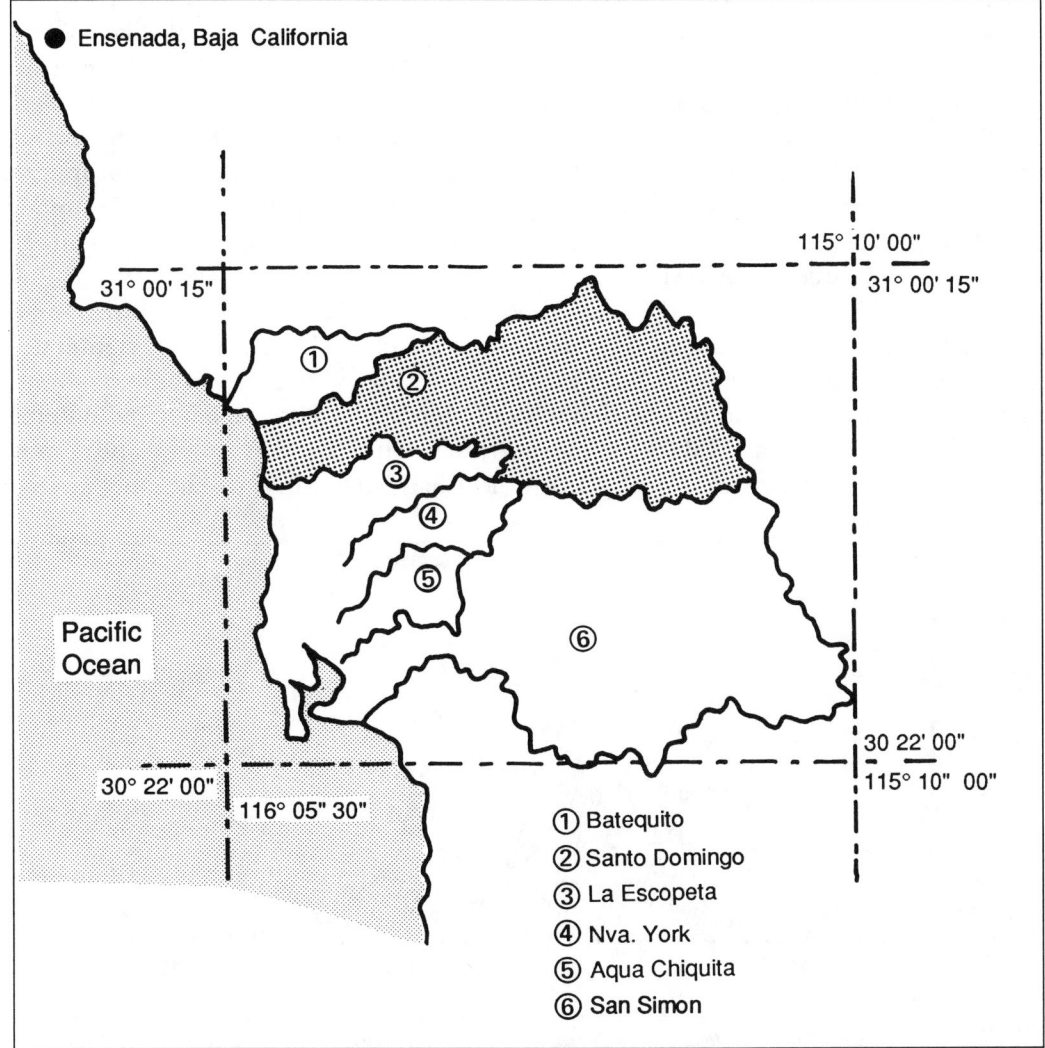

FIGURE 3. Basins adjacent to the Santo Domingo in northwestern Baja California.

latitude north and 115°10'00", 116°05'30" longitude west from the Greenwich Meridian). The La Mesa de San Jacinto and Bramadero ejidos (public lands) occur north of the basin, the San Pedro Mártir Park and Bramadero ejidos occur to the east, and the Pacific Ocean borders the west. Five other separate basins occur in the same general area.

The Sierra San Pedro Mártir and Sierra Juarez form the backbone of the BC Peninsula. Both ranges are a continuation of the peninsular extensions of southern California in the United States, and they bear a strong geological resemblance to the Sierra Nevada of California (Robinson 1967). The physical characteristics of the Santo Domingo Basin are a result of geological processes and block movements that created the mountain ranges, steep inland escarpments, and abrupt shorelines of the BC Peninsula.

The sierras (mountain ranges) of the BC Peninsula have two distinct slopes extending east and west (Figure 4). The more gradual western slope is made up primarily of two or three parallel escarpments, which contain bench-like valleys nestled between them. The more abrupt eastern slope consists of spectacular escarpments that drop sharply from cool, pine-shaded plateaus to the San Felipe desert 8,000 feet below; a low range, the San Felipe Mountains, lies between the desert and Gulf of California. The timbered area of the San Pedro Mártir Park is a unique boreal pine forest protected by law since 1951 and remaining largely pristine under the National Park Service of Mexico, administered by the Secretaria de Agricultura y Recursos Hidraulicos (SARH).

The San Pedro Mártir Park covers about 69,000 hectare. A granite peak, 3,050 m (10,000 ft) high and known as "Picacho del Diablo" (Devil's Peak), dominates the adjacent mountain range. A series of deep canyons extend toward the lower meadows on the west side of the sierras from this peak. Three main river systems drain the park, the San Rafael to the northwest and the San Telmo and Santo Domingo to the southwest (Figure 5). The Rio Santo Domingo is the only system

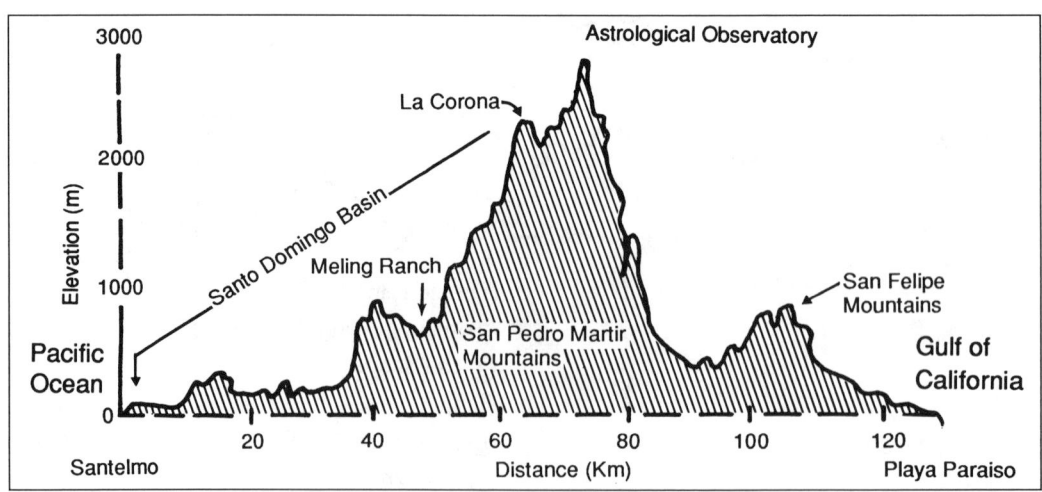

FIGURE 4. Cross-section of western and eastern slopes of the San Pedro Mártir Mountains in Baja California, Norte.

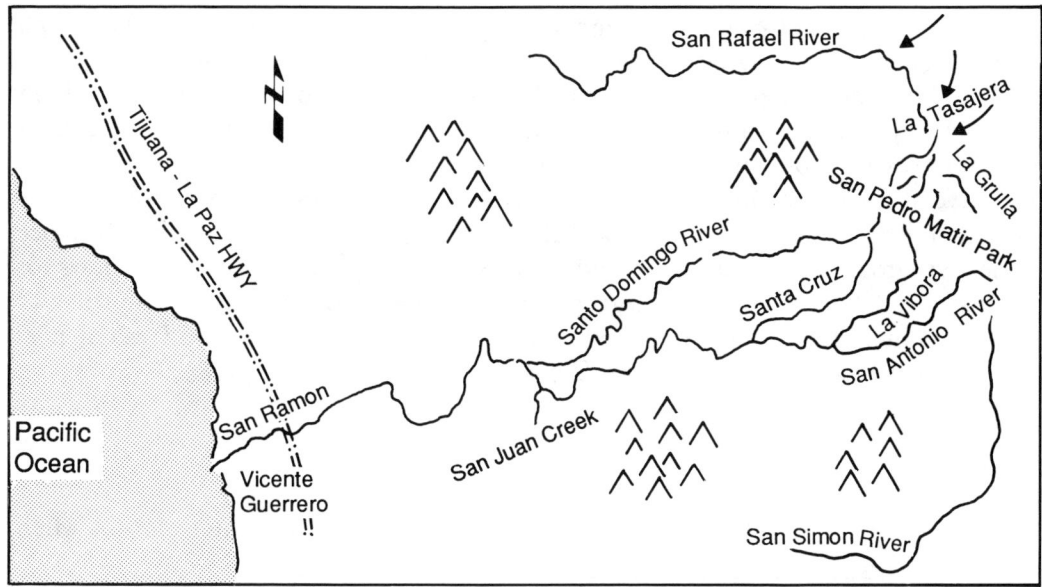

FIGURE 5. Major rivers and tributaries in the Santo Domingo Basin.

that receives sufficient precipitation and has sufficient storage in its headwater aquifers to remain flowing to the Pacific Ocean throughout the year.

The Rio Santo Domingo originates near La Corona in the mountains of the San Pedro Mártir Park about 2,800 m above sea level. The creeks of Valladares and Santa Cruz (at Valladares) and La Zanja, San Antonio, El Conejo, and El Caballo Falls (at San Juan) are the major tributaries entering the main-stem Santo Domingo River. Contributing inflows rise rapidly and are highest during periods of rainfall.

The San Simón River, also originating at San Pedro Mártir Park but at a lower elevation of about 2,300 m, is of minor importance locally. This river is also known as the Santa Eulalia. Place names typically change as the rivers flow toward the Pacific Ocean and into Bahia de San Quintin. Many different names, such as El Represo, El Rosario, El Morro, and San Pablo, are used for these streams by local residents. After heavy rainstorms in 1976, the lower section of the San Simon River near Bahia de San Quintin was channelized to limit future flood damage to the town and crops of the San Quintin Valley.

Water Quality Characteristics

The Santo Domingo River is about 170 km long. No major point-source effluents contaminate the headwaters. Soils along the river bed are mostly fluvial deposits with a maximum depth of 5 m and including gravels, sands, rock boulders, and some igneous fractured intrusive rocks (granioritic type) that range in color from light to dark grey. The river bed, no more than 30 m wide in its upper reaches and 100 m wide near the coast, typically has steep slopes shaped like a "V" (Marquez 1986).

Since granitic soils occur throughout almost all of the upper Santo Domingo Basin, excess flows pass rapidly downstream and erode soil in the valley below. From 1978 to 1985, annual discharge in the Rio Santo Domingo averaged 180 million m^3/y. However, during preceding drought years from 1950 to 1977, the annual discharge was no more than 10 million m^3/y (SARH 1983).

Several small streams arise near the headwaters of the Rio Santo Domingo. The water chemistry in these streams show differences (Figure 6) that reflect the influence of heated groundwater, springs, and near surface aquifers in the meadows of La Grulla and La Vibora. Recent geophysical studies in the San Pedro Mártir and San Felipe mountains suggest the existence of brackish or salty ground water, particularly in the northern part of the San Pedro Valley (Espinoza 1983).

Climate

The rain season in the BC Peninsula is short and limited to the winter months when

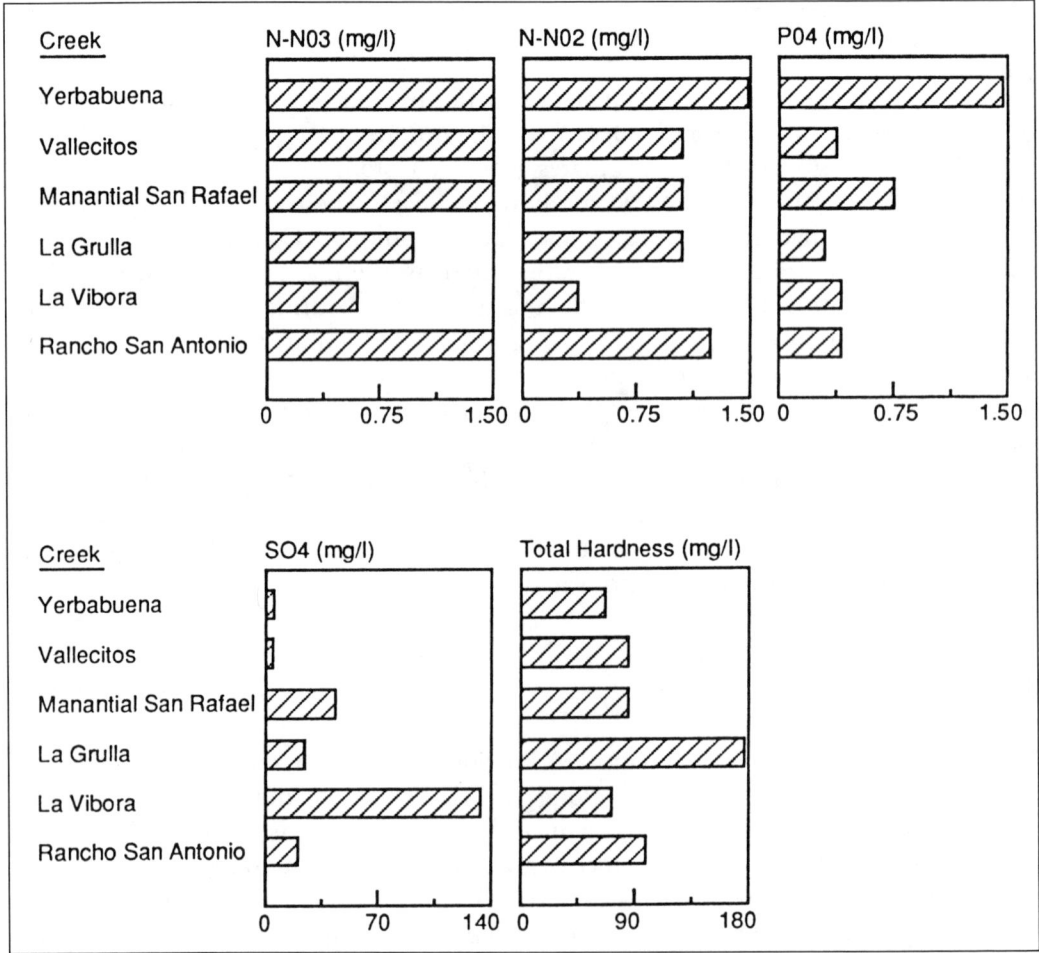

FIGURE 6. Some water quality features of six tributaries in the Santo Domingo Basin.

snow sometimes falls at higher elevations. Generally, precipitation in northern Baja California has always been sparse, averaging no more than 30 cm annually (Minnich 1983). Much humidity is caused by condensation of moisture where ever hot fronts from the land contact cool currents of the Pacific Ocean. Cold weather conditions can last from 40 to 60 days. However, occasional frosts or "santanas" (warm periods) may occur during winter.

Data from weather stations at Santa Cruz, San Pedro Mártir, and Vincente Guerrero suggest that the climate in local basins change with geographic characteristics and altitude (Hastings and Humphrey 1969). Five climatic regimes have been identified in the Santo Domingo Basin (Figure 7A). Areas 1 and 2 are the most dry and have the hottest summers. Area 3 is semi-dry and also has hot summers. Area 4 is subhumid and more temperate. Area 5, high in the sierras, is subhumid and relatively cold.

Generally, rain in the Santo Domingo Basin is most frequent during the winter. Annual precipitation ranges from 100 to 200 mm near the coast, increasing to 400 to 500 mm in the high sierras (Figure 7B). Evaporation is relatively high throughout the basin, averaging 1,800 to 2,000 mm/year; relative humidity averages about 80% near the coast and 70% inland (Figure 7C).

The annual air temperature of the Santo Domingo Basin averages 16°C. The minimum can fall as low as −12°C and the maximum can rise to 48°C (Figure 7D). Daily variations in air temperature, in many cases, are reflected by a wide diel

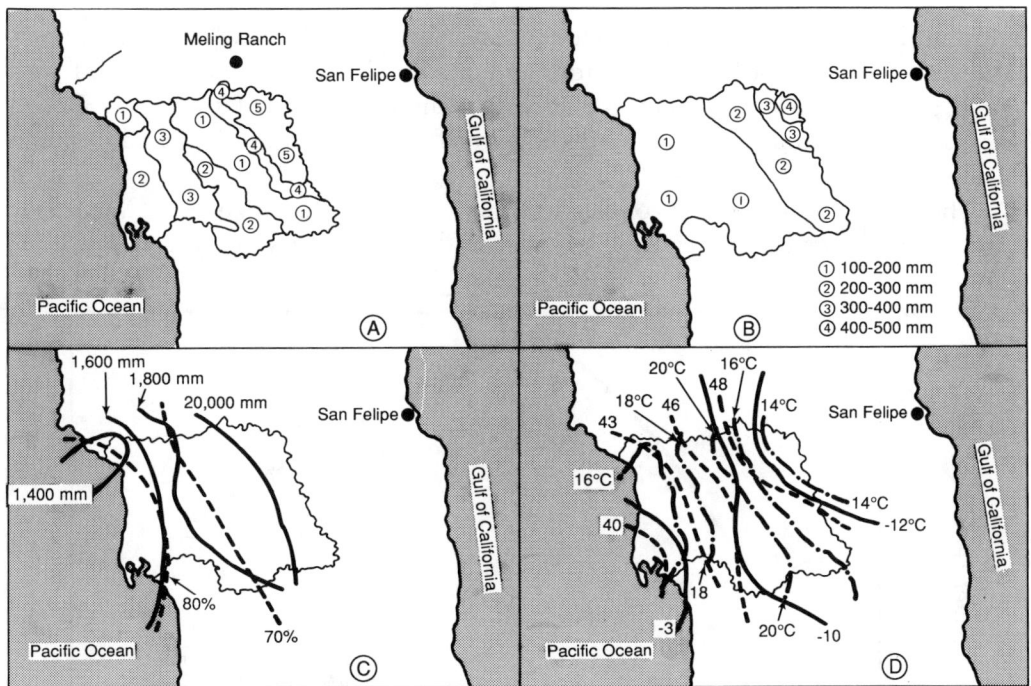

FIGURE 7. Climatological features of the Santo Domingo Basin. (Modified from SARH 1988.) A - Five climatic regimes (explained in text); B - differences in mean annual rainfall over the basin; C - differences in mean annual evaporation (mm) and relative humidity (%) over the basin; and D - Mean annual isotherms over the basin, where solid lines represent extreme minimum temperatures and broken lines represent extreme maximum temperatures.

range in surface water temperatures (Figure 8) to which aquatic organisms in basin streams must adapt in order to survive.

Ecology

Vegetation

Major plant communities form broad zonal belts that increase in elevation southward into Baja California (Minnich 1983). Grasslands and coastal sage scrub in lower coastal valleys are replaced by chaparral on mesic coastal slopes of the mountains. Mixed evergreen forest and mixed conifer forest occupy the highest mountains. These forests grade into pinyon and juniper forest and scrub communities on the east slope of the mountains adjoining the Sonoran and Mojave deserts (Figure 9, Table 1).

Generally the pine forest and adjacent regions are sanctuaries, not only for various indigenous life forms but for migratory waterfowl that stop during their migration to more southern nesting areas.

Fauna

Communities of land animals are abundant and they include many endemic species. Groups of insects, reptiles, fish, birds, and mammals include many species with limited geographical range. Such is the case of the San Pedro Mártir trout, *Oncorhynchus mykiss nelsoni*, the only native endemic fish in the basin (Ruiz and Conteras 1985).

The native trout was first collected in the Rio Santo Domingo by E. W. Nelson of the U. S. Biological Survey in 1905; it was identified a few years later by B. W.

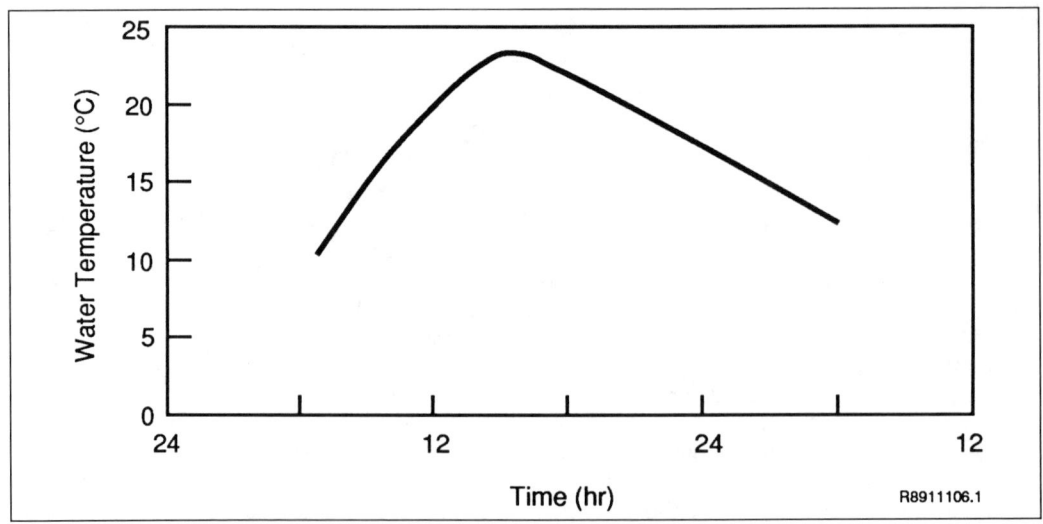

FIGURE 8. Diurnal variation in water temperature of the lower San Rafael River, June 1987.

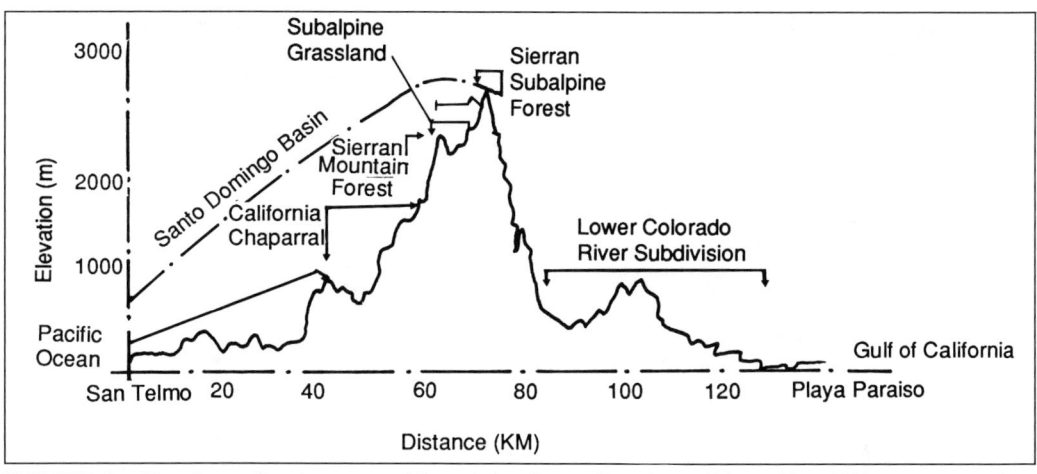

FIGURE 9. Dominant floral communities of the Santo Domingo Basin, East-West cross section.

Evermann as *Salmo nelsoni* (Evermann 1908). Subsequently, J. O. Snyder also collected "the Nelson trout" in the Rio Santo Domingo and concluded that it was a geographically isolated form of rainbow trout (Snyder 1926). P. R. Needham and R. Gard later confirmed Snyder's observations; they concluded that "*S. nelsoni*" resembled the typical rainbow in general appearance and definitely belonged to the rainbow series (Needham and Gard 1959).

According to a 1905 report by Nelson, the native trout was only found in "the Rio Santo Domingo and at the base of the mountains." However, presently the San Pedro Mártir trout can be found in both the upper tributaries of the Rio San Rafael and the Rio Santo Domingo because of the efforts of C. E. Utt. From 1929 until 1941, Utt captured and transplanted the native trout by placing them in six 5-gal cans fitted in special cases on both sides of pack mules (Yruretagoyena 1985).

The San Pedro Mártir trout consumes predominantly insects (78%), of which Trichoptera (27.5%), Odonata (16.2%), and Coleoptera (15.8%) are the main food groups. No significant qualitative difference has appeared in the food groups (diet) in trout of different size or age groups (Cirilo and Ruiz 1987). Erosive forces in the Rio Santo Domingo tend to create gravel deposits in relatively shallow areas. Placement of redds in shallow areas increases the vulnerability of developing eggs and embryos to scouring during flash floods. Scouring also reduces bottom-dwelling stream invertebrates (Table 2).

Electrophoretic analysis has revealed a distinctive genetic profile in the San Pedro Mártir trout, confirming the view of Needham and Gard (1959) that there has been little or no influence of stocking from hatcheries or from introduction of rainbow trout strains taken from streams along coastal California.

Preliminary studies indicate that, while the San Pedro Mártir trout is genetically similar to other populations of coastal rainbow trout, it represents a distinctive form. Because it is one of the few undisturbed forms of trout in the western part of Mesoamerica, it should be managed as a distinct entity and be protected from outside introductions (R. C. Smith, Davis, California; personal communication).

TABLE 1. Representative floral communities in the upper northeast part of the Santo Domingo Basin, Baja California Norte.

Pine Forest Species:
    Coulter pine - *Pinus coulteri*
    Lodgepole pine - *P. contorta*
    Jeffrey pine - *P. jeffrey*
    Sugar pine - *P. lambertiana* and *P. muricata*
    Four needled pinyon - *P. quadrifolia*
    White fir - *Abies concolor*
    Incense cedar - *Calocednus decurrens*
    Arizona cypress - *Cupressus arizonica*
    Guadalupe cypress - *C. guadalupensis*
    Black oak - *Quercus agricola*
Chaparral species:
    Manzanita - *Adenostoma* spp.
    Chaparral - *Arctostaphylos glandulosa*, *A. glauca*, and *A. peninsularis*
Grassland species:
    Wild oats - *Avena barbata*
              Brassica nigra
              Bromus ciliathus and *B. mollis*
              Erodium cicutarium and *E. botrys*

Human Influences

Due to its remote location, the Santo Domingo Basin now provides a refuge for many animals that live in the Sierra San Pedro Mártir. Thus, the basin provides a ideal site to study wildlife of Baja California in their natural, largely undisturbed habitat. Other parts of the region are subject to human intrusion as the population of the area continues to increase, a phenomenon common throughout Mexico.

In the late 1890's, early settlers introduced sheep and cattle to graze the upper meadows in the Santo Domingo Basin. Today, such actions as overgrazing and improper management of rangeland, and natural catastrophes such as forest fires, have increased the vulnerability of 70,000 hectare in the basin to erosion and other destructive forces of nature. Also, tenant property near the margin of the San Pedro Mártir Park and overstocking of cattle from spring to fall has destroyed much riparian vegetation, eroded river banks, and silted the gravel of many spawning areas used by the San Pedro Mártir trout.

Fortunately, the absence of dense human settlements, poor road maintenance, and lack of tourist services have, until now, limited ecological changes on the eastern slopes of the Santo Domingo Basin. The situation is quite different on the western side in the lower valleys of the San Quintin, Camalu, and Vicente Guerrero. Establishment of large farm communes and rotation of crops has been followed by accumulation of inorganic salts, fertilizers, and pesticides in the soil. Further, such actions as the discharge of urban and domestic septic effluents, the overuse of near-surface aquifers, and the burning of plastic materials used on farms have had a great impact on local environments.

TABLE 2. Some groups of aquatic organisms in the San Rafael and Santo Domingo rivers, northeastern part of the Santo Domingo Basin.

Order	Stream			
	La Grulla	La Vibora	Yerbabuena	San Rafael-1
Coleoptera		X	X	
Dipthera	X	X	X	
Ephemeroptera	X	X	X	
Hemiptera	X	X	X	
Hymenoptera			X	
Lepidoptera			X	
Neuroptera			X	
Odonata	X	X	X	
Plecoptera			X	
Trichoptera	X	X	X	
Amphipoda			X	
Aranea	X		X	
Crustacea		X		
Gasteropoda	X		X	X
Ostracoda			X	
Pisces	X	X		X

Mercury is now biologically available in coastal waters and DDT occurs in oysters, *Crassostrea gigas*, cultured in San Quintin Bay. The availability of mercury is associated with upwelling waters. However, both mercury and DDT in aquatic organisms remain below levels that present a hazard for human consumers (Gutierrez and Munoz 1984, 1986).

Water Impoundment Projects

Water in the arid Santo Domingo Basin is an extremely important commodity. However, water to irrigate crops is now in short supply, and sea water is infiltrating underground to contaminate the aquifer in the lower part of the basin (SARH 1988). As a solution to the region's need for water and to replenish the aquifer, SARH developed plans to construct 20 water storage and regulator dikes on streams throughout the basin. While many companies considered building water dams in the region from the late 1880's until 1949, they could foresee no economic recovery of costs and cancelled their plans (Memoria Administrativa 1924–1927; Pinera 1983). Phase I of SARH's project for developing the basin was scheduled to start in 1989 with the construction of a dike at La Boquilla and a water dam 18 km downstream (Table 3).

The physical characteristics of dams and dikes planned under the SARH project are in Table 4; their intended locations are in Figure 10. (Note: The Mexican government has apparently postponed the start of this project.)

Potential Problems

Generally speaking, the Santo Domingo project has three main objectives: 1) to make better use of water resources in Baja California; 2) to assure, consolidate, and

TABLE 3. Physical characteristics of the La Boquilla dam, lower Santo Domingo River.

Main Dam:	
Basin surface area	975.0 km²
Total storage capacity	41.9 M m³
Useful capacity	40.0 M m³
Sediment storage	1.9 M m³
Dam height	395.0 m
Dam width	10.0 m
Maximum level height	58.0 m
Flooded area	192.3 hectares
Dike:	
Length	123.0 m
River base width	92.0 m
Maximum level height	5.0 m
Distance, dike to dam	17.5 km
Distance to farm lands	8.0 m

increase crop production in the lower basin; and 3) to provide other economical means of producing food. When the many proposed dikes and dams are installed, the shortage of water will be temporarily alleviated for many farms on the western side of the San Pedro Mártir Mountains. Unfortunately, the development plan did not contemplate adverse environmental effects that might result from dam and dike construction. Yet, one analysis stated "Any modification of the environment can alternate or unbalance the ecological equilibrium that eventually may destroy the natural resources" (Conteras 1985).

The most likely problems to follow impoundment of water in the Santo Domingo Basin are:
1. Physical changes such as soil erosion, stream sedimentation, eutrophication behind dams, and changes in water flow patterns.
2. Biological changes such as unusual growth of forest and other plants, appearance of non-native species, and a local increase in wildlife populations.
3. New and increased use of the area by people for aquaculture, cattle grazing, farming, fishing, hunting, tourism, and urban development.
4. An increase in pollutant sources such as cattle wastes, herbicides, pesticides, and septic and urban effluents.
5. An increase in public and private interest in the area for social, political, and economic gain, leading to increased occupancy of land by laborers, agriculture, forestry, industry, and road construction.

Development of water resources in the Santo Domingo Basin may lead to the introduction of exotic fishes to provide food in a largely protein-deficient society. The new impoundments will provide habitat that, in many cases, may be suitable for stocking warm water species. Such introductions could put the endemic San Pedro Mártir trout at greater risk.

To biologists concerned with reductions and potential extinctions of native fishes, particularly of endemic species, introduction of exotic species may appear incongruous. Nevertheless, when social and economic conditions are considered, introductions

TABLE 4. Physical characteristics of dikes planned for the Santo Domingo Basin. (Locations are shown in Figure 10.)

Dike	Drainage basin (km²)	Mean annual flow (M m³)	Maximum flow (m³/sec)	Useful capacity (M m³)
Storage Dikes (downstream)				
El Potrero	136.0	4.02	265.6	0.22
La Grulla	85.4	0.64	166.8	0.64
El Alcatraz	50.5	0.24	99.0	0.24
El Horno	200.0	0.40	391.0	0.40
Regulator dikes (upstream)				
La Encantada	51.2	1.40	198.0	—
Botella Azul	3.0	0.10	224.0	—
Pico del Diablo	13.4	0.40	76.3	—
Parque Nacional	10.9	0.30	64.9	—
San Pedro	16.0	0.50	87.1	—
Rancho Viejo	29.4	0.50	135.5	—
La Puerta	4.5	0.10	32.0	—
La Tasajera III	15.0	0.20	86.1	—
El Cajon	7.4	0.20	47.9	—
La Piedra	18.2	0.50	95.8	—
San Ramon I	17.4	0.40	92.7	—
San Ramon II	23.6	0.60	116.7	—
Campo del Oso	14.5	0.30	80.8	—
San Ramon III	28.2	0.10	131.6	—
La Mision	41.9	0.90	173.3	—
La Tasajera III	8.0	0.20	50.9	—

can be viewed as justifiable, particularly by government agencies (Contreras and Escalante 1984).

Future Prospects

To preserve biological and ecological values, a realistic and effective management program must be developed for the Santo Domingo Basin and set in action by the government, conservation groups, and scientific institutions. To achieve success, three interrelated steps are required.
1. The legal status of the Forest Reserve concept must be reviewed and reinforced.
2. A multidisciplinary research program for evaluating the stream, riparian, and biotic conditions in the basin must be put into action, and
3. Public awareness and cooperation must be obtained so that the pristine conditions of the basin can be preserved to and beyond the next century.

Acknowledgments

I thank C. Dale Becker, Duane A. Neitzel, and E. P. Pister for reviewing this manuscript; and Oc. Carlos Arias and Ing. Jose Issac Orozco, Unidad de Hidraulica,

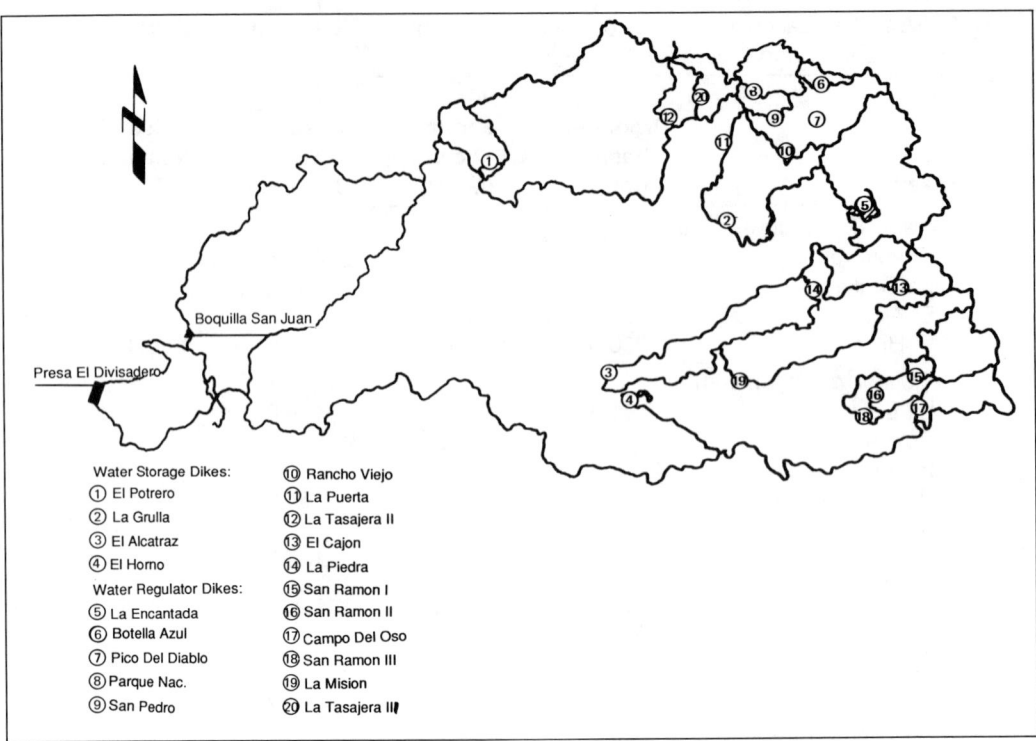

FIGURE 10. Existing and proposed dikes for storage and regulation of water in the Santo Domingo Basin.

SARH, Ensenada, for providing information on dam projects planned for the Santo Domingo Basin. Oc. Luis Aquilar Rosas kindly allowed me to use his computer, and Mrs. Sachiko Nishikawa considerately typed initial drafts of this manuscript.

References

Cirilo, S. H., and G. Ruiz Campos. 1987. Habitos alimentićios de la trucha de San Petro Maŕtir, *Salmo nelsoni* (Evermann). Proceedings of the Desert Fishes Council, Vol. 19. University of Nevada Press, Las Vegas.

Contreras Balderas, S. 1985. Impacto ambiental de obras hidraulicas. Universdad Autónoma de Nuevo León, Monterrey, N. L., México.

Contrearas Balderas, S., and M. A. Escalante-C. 1984. Distribution and known impacts of exotic fishes in Mexico. Pages 102–130, *in* W. R. Courtenay, Jr., and J. R. Stauffer, Jr., editors. Distribution, biology, and management of exotic fishes. The John Hopkins University Press, Baltimore, Maryland.

Espinoza Barreras, P. 1983. Estudio de resistividad en los valles de San Pedro Mártir y valle Chico Baja California. Tesis de Maestria, Centro de Investigacion Cientifica y de Educacion Superior de Ensenada (CICESE), Ensenada, B. C.

Evermann, B. W. 1908. Description of a new species of trout (*Salmo nelsoni*) and a new cyprinodontid (*Fundulus meeki*) with notes on other fishes from lower California. Proceedings of the Biological Society of Washington 21:19–30.

Gutierrez Galindo, E. A., and G. F. Munoz. 1984. Concentraciones de pesticídas y de mercurio en algunos vivalvos de estero de San Quintin, Baja California, México. Ciencias Marinas 10:17–30, Ensenada, B. C.

Gutierrez Galindo, E. A., and G. F. Munoz. 1986. Biological availability of mercury in coastal seawaters of northern Baja California. Ciencias Marinas 12:85–98, Ensenada, B. C.

Hastings, J. R., and R. R. Humphrey. 1969. Climatological data and statistics for Baja California. Technical Report No. 18. Institute of Atmospheric Physics, University of Arizona, Tucson.

Lindsay, G. 1970. Some natural values of Baja California. Pacific Discovery 23:1–10.

Marquez Landavazo, J. 1986. Informe geológico definitivo de La Boquilla San Juan, sobre el arroyo del Santo Domingo, en el estado de Baja California. Secretaria de Agricultura y Recursos Hidraulicos (SARH), Residencia General de Estudios en Ensenada, Ensenada, B. C.

Memoria Administativa. 1924–1927. Gobierno del distrito norte del territoria de Baja California, informe de avance de obras del periodo 1924–1927. Prensa del Gobierno del Distrito Norte, Mexicali, B. C.

Minnich, A. R. 1983. Fire mosaics in southern California and northern Baja California. Science 2119:1287–1294.

Murphy, R. W. 1983. Paleobiogeography and genetic differentiation of the Baja California herpetofauna. Occasional Papers of the California Academy of Science. 137: 1–43.

Needham, P. R., and R. Gard. 1959. Rainbow trout in Mexico and California. University of California Publications in Zoology 67:1–67.

Paredes, E. 1984. Importancia de los recursos hidraulicos de Baja California. Boletin Informatívo Hydrológico No. 8, Subdireccion de Planeacion SARH, Residencia Estatal, Mexicali, B. C.

Pinera Ramirez, D. 1983. Panarama histórico de Baja California. Centro de Investigacíones Históricas, UNAM-UABC, Revista Trimetral de la Direccion General de Asuntos Académicos 4:33–39, Mexicali, B. C.

Robinson, W. J. 1967. Camping and climbing in Baja. La Siesta Press, San Diego, California.

Ruiz Campos, G., and S. Contreras Balderas. 1985. Ecological and geographic checklist of inland fishes from the Baja California Peninsula, Mexico. Proceedings of the Desert Fishes Council, Vol. 17. University of Nevada Press, Las Vegas.

SARH (Secretaria de Agricultura y Recursos Hidraulicos). 1983. Informe geologico y geohydrologicos de diversas zonas de Baja California. Subdireccion de Planeación, Residencia Estatal, Mexicali, B. C.

SARH (Secretaria de Agricultura y Recursos Hidraulicos). 1988. Diagnostico del proyecto de desarrollo regional Santo Domingo-San Simón; Caracteristícas físicas y naturales. Subsecretaria de Infraestructura Hidraulica, Del. B. C., Residencia Gral. de Estud., Ofic. Ensenada, B. C.

Snyder, J. O. 1926. The trout of the Sierra San Pedro Mártir, lower California. University of California Proceedings in Zoology 21:419–426.

Wiggins, L. I. 1980. The origin and relationships of the land flora. Terrestrial and fresh water biota, Part II. Symposium on the Biogeography of Baja California and Adjacent Seas. Systematic Zoology, Vol. 9 (Sept.–Dec. 3 and 4).

Yruretagoyena U., C. 1985. Reseña sobre los estudios de los peces de las aguas continentales en Baja California con especial referencia en la trucha de San Pedro Mártir (*Salmo nelsoni*). Proceedings of the Desert Fishes Council, Vol. 16. University of Nevada Press, Las Vegas.

The Missouri River—Great Plains Thread of Life

JAMES C. SCHMULBACH
Department of Biology, University of South Dakota
Vermillion, South Dakota 57069, USA

LARRY W. HESSE
Nebraska Game and Parks Commission
Post Office Box 934 Norfolk, Nebraska 68701, USA

JANE E. BUSH
Missouri Department of Conservation
1110 College Avenue, Columbia, Missouri 65201, USA

ABSTRACT. *After the arrival of Caucasian settlers, the Missouri River endured severe physical alterations for human benefits with scant regard for aquatic and riparian biota. The current length of the main stem consists of approximately equal distances (i.e., one third) of channelized, impounded, and "free-flowing" reaches. Attendant water quality changes reflect an altered hydrologic cycle. Much of the river's suspended solids are now sedimented in reservoirs. The river's present annual sediment load at its mouth is less that 50% of that observed prior to the closure of main stem and tributary dams. Sediment-free water released from the main-stem dams renews its suspended sediment load in downstream reaches causing the stream bed to degrade. Also, humans use the reservoirs to control flooding, erosion, and meandering of the channel, thus reducing the influx of allochthonous organic matter. In the lower Missouri River, chemicals such as chlordane, dieldrin, and polychlorinated biphenyls from nonpoint sources have accumulated in fish at concentrations that exceed safe human consumption standards. Fish in some tributaries of the middle Missouri River were once contaminated with mercury from gold mining wastes. This problem was alleviated by eliminating mercury as a gold amalgam, but many tributary sediments still contain large amounts of mercury. Consumptive water uses, primarily irrigation, along with evaporation presently take more than 8.6% of the river's total annual discharge ($6.0 \times 10^9 m^3$). A limited water supply in the drainage basin will influence future water usage and priorities.*

The Missouri River has served as a travel corridor for migrating fishes, birds, mammals, and humans since the Pleistocene. Prior to the expedition of Lewis and Clark (1804–1806), the river was a major route used by aboriginal Americans to conduct trade and cultural exchanges. Most permanent Indian villages in the upper Great Plains were located either along the banks of the Missouri River or the banks of major tributaries. The river and its floodplain provided a reliable source of water, cropland, shelter, and fuel in an otherwise expansive and semiarid grassland, while fish, shellfish, and wildlife supplied food. The first Caucasian settlers effected little change in the river's role in human lives because all arrivals remained tied to the river for survival. For the settlers, the river also provided an essential but tenuous connection with civilization.

The Missouri River was once euphemistically called the "Big Muddy." After the construction of main-stem dams, it was converted from a sediment-laden river to a series of deep reservoirs with alternating aggrading and degrading reaches. Much of the alluvium previously transported downstream is now collected in reservoirs and the water discharged through the dams is relatively free of sediment. In general, the suspended sediment load at the river's mouth is now about 50% of the load observed prior to closure of main-stem and tributary dams.

Today, the Missouri River influences the lives and life styles of more humans than ever before, but its effects are less pervasive. This paper addresses how human use of the Missouri River has affected aquatic and riparian communities, particularly where water quality drives the forces of change. Three contemporary concerns are discussed, viz. physical-chemical changes caused by an altered hydrologic cycle, industrial and municipal pollution in the lower river, and problems with heavy metals.

Morphometry and Hydrology

The Missouri River is formed near Three Forks, Montana, at the confluence of the Gallatin, Madison, and Jefferson rivers. It is the longest river in the United States, and it flows generally east and south for 3,768 km until it joins the Mississippi River near St. Louis, Missouri (Figure 1). Prior to Pleistocene glaciation, the river flowed northward to Hudson Bay. It survived the advancing ice sheet by turning southward, gaining tributaries, and increasing in size. The present course of the river generally outlines the most southern extension of the Pleistocene glaciers (Flint 1955).

The drainage basin of the Missouri River encompasses 137 million hectares and includes three major physiographic divisions: Rocky Mountains (10.6%), Interior Plains (87.6%), and Interior Highlands (1.8%). The surface mantle and topography south and west of the present river channel were developed by erosion of a fluvial plain that extended east of the Rocky Mountains. Topography of the basin north and east of the Missouri River was affected by continental glaciation, and the present condition resulted from erosion of glacial drift and till. Consequently, all but two of the extant major tributaries in the upper and middle basin enter the main channel along the river's right bank.

The river bed over most of its distance originally consisted of braided and shifting channels, eroding banks, numerous sandbars, and many islands. In the most upstream portion of the basin, the Missouri River is a clear stream traversing canyons

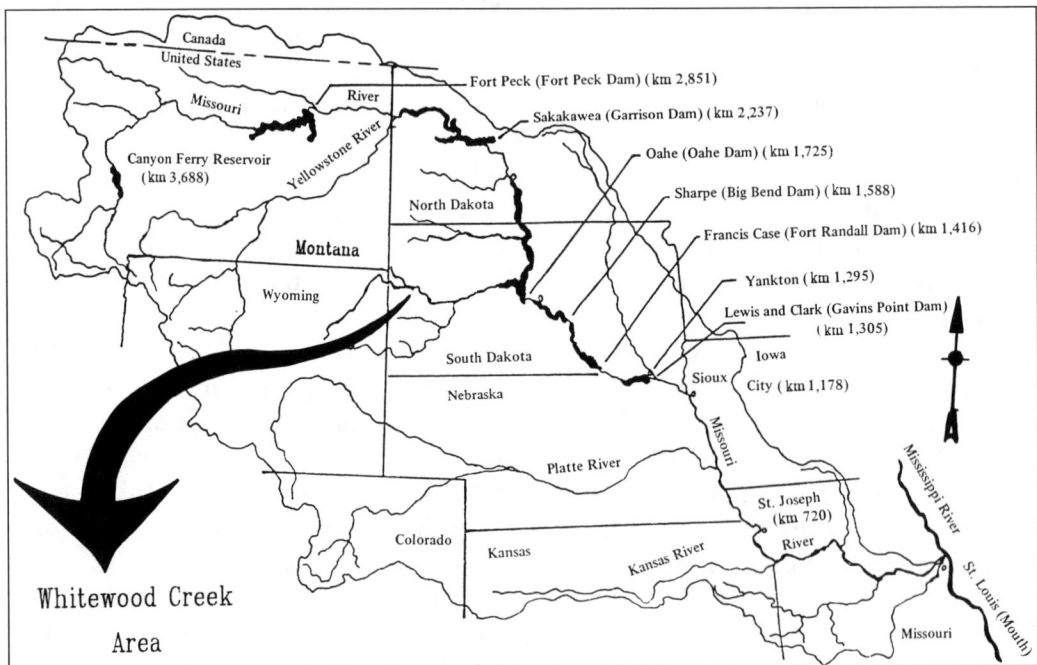

FIGURE 1. Missouri River drainage basin with location of Whitewood Creek area and the main-stem reservoirs. The main-stem dams are indicated by distances measured along the thalweg upstream from the confluence of the Missouri and Mississippi rivers based on 1960 data. Whitewood Creek was the site of mercury contamination as the result of gold mining activities.

and rugged mountain terrain. Near its source, the valley width is 240 m. As the river flows downstream, it is joined by many tributaries; those that drain the fluvial plain flow through highly erodible, unglaciated soils, which add a heavy sediment load. In the lower basin floodplain, width varys from 2.4 to 27.4 km and averages 8.1 km. The gradient of the Missouri River varies from 38 to less than 0.17 m/km (USACE 1985). Annual precipitation averages less than 41 cm in about 50% of the basin. About 70% of the precipitation falls as rain during the growing season (mid-May through mid-September) and 80% of the basin's annual water supply is received before August 1. The mean annual discharge to the Mississippi River is $7.0 \times 10^{10} m^3$, a modest volume for a long river with such an extensive drainage basin.

The natural hydrologic cycle of the Missouri River was once characterized by two floods each spring. Water levels and discharges were low in the fall and winter. Temporally, the first flood or "March rise" was caused by melting snow on the plains and breakup of ice in the main channel and tributaries. The crest of this flood usually flattened as it progressed downstream. The second flood or "June rise" was produced by the combined runoff from melting snow in the Rocky Mountains augmented by rainfall throughout the basin. This flood generally was the larger of the two.

Shortly after the first settlers entered the Missouri River basin in the 1800s, development of water resources began in response to a need for a dependable water supply for irrigation, navigation, and mining. The U. S. Reclamation Act of 1902 fostered irrigation and settlement of land in the western states, and the Rivers and Harbors Act of 1912 authorized a 1.8-m-deep navigation channel from the river's mouth to Kansas City, Missouri. Tributary development and levee construction along the main stem in the lower basin were authorized in the Flood Control Act of 1941. Presently, the navigation channel is about 2.7 m deep and 91.4 m wide. These dimensions were established by the Rivers and Harbors Act of 1945, which extended the navigation channel to Sioux City, Iowa. However, the greatest man-made change in the basin was effected by the Flood Control Act of 1944, which included a comprehensive plan for the development of the entire Missouri River basin called the Pick-Sloan (P-S) Plan. The P-S plan embraced the multiple-use concept; included among its purposes were irrigation, navigation, hydroelectric power, flood control, water quality, water supply, fish and wildlife, and recreation.

The most prominent features resulting from the P-S plan were the construction of six main-stem reservoirs, Ft. Peck, Sakakawea, Oahe, Sharpe, Francis Case, and Lewis and Clark (Table 1). These reservoirs can collectively store $9.26 \times 10^{10} m^3$ of water, which exceeds the river's mean annual total discharge at its mouth.

Other large storage reservoirs, more than 1,300 smaller reservoirs, and numerous farm ponds were built on the main stem and tributaries by federal agencies, conservation districts, and private entities. These structures also control runoff and limit the movement of sediment downstream.

Relatively sediment-free water leaving the main-stem dams has degraded the river bed and little permanent armoring of the river bed has occurred even three decades after the dams were closed. Armoring is a phenomenon whereby a residual layer of bedrock, boulders, or rubble prevents erosion of the stream bed by water

TABLE 1. Characteristics of main-stem Missouri River reservoirs in the Pick-Sloan Plan.

Dam	Location (RKm)[a]	Reservoir	Length (km)	Total volume (10^{10}m^3)	Mean annual discharge (10^9m^3)/yr	Storage ratio	Flushing rate
Fort Peck	2,851	Fort Peck	216	2.33	7.78	3.00	1095.0
Garrison	2,237	Sakakawea	286	2.95	21.28	1.40	511.0
Oahe	1,725	Oahe	372	2.88	22.80	1.00	365.0
Big Bend	1,588	Sharpe	129	0.23	19.37	0.12	43.8
Fort Randall	1,416	Francis Case	172	0.69	13.76	0.50	145.6
Gavings Point	1,305	Lewis & Clark	40	0.01	15.55	0.04	14.6

[a]RKm = river kilometers; distance along the thalweg upstream of the confluence of the Missouri and Mississippi rivers based on 1960 data.

that has the competence (capacity) to transport a larger quantity of bed material. Degradation is especially severe in the reach below the most downstream dam. At a gaging station 9.6 km downstream from Gavins Point Dam, the river bed has degraded 2.26 m. Degradation continues from Gavins Point Dam to the mouth of the Platte River. Conversely, there are varying degrees of aggradation from the Platte-Missouri confluence to the mouth of the Missouri River, causing the bed elevation to rise (USACE 1980).

Degradation and aggradation in rivers cause numerous problems to humans and also deteriorates fish and wildlife habitat (Sayre and Kennedy 1978). Degradation is directly implicated in the draining of many backwaters and oxbows because they are hydraulically connected with the river. Lowering the channel bed causes "cut-off" lakes to lose water and, through succession, become terrestrial habitats. Even oxbows deep enough to retain water are no longer connected to the main channel, and lie perched and isolated several meters above the deepening channel bed. The energy contained in dissolved and particulate organic matter produced in the cut-off lakes formerly flowed through the river's trophic levels. This no longer occurs.

The difference in water storage in the six main-stem reservoirs between the minimum for flood control and maximum pool levels is 2.13×10^{10}m^3. Storage is reduced to near the minimum for flood control by March 1. During spring runoff, water is stored primarily in the three uppermost main-stem reservoirs (Fort Peck, Sakakawea, and Oahe) and released slowly to prevent floods but still supply enough water for navigation and power production, and to maintain water quality standards (USACE 1985). Open water releases from April 1 through November 30 at Sioux City, Iowa, range between 708 and 990 m^3/s daily whereas releases during the nonnavigation season in winter range between 170 and 566 m^3/s (Slizeski et al. 1982).

Downstream discharge rates for water quality were established by the Federal Water Pollution Control Authority in 1969, and reapproved by the U. S. Environmental Protection Agency (USEPA). For example, based on Public Law 92–500 programs for managing both point and nonpoint pollution sources that may discharge to the river and the federal provision for a minimum dissolved oxygen content of 5 mg/L, the required minimum daily flows for the Missouri River to Omaha, Nebraska, are as follows: December through February and in May, 127 m^3/s; March, April, October, and November, 96 m^3/s; June to September, 212 m^3/s (Slizeski et al. 1982). Discharge volumes have always exceeded the minimum requirements.

Pollution in the Missouri River basin is mostly confined to the downstream reaches near population centers or adjacent to industrial activities. Conversely, structural alterations of the Missouri River has been extensive in all reaches. About 35% (1,316 km) of the river's length consists of reservoirs and 32% (1,212 km) is channelized or stabilized. All of the lower Missouri River from Sioux City, Iowa, to its mouth is channelized and consists mostly of a single channel with rock-lined banks. Water volumes, velocities, and levels are regulated by releases from mainstem reservoirs. Even during flooding, the river is confined to a narrow floodplain by high agricultural levees. The present width between levees ranges from 183 to 335 m, about 10% of the original floodplain width. Remnant unchannelized and unimpounded reaches of the Missouri River are termed "free-flowing" and constitute 33% (1,241 km) of the river's length. Recently 240 km of "free-flowing" river in Montana and 93 km below Gavins Point Dam were incorporated into the National Wild and Scenic Rivers System. After the Bureau of Reclamation constructed Canyon Ferry Reservoir in Montana, only 1% (40 km) of the river's length near the headwaters has truly uncontrolled releases of water. Alterations of the natural hydrologic cycle have profoundly affected water quality and the river's biota.

Physical and Chemical Characteristics

> "There is only one river with a personality, habits, dissipations, and a sense of humor.... a river that goes travelling sidewise, that interferes in politics, rearranges geography, and dabbles in real estate; a river that plays hide and seek with you today and tomorrow follows you around like a pet dog with a dynamite cracker tied to its tail.... It cuts corners, runs around at night, lunches on levees, and swallows islands and small villages for dessert. Its perpetual dissatisfaction with its bed is the greatest peculiarity of the Missouri.... It makes farming as fascinating as gambling. You never know whether you are going to harvest corn or catfish" (Fitch 1907).

This early description of the Missouri River aptly describes the physical wanderings that contribute to the river's chemical constitution. The shifting of the channel with attendant erosion and accretion was legendary. As much as 10 hectares have been eroded from a particular bend in 24 h (Duncanson 1909).

The eroding soils were developed primarily under forest or prairie conditions and contained much organic material in various stages of decay. Even today, the Missouri River carries an annual organic carbon load of 725,000 tonnes to the Mississippi River (Malcolm and Durum 1976). This load constitutes 33% of the organic carbon carried by the Mississippi River even though the Missouri River accounts for only 10% of the Mississippi's volume. But the organic carbon now carried by the Missouri River is less than 20% of the amount carried prior to the construction of reservoirs in the basin (Hesse et al. 1988).

The main-stem reservoirs have significantly changed physical and chemical features in large reaches of the Missouri River, including free-flowing sections downstream of the reservoirs. As previously noted, the Missouri River once experienced two spring floods. Today, main-stem reservoirs store much of the spring run-off, which is released gradually throughout the summer, fall, and winter. Compared to pre-reservoir conditions, the water volume flowing past any point below the first main-stem reservoir is lower in the spring and higher during the fall and winter.

Temporally controlled discharges from main-stem dams alter the quality and quantity of water below them. For example, mean flow rates today average 19.65 m^3/s in the Milk River, a turbid tributary of the Missouri River in Montana. These rates are nearly the same as flows before the main-stem reservoirs were constructed and they contribute less than 5% to the total volume of the Missouri River. However, during April, May, and June, flows from the Milk River, which enter the main stem downstream of Fort Peck Dam, average 23.4 m^3/s and they often constitute more than 10% of the Missouri River's total volume (Gardner and Stewart 1987). Consequently, tributary streams can seasonally exert a great influence on water quality in the Missouri River.

Water temperatures also have been affected by the reservoirs and are especially noticeable in reaches above and below Fort Peck Reservoir. The Missouri River downstream of Fort Peck Dam in Montana is clear and cold, very much unlike the warm and turbid Missouri River that flows into Fort Peck Reservoir 216 km upstream. Indeed, the water temperatures 243 km downstream from Fort Peck Dam remain about 4°C colder than at the headwaters of Fort Peck Reservoir (Gardner and Stewart 1987). Releases at Fort Peck Dam are from the hypolimnion, 56.5 m below the reservoir's surface at full pool and, hence, are much colder than at the surface.

Water quality of the Missouri River downstream from Fort Peck Dam in 1982 was as follows: specific conductance, 670 micromhos (μS)/cm; pH, 8.4; dissolved oxygen (DO), 11.5 mg/L; total alkalinity, 160 mg/L; sulfate ions, 198 mg/L; and total dissolved solids, 430 mg/L (Gardner and Stewart 1987). Some of these features were measured upstream from Fort Peck Reservoir during 1978 and 1979, a similar water year to 1982. Berg (1981) reported the following values: specific conductance, 629 μS/cm; total alkalinity, 152 mg/L; sulfate ions, 158 mg/L; and total dissolved solids, 472 mg/L.

Before flows in the Missouri River were controlled, the river transported a mean annual sediment load that increased from 25 million tonnes at Fort Peck, Montana, to 150 million tonnes at Yankton, South Dakota; 175 million tonnes at Omaha, Nebraska; and 250 million tonnes at its confluence with the Mississippi River (USACE 1979). After closure of the main-stem dams, suspended sediments at Omaha and the river's mouth averaged 25 and 125 million tonnes per year, respectively (USACE 1979). Since the river's banks between Sioux City, Iowa, and its mouth near St. Louis, Missouri, have been armored, the new sediment load must be obtained from the river bed after the flow passes Gavins Point Dam at Yankton, South Dakota, nearly sediment-free. The imbalance between degradation and aggradation has caused severe environmental deterioration for more than 346 km downstream of Gavins Point Dam (Hesse et al. 1989).

One of the earliest scientific accounts of the physical and chemical characteristics of the Missouri River was made in 1945 (Berner 1951). Berner worked on the lower river from its mouth to the Iowa-Missouri state boundary at river kilometer (RKm) 890. (River kilometer designations refer to distance upstream from the Mississippi River along the thalweg based upon 1960 measurements.) He noted that gravel and coarse sand covered most of the river bottom in this reach. Gravel substrates are scarce now in the lower river and the bottom is more uniformly covered by fine sands, a result of reservoir construction and channelization. After closure of Gavins Point Dam, the sand fraction in suspended sediments increased by a factor of 2.6 (Slizeski et al. 1982).

In 1945, turbidity in the lower river averaged 1,700 mg/L after August 15. The highest turbidity during spring and summer was 8,000 mg/L and the Secchi disk transparency was less than 13 mm (Berner 1951). Turbidity in the lower river at RKm 857 averaged 60 nephelometer turbidity units (NTU) for the period 1971 to 1975 (Todd and Bender 1982).

In 1945, with only rare exceptions, pH ranged between 7.4 and 7.8, and DO between 3.5 and 9.9 mg/L; oxygen saturation values of less than 50% were common during summer (Berner 1951). In the early 1970's, pH ranged from 7.1 and 8.3 at RKm 857 (Todd and Bender 1982). However, after a nuclear power station begin operating in 1974, the pH ranged from 7.0 to 9.6. Dissolved oxygen ranged between 6.8 and 12.3 mg/L and always remained above 77% saturation before the nuclear station started up. After startup, DO was measured as low as 21% saturation (Todd and Bender 1982).

No carbonate concentrations were measured in 1945 but bicarbonates ranged from 83 to 185 mg/L (Berner 1951). In the early 1970's, total alkalinity ranged between 163 and 187 mg/L and hardness averaged 242 mg/L (Todd and Bender 1982).

After impoundment of the middle Missouri River, total hardness and alkalinity relationships in some main-stem reservoirs reflected the use of nutrients by primary producers. In the early stages of impoundment of Lake Francis Case, 1953 through 1957, photosynthesis reduced alkalinity and hardness of water entering the reservoir (Neel 1967). Conversely, in other reservoirs during the early stages of impoundment (Lake Sakakawea and Lewis and Clark), water released through the dams exhibited gains in total alkalinity and hardness over inflows in most years. The increases above amounts contained in inflowing water represented amounts leached from the reservoir floor because dilution and photosynthesis normally reduce minerals in reservoirs (Neel 1967).

Surface waters in the drainage basin of the Missouri River are characterized by high sulfate-ion concentrations (Eddy 1963). Missouri River water follows this convention with sulfate-ion concentrations that range from 150 to 230 mg/L. Concentrations are relatively uniform throughout the river's length although they vary temporally. Sulfate-ion concentrations varied from 160 to 200 mg/L below Fort Peck

Dam (Gardner and Steward 1987) and averaged 150 to 220 mg/L in the channelized Missouri River at RKm 857 and 1040 (Todd and Bender 1982). Moreover, the mainstem impoundments apparently had little effect on sulfate-ion concentrations because preimpoundment values on the middle Missouri River varied from 172 to 268 mg/L (Damann 1951; Gastler and Moxon 1948).

Selected water quality variables during recent times can be compared between the channelized Missouri River at RKm 890 and an unchannelized reach at RKm 1,391 (Table 2). Most striking are the differences in nitrate and phosphate concentrations. Water leaving the reservoirs to the unchannelized reach is nearly devoid of these essential elements. Turbidity is higher downstream in the channelized reach, which is expected, but also demonstrates the influence of the reservoirs as sediment traps. Because dissolved organic carbon measures nearly the same at the two locations, controlling meander, erosion, and flooding downstream from the dams has apparently reduced the input of organic matter from the floodplain and watershed.

Biota and Abiotic Environment of the Missouri River

Plankton densities in the main channel of the lower Missouri River in 1945 were low, averaging 44 phytoplankton and 24 zooplankters per liter. The low densities were apparently caused by high turbidity, high current velocity, and lack of adjoining lentic habitat (Berner 1951). Since 1945, turbidity has been greatly reduced and 55 reservoirs under the P-S plan have created 10.4×10^{10} m^3 of lentic habitat (Missouri Basin States Association 1986). In 1974, phytoplankton density in the channelized Missouri River ranged between 1,000 and 25,000 units/L (Reetz 1982). A unit

TABLE 2. Comparison of selected physicochemical variables from two reaches of Missouri River, Nebraska. Channelized reach = RKm 890; unchannelized reach = RKm 1,391.

Variable	Channelized	Unchannelized
Specific conductance (s/cm)	709.0[e]	638.0[a]
pH	7.9[e]	8.6[a]
Total alkalinity (mg/L as Ca/CO$_3$)	169.0[e]	163.0[a]
Turbidity (Nephelometer turbidity units)	58.0[e]	18.0[b]
Total dissolved solids (mg/L)	475.0[e]	452.0[b]
Biochemical oxygen demand (mg/L)	3.8[e]	2.3[b]
Chemical oxygen demand (mg/L)	54.3[e]	126.8[b]
Total phosphate (mg/L)	0.3[e]	0.1[a]
Total nitrate (mg/L)	0.4[e]	0.1[b]
Chlorophyll a (mg/m^3)	18.6[d]	5.0[a]
Phytoplankton, carbon fixation rate (mg C/m^3.h)	124.9[d]	
Periphyton, carbon fixation rate (mg C/m^3.h)		1.2[a]
Dissolved organic carbon (mg/L)	4.6[c]	4.8[a]

[a]Hergenrader and Carr (1986)
[b]Hesse and Mestle (1987)
[c]Malcom and Durum (1976)
[d]Reetz (1982)
[e]Todd and Bender (1982)

consisted of one cell for unicellular species, a 100 millimicron length for filamentous forms and four cells for colonial species. Zooplankton density was measured at 1.7/L and these zooplankters apparently originated in the main-stem reservoirs (Repsys and Rogers 1982). Cladoceran and copepod densities in Lewis and Clark Reservoir were dependent upon discharges from an upstream reservoir, Lake Francis Case (Cowell 1970). Downstream of the main-stem reservoirs, the loss of localized off-channel areas through channelization and degradation has reduced the availability of zooplankton; even the substantial water volume held in storage upstream has not offset this loss. The gain in autochthonous primary production (phytoplankton) in the reservoirs is probably overshadowed by reduced input of allochthonous organic matter from the floodplain and basin, which is no longer available to aquatic communities due to a cessation of flooding and meander.

In 1945, the biomass (wet weight) of benthic invertebrates in the middle and lower reaches of the Missouri River averaged 0.73 kg/hectare (Berner 1951) and, two decades later, the mean benthic biomass (wet weight) in channelized locations was estimated as 0.69 kg/hectare (Morris et al. 1968). Both studies attributed low benthic standing crops to shifting bottom substrates, siltation, fluctuating water levels, swift current, and the absence of aquatic vegetation. The most important habitat lost to channelization was the large area of backwaters, none of which were surveyed by Berner (1951) or Morris et al. (1968). More recently, invertebrates in remnant backwaters in an unchannelized reach produced about 4.8 g/m^{2}yr (dry weight) (Mestl and Hesse 1987). Annual production by aufwuchs was estimated as more than 70 g/m^2 (dry weight) in chute habitat of the unchannelized river (Dixon 1986). Although Berner (1951) noted that benthic production in the lower Missouri River was low, about 0.14 kg/hectare, he estimated commercial fish yield to be 1.82 kg/hectare. Benthic sampling methods were probably inadequate to estimate secondary production accurately.

Organisms colonizing the Missouri River, especially the middle and lower reaches, during the pre-control river era were probably adapted to high turbidity. Turbidity was an important indicator of erosion and flooding. The latter carried large amounts of allochthonous organic matter to stream inhabitants, which surely contributed substantially to an estimated commercial harvest of over 800,000 kg of fish in Missouri in 1947 (Funk and Robinson 1974).

Algal growth was not believed to be limited by nutrients in the channelized reach investigated by Todd and Bender (1982). Nitrate and phosphate concentrations are at least four times less in the unchannelized river than in the channelized river, but existing data do not explain a cause/effect relationship (Table 2). Conversely, there is evidence that reservoirs on the main-stem Missouri River are nutrient limited. Phytoplankton standing crop and primary productivity were limited by phosphorus in Lake Francis Case and, to a lesser extent, in Lewis and Clark Lake (Martin and Novotny 1975).

Quantitative fish surveys were not completed in the lower one third of the Missouri River before work on the main-stem dams and channel was completed. The following species of fish are now abundant in unchannelized and channelized portions of the river: shovelnose sturgeon *Scaphirhynchus platorynchus*, shortnost gar *Lepisosteus platostomus*, white bass *Morone chrysops*, sauger *Stizostedion canadense*, freshwater drum *Aplodinotus grunniens*, gizzard shad *Dorosoma cepedianum*, goldeye *Hiodon alosoides*, common carp *Cyprinus carpio*, emerald shiner *Notropis atherinoides*, red shiner *Notropis lutrensis*, sand shiner *Notropis stramineus*, river carpsucker *Carpiodes carpio*, shorthead redhorse *Moxostoma macrolepidotum*, smallmouth buffalo *Ictiobus bubalus*, and channel catfish *Ictalurus punctatus* (Hesse et al. 1989).

Industrial and Municipal Pollution

Water quality problems have occurred on the lower Missouri River since at least 1910, when a notable increase in human deaths caused by typhoid was documented in bordering towns. The United States Public Health Service (USPHS) identified sewage pollution as a major contributor to the high incidence of typhoid (Ford 1982). This observation marked the first documented water pollution in the lower river. Many pollution problems stemmed from the public view that waste disposal was one valuable use of the Missouri River. For the next 50 to 60 years, pollution from untreated human and animal wastes dominated water quality concerns. From 1920 to 1958, levels of bacterial contamination rose while treatment of water supplies for drinking water failed to meet USPHS standards. Not only had the human population grown in areas such as Kansas City, but so had the meatpacking industry and stockyards (Ford 1982).

Low dissolved oxygen levels were related to how waste was disposed. In May 1964, this was dramatically demonstrated with a major fish kill extending from Kansas City to at least 161 km downstream. Heavy thunderstorms washed large amounts of organic material into the river, resulting in oxygen depletion. Another fish kill, in June 1967 near Rulo, Nebraska, was also caused by low dissolved oxygen levels following a heavy rain (Missouri Department of Conservation, unpublished).

From 1947 to 1959 commercial fishing around Kansas City declined due to several factors, including water quality impairment associated with petroleum spills. The Missouri Department of Conservation received numerous complaints from citizens and fishermen of oil contamination that killed fish, caused odor problems, rotted nets, and altered the taste of fish. In addition, black, "foul" waters and dying fish were noted in 1966 at the confluence of the Big Blue and Missouri rivers directly downstream from Kansas City (Missouri Department of Conservation, unpublished).

Flavor tests in 1969 revealed unacceptable tastes in fish taken 1.6 km downstream of Sioux City, 4 km downstream of Omaha, and up to 35 km downstream of the Kansas City metropolitan area (Ford 1982). This problem was attributed to releases of municipal and slaughter house wastes in the Sioux City and Omaha areas, and of oil refining and chemical industry wastes in the Kansas City area.

Data collected by the Federal Water Quality Administration from 1968 to 1970 revealed another serious pollution problem. Nineteen *Salmonella* serotypes were isolated from samples of Missouri River water, many of which were pathogenic human strains. Results of investigations in 1969 and 1970 showed that the water was of acceptable quality upstream of Sioux City. However, fecal coliform densities were high and *Salmonella* was present downstream of Sioux City (USEPA 1971a). These studies also demonstrated that viruses could survive in Missouri River water for 25 h or longer, indicating that they could survive long enough to enter most water supply intakes (USEPA 1971a). The investigators concluded that, at a minimum, secondary sewage treatment was needed to protect other water users. In 1971, the USEPA believed the lower Missouri River represented a potential hazard to anyone using it as a source of drinking water or for recreation.

By 1974, water quality in the Missouri River had improved somewhat as most cities had at least primary treatment systems to remove floating materials and, thus, to prevent sludge depositions along banks. Some cities also had installed secondary treatment systems (Whitley and Campbell 1974). However, the Missouri Department of Natural Resources reported that oil, floating debris, and the presence of materials that caused objectionable tastes and odors in drinking water and fish were still problems in 1974.

In the 1970's, attention turned to the presence of chlorinated hydrocarbons, other pesticides, polychlorinated biphenyls (PCBs), and mercury in fish and water. Early sampling for these contaminants was infrequent and nonsystematic. However, analyses of water from the lower Missouri River during the 1970s revealed ten heavy metals, three pesticides, and two volatile organics (Ford 1982). In 1980, 37 pollutants (13 metals, 23 organic compounds, and cyanide), most of which were common industrial solvents, were detected at four water treatment plans in St. Louis (Ford 1982).

Monitoring contaminants in fish flesh was infrequent in early years and often not well documented. Iowa biologists took samples at Council Bluffs, Iowa, from 1968 to 1976 and found that DDE levels in fish flesh violated standards 33% of the time, dieldrin 13%, and DDT 9% (Ford 1982). The National Pesticide Monitoring Program (NPMP) also sampled fish in the fall of 1969 on the lower Missouri River. Dieldrin, PCBs, and DDT were detected in common carp, channel catfish, goldeye, and white crappie *Pomoxis annularis*. No concentrations exceeded the action levels established by the United States Food and Drug Administration at that time. (Action levels are concentrations that assume a potential threat and call for steps to protect human health.) However, PCBs in common carp were recorded at 4.58 mg/kg, which exceeded the current action level of 2.0 mg/kg (Henderson et al. 1971). At Hermann, Missouri, from 1970 to 1974, PCBs, aldrin, and dieldrin in fish equaled or exceeded criteria set by the National Academy of Science and National Academy of Engineers (NAS-NAE), which were greater than 0.5 mg/kg for PCBs, and greater than 0.1 mg/kg for aldrin and dieldrin (Schmitt et al. 1981). Samples during 1976 to 1978 equalled or exceeded these criteria and some samples also exceeded the more than 0.1 mg/kg limit for chlordane (Schmitt et al. 1983). The NPMP residue values represented whole-body concentration of contaminants in fish and, thus, were not directly comparable to samples of only edible portions.

Sampling by the Missouri Division of Health (MDH) in 1970 to 1971 revealed the limit for chlordane (0.3 mg/kg) was occasionally exceeded in the flesh of channel catfish, with values that ranged from 0.16 to 0.41 mg/kg (Missouri Department of Conservation, unpublished). Further sampling in 1978 to 1979 showed that the limit for chlordane was exceeded in the flesh of several shovelnose sturgeon from the Rockport to Easley, Missouri reach (Missouri Department of Conservation, unpublished).

Missouri River fish were also examined for contaminants in the mid–1970s by the Missouri Department of Conservation. Dieldrin levels were high in catfish at several locations. As a result, the department issued a public statement warning people of the problem. The commercial fishery was impacted because the public no longer purchased fish. Analyses were inconclusive for chlordane because of interference from PCBs but the pesticide was present (J. R. Whitley, Missouri Department Conservation, personal communication).

The USEPA in 1976 sampled fish from three sites on the Missouri River, viz. St. Joseph, Missouri, US 291 Bridge (RKm 565), and Jefferson City, Missouri. Low levels of PCBs, DDT, DDE, DDD, chlordane, dieldrin, and BHC were detected in goldeye, shortnose gar, river carpsucker, buffalo, common carp, and shorthead redhorse. The wide variety of pesticides detected at some sites was considered strong evidence of non-point source loading (Lorenz 1977).

Legislation was enacted in the 1970s to attempt to control chemical contamination of the Missouri River. In 1974, the registration of aldrin and dieldrin was suspended on the basis of adverse health effects in rodents. Use of these chemicals was still permitted for termite treatment by direct application to the soil and for treating non-food nursery plants. The Toxic Substance Control Act (TSCA), Public Law 94-469, was signed into law in October 1976. This act restricted the use, manufacture, sale, and distribution of PCBs. Use was restricted to sealed systems as of October 11, 1977, manufacture was banned on January 1, 1979, and all processing and distribution ceased July 1, 1979. Use of chlordane was also restricted in the 1970s. In 1978, the USEPA prohibited agricultural use of chlordane but allowed it for subsurface termite control and treatment of non-food nursery plants.

The Missouri Department of Conservation began a more comprehensive study of contaminants in the Missouri River in 1984. Its goal was to determine whether contaminants were present in fish flesh at levels hazardous to human health. Chlordane, dieldrin, and PCBs were still the contaminants recorded at the highest levels in fish flesh (Table 3). All samples analyzed were skinless fillets, unlike whole fish samples of some previous studies. In 1985, Missouri residents were advised not to eat shovelnose sturgeon taken from the Missouri River in the state due to PCB contamination and, in 1987, not to eat common carp, buffalo, catfish, river carpsucker, freshwater drum, and shovelnose sturgeon from the reach between Kansas City and St. Louis because of chlordane. The advisory issued in 1987 had great impact on the public and substantially reduced sales of commercially caught fish.

Problems with Heavy Metals

The Missouri River basin varies widely in climate, topography, geography, and soils, and these variables determine human land and water use. Water quality and biota

TABLE 3. Summary of contaminant analysis for flesh samples from Missouri River, 1984 to 1986 (Bush and Grace 1987). Numbers in parenthesis represent sample size on which results were based.

Site, year, and species	Sample size	Contaminants[a] Dieldrin (mg/kg)	Chlordane (mg/kg)	PCBs (mg/kg)
Howell Island, Missouri 1984, RKm 76				
Common carp	(20)	0.073	0.143	0.043
Easley, Missouri 1984, RKm 272				
Channel catfish	(20)	0.094	0.205	0.056
Common carp	(20)	0.065	0.149	0.038
Shovelnose sturgeon (flesh)	(20)	0.126	0.210	2.518
River carpsuckers	(20)	0.043	0.152	0.267
Flathead catfish	(20)	0.080	0.160	0.080
Easley, Missouri 1986				
Channel catfish	(5)	0.002	0.518	0.075
Common carp	(5)	0.004	0.118	0.075
Shovelnose sturgeon (flesh)	(5)	0.062	0.860	0.641
Shovelnose sturgeon (eggs)	(5)	0.194	0.921	1.572
River carpsuckers	(5)	0.069	0.921	0.356
Freshwater drum	(5)	0.033	0.139	0.123
White bass	(2)	0.027	0.124	0.075
Bigmouth buffalo	(2)	0.142	0.423	0.075
Lexington, Missouri 1984, RKm 517				
Common carp	(20)	0.203	0.126	0.045
Kansas City, Missouri RKm 596				
Channel catfish	(5)	0.112	0.777	0.196
Common carp	(5)	0.025	0.548	0.075
Shovelnose sturgeon (flesh)	(1)	0.004	0.146	0.153
Shovelnose sturgeon (eggs)	(1)	0.252	6.735	3.020
Atchison, Kansas 1984, RKm 704				
Common carp	(20)	0.065	0.058	0.034
Rulo, Nebraska 1984, RKm 802				
Common carp	(20)	0.098	0.058	0.027

[a]Food and Drug Administration action levels: Dieldrin 0.3 mg/kg; Chlordane 0.3 mg/kg; PCBs 2.0 mg/kg.

in streams reflect current and previous land use patterns. In the Great Plains, many soils are derived from shales deposited in the Cretaceous and Tertiary periods under marine and lacustrine conditions. These soils contain soluble salts with heavy metals that occasionally result in water quality problems. In South Dakota, 8% of the water samples may contain more than 10 μg/L of selenium, the USEPA allowable limit for drinking water (Stach 1978). Dietary selenium is periodically toxic to livestock in South Dakota.

Perhaps the heavy metal of greatest concern in the Missouri River basin, however, is mercury. Natural mercury is emitted from nonpoint sources, the black shales, and from industrial point sources (gold mines) in the Black Hills. For more than 100 years, mercury was used to form amalgams to recover gold from ore and considerable mercury was lost to streams draining the Black Hills.

Methylmercury is the common form lethal to aquatic organisms. Moreover, sublethal concentrations can impair reproduction and recruitment, thus, subtly changing community structure. Methylmercury in polluted ecosystems bioaccumulates in fish and fish-eating birds. Humans are usually exposed to methylmercury by eating contaminated fish.

The Homestake Mining Company, established in 1878, was the only commercial gold mine in the Black Hills and a major economic force in western South Dakota for years. Mine effluents were released into Whitewood Creek, a small stream whose waters eventually enter the Missouri River about 70 km upstream of Pierre, South Dakota, via the Belle Fourche and Cheyenne rivers (Figure 1). The effluents had a pronounced effect on the biota of Whitewood Creek and portions of the Belle Fource and Cheyenne rivers. These streams were also polluted by untreated human sewage from Lead and Deadwood, South Dakota, population 12,000. For decades Whitewood Creek harbored no fish and riparian vegetation was absent 58 km downstream of the mine effluent. Bottom organisms were eliminated from Whitewood Creek and from 97 km of the Belle Fourche River downstream of its confluence with Whitewood Creek. Even in the lower Belle Fourche and the Cheyenne rivers, the species richness and densities of benthos never attained the levels present upstream of Whitewood Creek (Thilenius 1965).

Analysis of mine effluent entering Whitewood Creek during June 1971 revealed that the Homestake Mine was discharging 142 kg of cyanide, 109 kg of zinc, 33 kg of copper, 5 to 18 kg of mercury, 9.65 metric tonnes of arsenic, and 2,779 metric tonnes of suspended solids daily (USEPA 1971b). Concentrations of cyanide, arsenic, mercury, and suspended solids were each sufficiently high, independently or in concert, to eliminate biota from Whitewood Creek. Estimates of the mercury discharged into Whitewood Creek after nearly 100 years vary between 3,962 and 15,850 metric tonnes. Most mercury was bound to mine tailings that were distributed as alluvial deposits downstream to the confluence of the Cheyenne and Missouri rivers, a distance of more than 300 km. Perhaps as much as 500 million metric tonnes of tailings exist in the system (Fox Consultants 1984). Alluvial sediments resembling mine tailings encompass 1,100 hectares of land on the floodplain of Whitewood Creek and the Belle Fourche River (USEPA 1973b). Deposits vary in depth from a few centimeters to greater than 3m.

All components necessary for continuous dispersal of mercury in its most hazardous form, methylmercury, exist in this portion of the Missouri River's drainage basin. Mercury lost during gold amalgamation was transported downstream sorbed to finely divided sediments. Dispersal was accompanied by organic matter in the form of raw sewage that, by volume, comprised 20% of the wastes entering Whitewood Creek (Thilenius 1965). Methylation and biomagnification of mercury created a potential human health problem in addition to severe environmental perturbations.

In the summer of 1970, analyses revealed that over 25% of all fish sampled from the Cheyenne River embayment of Lake Oahe, a main-stem Missouri River reservoir, contained more than 0.5 mg/kg of methylmercury. Concentrations above this level are considered hazardous to human health by the U. S. Food and Drug Administration. Subsequent studies in 1971 verified these results. Federal and state concerns eventually forced stringent pollution abatement efforts by the Homestake Mining Company. Almost a century of gross pollution was not a culpable act solely attributable to the mining company. Impunity was accorded the company by the state with tacit consent from an apathetic public.

The Homestake Mining Company discontinued the use of mercury to amalgamate gold in December 1970 (USEPA 1971b). This was the first step towards alleviating mercury pollution in the drainage basin although contaminated effluent from mine tailings and untreated sewage still entered Whitewood Creek. Surface waters, sediments, and fish flesh from the polluted system contained mercury, cyanide, and arsenic at concentrations above those in adjacent streams (Table 4). Some sport fishes exhibited mercury levels as high as 2.0 mg/kg, prompting the state of South Dakota to post notices recommending restricted consumption of fish from the Cheyenne River embayment of Lake Oahe. The hair of humans who ate contaminated fish from the Cheyenne embayment two or more times per week contained significantly higher mercury levels ($\bar{X}$ = 5.65 mg/kg, $P < 0.01$) than did non-consumers or those who ate fish only twice a month ($\bar{X}$ = 1.82 mg/kg). A few people acquired mercury in excess of the "safe" level of 6 mg/kg but none exhibited clinical symptoms of mercury toxicity (Heisinger et al. 1974).

The second major step to alleviate mercury pollution was taken when the mining company constructed a permanent tailings pond in 1977. The mine effluent immediately cleared. It still contained high levels of cyanide, arsenic, and other heavy metals but little mercury. In 1979, the Lead-Deadwood wastewater treatment plant, an advanced secondary treatment facility, became operational and untreated sewage no longer entered Whitewood Creek. Subsequent analyses of mercury in surface waters, sediments, and fish flesh reflect the success of initial efforts to control mercury emissions and mobility (Table 4). No samples of fish, ground water, surface water, or sediment collected after 1980 had mercury levels in excess of safe drinking-water or food standards (USFWS 1985). Whitewood Creek still had no biota but the danger from mercury toxicity was lessened.

Whitewood Creek was declared a "superfund" site in 1981 under the Comprehensive Environmental Response Compensation and Liability Act of 1980. Subsequent studies generally concluded that water quality had improved since disposal of tailing into Whitewood Creek ceased and there were no significant water quality problems in surface waters caused by mine waste disposal (Fox Consultants 1984). No superfund money was used to recover mercury distributed throughout alluvial deposits. Natural pathways appeared to be gradually ameliorating the threat that mercury and other heavy metals posed to public health and the environment. The principal regulatory concern shifted to arsenic in ground and surface waters and in riparian vegetation (Fox Consultants 1984).

The final phase in restoring Whitewood Creek occurred when the wastewater treatment plant for the Homestake Mine effluent was completed in January 1985. This plant removes cyanide, ammonia, and other contaminants. Since the late 1970s, surface water quality has improved remarkably and mercury levels in Whitewood Creek are now similar to those of adjacent streams (Table 4).

TABLE 4. Mean concentrations of mercury (Hg), total cyanide (Cn), and total arsenic (As) from the Cheyenne River drainage, Missouri River basin. * = Exceeds USEPA or USFDA standards for human consumption; ND = not detectable. All values are µg/kg.

Station and Source	Period I (1970–1972) No Pollution controls			Period II (1982–1984) Some pollution controls			Period III (1985–1987) Advanced pollution control		
	Hg	Cn	As	Hg	Cn	As	Hg	Cn	As
Surface water									
Whitewood Creek I (above effluent)	<1.00	<20	12.5[a]	1.8	<0.1	11[c]	<0.4	<10	12[d]
Whitewood Creek II (below effluent)	*69.00	2600	*3,650.0[a]	*2.4	1,170.0	*425[c]	<0.5	46	20d
Belle Fourche River (above effluent)	*3.00	ND	ND[a]	1.0	238.0	46[c]	<1.0	53	28[d]
Cheyenne River (below effluent)	1.80	ND	ND[a]	—	—	—	—	—	—
Ground water									
Whitewood Creek I (above effluent)	<0.20	<20	<10.0[a]	0.3	<10.0	3[c]			
Whitewood Creek II (below effluent)	0.25	<20	<10.0[a]	ND	—	*218[c]			
Belle Fourche River (above effluent)	0.33	<20	*462.0[a]	ND	—	24[c]			
Cheyenne River (below effluent)	—	—	—	—	—	—			

TABLE 4. Continued.

Station and Source	Period I (1970–1972) No Pollution controls			Period II (1982–1984) Some pollution controls			Period III (1985–1987) Advanced pollution control		
	Hg	Cn	As	Hg	Cn	As	Hg	Cn	As
Sediments									
Background	<100.00	—	10,217.0[a]						
Whitewood Creek II (below)	855.00	—	3,100,000.0[a]						
Belle Fourche River (above effluent)	569.00	—	1,621,519.0[a]						
Cheyenne River (below effluent)	333.00	—	393,333.0[a]	10,550.0	—	15,795[b]			
Fish Flesh									
Whitewood Creek I (above effluent)	<40.00[a]	—	—	—	—	—			
Whitewood Creek II (below effluent)	No fish present			—	—	—			
Belle Fourche River (above effluent)	—	—	—	123.0[c]	—	—			
Cheyenne River (below effluent)	*500.00[a]	—	—	199.0[b]	—	—			

[a]U.S. Environmental Protection Agency (1971b, 1973a, and 1973b)
[b]U.S. Fish and Wildlife Service (1985)
[c]Fox Consultants (1984)
[d]J. Bowar, South Dakota Department of Water and Natural Resources, personal communication.

In the early 1980s, insects, mosses, and algae reappeared in Whitewood Creek and riparian vegetation returned to the denuded 58 km reach (Glover 1982). Trout were successfully introduced to the reach formerly decimated by mining and municipal pollutants.

Whitewood Creek is an example of gross environmental degradation tacitly condoned by public apathy that was halted and then ameliorated by a substantial pollution-control effort. Once pollutants were no longer discharged, the ecosystem repaired itself, a tribute to its resilience. This story has not reached its conclusion, however, as unresolved questions remain concerning arsenic. Moreover, more than 15,000 metric tonnes of mercury may be distributed throughout the watercourse. The potential for future problems with heavy metal toxicity are real and a constant monitoring program is necessary, not only in the Cheyenne River system but in the main-stem Missouri River as well.

Algae, macroinvertebrates, and fish flesh were recently analyzed for copper, chromium, lead, and zinc at two sites, one upstream and one downstream of Sioux City, Iowa (Stevens and Tondreu 1986). Algae and macroinvertebrates contained relatively high levels of heavy metals, suggesting bioaccumulation from the sediments. Additionally, heavy metals were considerably higher at the site downstream of Sioux City than upstream. Industrial use of these metals was discontinued in Sioux City decades ago. Perhaps heavy metals in the sediments represent a "time bomb" that will require future attention.

Future Prospects

The Missouri River's future quality and quantity are clouded by potential climatic changes, past human usage, and future plans for the resource. The principal industry of the basin is agriculture, primarily grazing and dryland farming, and about 95% of the basin's area is devoted to this use. Surface water constitutes only 1.2% of the drainage basin but it plays a disproportionatelly larger role in economics of the area (USACE 1979). Presently about 2 to 3% of the basin's farmland is irrigated but this consumptive water use is increasing. Indeed, if all irrigation units of the P-S Plan authorized by Congress were developed, they would use 30% of the average annual discharge at the river's mouth (Missouri Basin States Association 1985).

The extensive developments envisioned in the Missouri River basin by the P-S Plan must one day deal with an insufficient water supply as the greatest threat to further irrigation, conservation of wildlife resources, navigation, and maintenance of water quality (Lord et al. 1975). By 1970, just 26 years after the P-S Plan was authorized, $6.0 \times 10^9 m^3$ of water (8.6% of the annual discharge) was consumed annually (USACE 1979). Evaporation removed about 37% of the depletions while irrigation accounted for 43%. Depletions are larger today. Since the basin is already deficient in water, as the river gives up more water to grow crops, new depletions will complicate already untenable fisheries management problems by intensifying reservoir fluctuations. Additionally, new depletions will probably increase temporal demands for water, reduce hydroelectric power production, alter the navigation season, destabilize aggradation and degradation processes, and create additional water quality problems. Degraded water quality could be severe locally if irrigation return flows contribute contaminated water to a system lacking capacity to dilute the contaminants to an acceptable level.

In final analysis, however, climatic changes effected by anthropogenic causes could be the most important factor influencing the water supply in the Missouri River basin. Continued reliance on fossil fuels as a primary energy source may increase atmospheric carbon dioxide, resulting in the "greenhouse effect." If scientific estimates are correct, CO_2-induced warming may account for about a 1°C rise in global temperatures by the early part of the 21st century. Not all parts of the world would be equally affected by this temperature change. In the United States, the Great Plains and, hence, the Missouri River basin, would be severely impacted. This area, already water deficient, would become much warmer and drier (Bernard 1980). Maintaining water quality would, presumably, be granted priority status but minimum stream flows for aquatic biota might not fare well.

More than 180 years have elapsed since Lewis and Clark encountered the abundant living resources of the Missouri River. The ecosystem has been severely altered in intervening years. Aquatic biologists have the responsibility to plan for the future and to insure that all users of Missouri River water are considered.

References

Berg, R. K. 1981. Fish populations of the wild and scenic Missouri River, Montana. Federal Aid in Fish Restoration, Project FW-3-R, Performance Report. Montana Department of Fish, Wildlife and Parks, Helena.

Bernard, H. W., Jr. 1980. The greenhouse effect. Ballinger, Cambridge, Massachusetts.

Berner, L. M. 1951. Limnology of the lower Missouri River. Ecology 32:1–12.

Bush, J., and T. Grace. 1987. Fish contaminant study of the Missouri and Mississippi rivers. Performance Report. Missouri Department of Conservation, Columbia.

Cowell, B. C. 1970. The influence of plankton discharges from an upstream reservoir on standing crops in a Missouri River reservoir. Limnology and Oceanography 15:427–441.

Damann, K. E. 1951. Missouri River basin plankton study, 1950. United States Public Health Service, Environmental Health Center, Cincinnati, Ohio.

Dixon, K. M. 1986. Secondary production of macroinvertebrates from selected habitats of the unchannelized Missouri River. Master's thesis. University of South Dakota, Vermillion.

Duncanson, H. H. 1909. Observations on the shifting of the channel of the Missouri River since 1883. Science 29:869–871.

Eddy, S. 1963. Minnesota and the Dakotas. Pages 301–315 *in* D. G. Frey, editor. Limnology in North America. University of Wisconsin Press, Madison.

Fitch, G. 1907. The Missouri River; its habits and eccentricities described by a personal friend. The American Magazine 63:637–650.

Flint, R. F. 1955. Pleistocene geology of eastern South Dakota. United States Geological Survey, Professional Paper 262, Washington, D. C.

Ford, J. C. 1982. Water quality of the lower Missouri River, Gavins Point Dam to the mouth. Missouri Department of Natural Resources, Jefferson City.

Fox Consultants. 1984. Whitewood Creek project, South Dakota project 21024.0. Report to South Dakota Department of Water and Natural Resources, Pierre. (3 Volumes with appendices)

Funk, J. L., and J. W. Robinson. 1974. Changes in the channel of the lower Missouri River and effects on fish and wildlife. Aquatic Series 11. Missouri Department of Conservation, Jefferson City.

Gardner, W. M., and P. A. Steward. 1987. The fishery of the lower Missouri River, Montana. Federal Aid in Fish Restoration, Project FW-2-R, Performance Report. Montana Department of Fish, Wildlife and Parks, Helena.

Gastler, G. F., and A. L. Moxon. 1948. Composition of Missouri River water samples taken at monthly intervals from May 15 to September 15, 1947. Proceedings South Dakota Academy of Science 27:32–35.

Glover, R. 1982. Whitewood Creek, a future trout system. South Dakota Conservation Digest 49:22–24.

Heisinger, J. F., F. Foss, and S. Dickout. 1974. Mercury in the hair of South Dakota fishermen. University of South Dakota, Biology Department, Vermillion (Mimeo.)

Henderson, C., A. Inglis, and W. Johnson. 1971. Residues in fish, wildlife and estuaries; organochlorine insecticide residues in fish-fall 1969. National Pesticide Monitoring Program. Pesticides Monitoring Journal 5:145–151.

Hergenrader, G. L., and J. M. Carr. 1986. Biology of attached algae in unchannelized sections of the Missouri River. Project Completion Report to United States Army Corps of Engineers, Contract DACW 45-83-C-0206. University of Nebraska, Lincoln.

Hesse, L. W., and G. E. Mestl. 1986. Ecology of unchannelized reaches. Federal Aid in Fish Restoration, Project F-75-R, Performance Report. Nebraska Game and Parks Commission, Norfolk.

Hesse, L. W., C. W. Wolfe, and N. K. Cole. 1988. Some aspects of energy flow in the Missouri River ecosystem and a rationale for recovery. Pages 13–29 *in* N. G. Benson, editor. The Missouri River-the resources, their uses, and values. North Central Division American Fisheries Society, Special Publication 8.

Hesse, L. W., and G. E. Mestl. 1987. Ecology of unchannelized reaches. Performance Report, Federal Aid in Fish Restoration, Project F-75-R. Nebraska Game and Parks Commission, Norfolk.

Hesse, L. W., J. C. Schmulbach, J. M. Carr, K. D. Keenlyne, D. G. Unkenholz, J. W. Robinson, and G. E. Mestl. 1989. Missouri River fishery resources in relation to past, present, and future stresses. Pages 352–371 *in* D. P. Dodge, editor. Proceedings of the international large rivers symposium (LARS). Canadian Special Publication of Fisheries and Aquatic Sciences 106.

Lord, W. D., S. K. Tublesing, and C. Althen. 1975. Fish and wildlife implications of upper Missouri River basin water allocation. Institute of Behavioral Science, Monography 22. University of Colorado, Boulder.

Lorenz, T. F. 1977. Summary report of the occurrence of PCB fish flesh contamination in the rivers and streams of Region VII. U. S. Environmental Protection Agency, Kansas City, Kansas.

Malcolm, R. L., and W. H. Durum. 1976. Organic carbon and nitrogen concentrations and annual organic carbon load of six selected rivers of the United States. Water Supply Paper 1817-F. U. S. Geological Survey, Washington, DC.

Martin, D. B., and J. F. Novotny. 1975. Nutrient limitation of summer phytoplankton growth in two Missouri River Reservoirs. Ecology 56:199–205.

Mestl, G. E., and L. W. Hesse. 1987. Investigation into the secondary productivity of an unchannelized Missouri River backwater. Final Report, Dingle-Johnson Project F-75-R-4. Nebraska Game and Parks Commission, Norfolk.

Missouri Basin States Association. 1985. Interim report on irrigation. MBSA, Omaha, Nebraska.

Missouri Basin States Association. 1986. Interim report on multipurpose program features and non-separable program effects. MBSA, Omaha, Nebraska.

Morris, L. A., R. N. Langemeier, T. R. Russell, and A. Witt, Jr. 1968. Effects of main-stem impoundments and channelization upon the limnology of the Missouri River, Nebraska. Transactions of the American Fisheries Society 97:380–388.

Neel, J. K. 1967. Reservoir eutrophication and dystrophication following impoundment. Pages 322–334 *in* Reservoir fishery resources symposium. American Fisheries Society, Southern Division, Bethesda.

Reetz, S. D. 1982. Phytoplankton studies in the Missouri River at Fort Calhoun station and Cooper nuclear station. Pages 71–84 *in* L. Hesse, G. Hergenrader, H. Lewis, S. Reetz and A. Schlesinger, editors. The middle Missouri River. Missouri River Study Group, Norfolk, Nebraska.

Repsys, A. J., and G. D. Rogers. 1982. Zooplankton studies in the channelized Missouri River. Pages 124–146 *in* L. Hesse, G. Hergenrader, H. Lewis, S. Reetz, and A. Schlesinger, editors. The middle Missouri River. Missouri River Study Group, Norfolk, Nebraska.

Sayre, W. W. and J. F. Kennedy. 1978. Degradation and aggradation of the Missouri River. University of Iowa, Iowa Institute of Hydraulic Research, Iowa City.

Schmitt, C. J., J. L. Ludke, and D. F. Wash. 1981. Organochlorine residues in fish: National Pesticide Monitoring Program 1970–74. Pesticide Monitoring Journal 14:136–206.

Schmitt, C. J., M. A. Ribick, J. L. Ludke, and T. W. May. 1983. National Pesticide Monitoring Program: Organochlorine residues in freshwater fish, 1976–79. Resource Publication 152, U. S. Fish and Wildlife Service. NTIS, Springfield, Virginia.

Slizeski, J. J., J. l. Andersen, and W. G. Dorough. 1982. Hydrologic setting, system operation, present and future stresses. Pages 15–38 *in* L. Hesse, G. Hergenrader, H. Lewis, S. Reetz, and A. Schlesinger, editors. The middle Missouri River. Missouri River Study Group, Norfolk, Nebraska.

Stach, R. L. 1978. Selenium in South Dakota ground and surface water. Job Completion Report, Water Resources Institute, Project A-006-SDAK. South Dakota State University, Brookings.

Stevens, M., and R. Tondreau. 1986. Biomagnification of heavy metals (Cu, Cr, Pb and Zn) in the Missouri River near Sioux City, Iowa. Morningside College, Sioux City, Iowa. (Mimeo.)

Thilenius, C. A. 1965. An evaluation of pollution in the Belle Fourche and Cheyenne rivers due to wastes carried by Whitewood Creek. South Dakota Department Game, Fish and Parks, Pierre. (Mimeo.)

Todd, R. D., and J. F. Bender. 1982. Water quality characteristics of the Missouri River near Fort Calhoun and Cooper Nuclear stations. Pages 39–70 *in* L. Hesse, G. Hergenrader, H. Lewis, S. Reetz, and A. Schlesinger, editors. The middle Missouri River. Missouri River Study Group, Norfolk, Nebraska.

USACE (U. S. Army Corps of Engineers). 1979. Master Manual, Omaha District, USACE, Omaha, Nebraska.

USACE (U. S. Army Corps of Engineers). 1980. Investigation of channel degradation. Missouri River, Gavins Point Dam to Platte River confluence. Engineering Division, Omaha District, USACE, Omaha, Nebraska.

USACE (U. S. Army Corps of Engineers). 1985. Missouri River main-stem reservoirs. Summary of actual 1984–85 operations and operating plan for 1985–86. USACE, Omaha, Nebraska.

USEPA (U. S. Environmental Protection Agency). 1971a. Everybody can't live upstream, a contemporary history of the water quality problems on the Missouri River. Region VII, USEPA, Kansas City, Missouri.

USEPA (U. S. Environmental Protection Agency). 1971b. Pollution affecting water quality of the Cheyenne River system, western South Dakota. Division of Field Investigations, Region VII, USEPA, Kansas City, Missouri and VIII, USEPA, Denver, Colorado.

USEPA (U. S. Environmental Protection Agency). 1973a. Mercury concentrations in fish in Lake Oahe, South Dakota. Report 8SA/TSB-20. Region VIII Technical Support Branch, USEPA, Denver, Colorado. (mimeo.)

USEPA (U. S. Environmental Protection Agency). 1973b. Mercury, zinc, copper, arsenic, selenium and cyanide content of selected waters and sediment along Whitewood Creek, the Belle Fourche River and the Cheyenne River in western South Dakota. Report SA/TSB-17. Region VIII, Technical Support Branch, USEPA, Denver, Colorado.

USFWS (U. S. Fish and Wildlife Service). 1985. Results of the South Dakota field office of the ecological services section, fish and wildlife service-1984. Report. Resource Contaminant Assessment Program, USFWS, Pierre, South Dakota. (mimeo.)

Whitley, J. R., and R. S. Campbell. 1974. Some aspects of water quality and biology of the Missouri River. Transactions of the Missouri Academy of Science 7-8:60–72.

Water Quality Changes and Their Relation to Fishery Resources in the Upper Mississippi River

LESLIE E. HOLLAND-BARTELS[1]
U. S. Fish and Wildlife Service
National Fisheries Research Center
P. O. Box 818, La Crosse, Wisconsin 54602, USA

---

1. Current Address: U.S. Fish and Wildlife Service, 750 Spring Street S. W., Atlanta, Georgia 30303, USA

ABSTRACT. Despite a long history of human manipulation, the most dramatic changes in the upper Mississippi River occurred in the 1930s with construction of a lock and dam system to facilitate the commercial transport of commodities. In 1988, barge traffic through the system ranged from 7,500 tows per year at Lock and Dam 26 (near Alton, Illinois) to 1,118 at Lock and Dam 1 (in Minneapolis/St. Paul). The low-head dam system created a diversity of lentic habitats, but it also changed the stage and sediment transport characteristics of the river. The principal fishery-related water quality issues of this modified system concern the effects of sediments and toxic contaminants from nonpoint sources. Between 42 and 99% of the streams in the five states of the Mississippi River basin fail to fully support their designated uses because of pollution, primarily from nonpoint sources (e.g., 73% in Minnesota, 98% in Wisconsin, 75% in Illinois). Annual sediment inputs into the upper Mississippi River basin range from minimal in the upper reaches to about 210,000 kg/hectare in the lower reaches. This sediment results in significant losses of fishery habitat. Although only 5 to 9% of the total open water area of many pools had been lost by 1975, those losses were in highly productive side channel and backwater areas. Under existing conditions, a loss of an additional 22 to 49% of existing lentic habitats is predicted within 50 years. In addition, toxic contaminants transported along with fine sediments have become more available to stream biota. Although significant interagency efforts have been made to evaluate the impacts on biotic communities of the river, present data are inadequate to determine how changes in water quality affect the fisheries. This lack of data undermines our ability to judge the success of programs initiated to control pollution from point and nonpoint sources.

> *"The Mississippi today does not lend itself to easy definition, for it is as vulnerable as any other geographic phenomenon to the consistent inconsistencies of human use and interpretation... Men see the river as they want to see it: For some it is the great road of commerce... the heartline for an industrial civilization. For others it is a river of memory, rich with the ghosts of... a simpler, more understanding, more human time. For yet others it is a recreational resource, an escape hatch from the pressures of a more complicated world... And for those who would preserve what has so far been left untouched, it is an abundant natural force whose existence can serve to remind us from time to time that we are, after all, natural creatures living in a natural world, no matter what we have tried to do to that world with plastic and concrete. The river is each of these things... and all of them."*
>
> T. H. Watkins in GREAT II (1981)

The upper Mississippi River is truly each and all of these things; it has been called the best example of a multiuse river in the USA (GREAT II 1980a). This 1370 km stretch of river, lying between Lake Itasca in north-central Minnesota and the confluence of the Ohio River at Cairo, Illinois, is also an example of the conflicts and compromises that are associated with multipurpose exploitation. The differing requirements of commercial transportation, recreation, federally mandated resource protection, agriculture, industry, and an expanding human population result in a conflict that is difficult for federal and state agencies to resolve under their different management agendas. However, the upper Mississippi River today is as much an example of ecological tolerance and survival as it is an example of human manipulation and system degradation. The upper Mississippi River is unique among the large inland rivers of North America in that the need to modify the river for commercial navigation not only conflicts with, but also provides opportunities to protect an aquatic resource of national significance.

Nearly 200 years of navigation development on the river have radically changed the physical and water quality characteristics of the system (UMRBC 1982). The most significant change occurred with the construction of low-head navigational dams along about 1,050 km of the upper river between 1930 and 1950 by the U. S. Army Corps of Engineers. The resultant impoundments changed the free-flowing nature of the river but also created valuable fisheries and wildlife habitat. Before the impoundments, a limited amount of aquatic habitat along the river was being brought under the protection of the federal refuge system (UMRNW & FR 1987). However, after impoundment, off-channel lands purchased by the Corps of Engineers in conjunction with the navigation project were placed under the management of the U. S. Fish and Wildlife Service. Today, about 63% of the length of the river, encompassing 92,000 hectares of habitat, is managed under the upper Mississippi River National Wildlife and Fish Refuge authorized in 1924 and the Mark Twain National Wildlife Refuge authorized in 1958. The Upper Mississippi River National Wildlife and Fish Refuge is the only federal refuge that specifically includes management of fish under its authorization. The abundant fish and wildlife resources in the navigation impoundments have encouraged heavy recreational use; 12 million user-days per year have been estimated for the upper 10 navigation pools alone (GREAT I 1980a). The estimated 3 million user-days on the Upper Mississippi River National Wildlife and Fish Refuge represent the largest annual public use of any unit of the National Wildlife Refuge System (UMRNW & FR 1987).

The upper Mississippi River is also unique in that three major congressionally authorized efforts have compiled information on its present condition and evaluated its future prospects for survival. The Upper Mississippi River Comprehensive Basin Study (UMRCBS) was initiated by the Corps of Engineers in 1962. This study presented a comprehensive framework for planning the efficient use of water and related land resources (UMRCBS 1972). The Great River Environmental Action Teams also examined a number of river resource needs, including the impacts of navigation on water quality, fish, and wildlife (GREAT I 1980a; GREAT II 1980a; GREAT III 1982). Finally, the Comprehensive Master Plan for Management of the Upper Mississippi River estimated the increased navigational activity that would occur following renovation of Lock and Dam 26 and evaluated potential impacts on the river's water quality and biota (UMRBC 1982). These studies produced more than 50 volumes of information on water quality and fish and wildlife related issues in the upper Mississippi River. Despite the volumes of data that exist on general fish, wildlife, and water quality patterns in the system, little of this provides the manager with the correlative information necessary to predict future fisheries changes related to water quality degradation or improvement.

I do not intend to present a complete synthesis of these and more recent materials, but hope to provide a general overview of changes in the water quality that have occurred in the impounded portions of the Mississippi River and to demonstrate the need for more comprehensive efforts to relate water quality patterns to trends in fish populations.

River Basin Characteristics

The upper Mississippi River system consists of a watershed of 489,510 km^2 that stretches from its headwaters in north-central Minnesota to the mouth of the Ohio River. Most of the watershed is in Minnesota, Wisconsin, Iowa, and Illinois (Figure 1). The upper Mississippi River includes 14 major river basins that range in area from 11,920 km^2 (St. Croix River basin) to 63,500 km^2 (Iowa-Skunk-Wapsipinicon River basin).

The upper Mississippi River basin is an area of rolling terrain with few major contrasts in orographic features. The topography and soils were shaped primarily by glaciation. However, little glacial modification occurred in the lower third of Wisconsin and a portion of Minnesota, known as the driftless area. These areas contain escarpments and bluffs that are not typical of the rest of the region. Soils of the basin are of glacial, fluvial, eolian, and lacustrine origin (Olson and Meyer 1976) and are well-suited for agriculture. About 70% of the total land area is in cropland or pasture (UMRBC 1980).

Nearly all the water runoff is from precipitation. Most of the 81-cm annual precipitation comes as rainfall during spring and summer (UMRCBS 1972). Snowfall, which plays a key role in the severity of spring floods, varies from a mean annual average of 122 cm in much of the upper basin to 20 cm in the southern part.

River Features

History of Modification

The upper Mississippi River, as seen by early explorers, was shallow, swift, and rock-obstructed. Although the first steamboat did not ascend the river to St. Paul until 1823, more than 1,000 steamboats per year were carrying migrants to the upper end of the system by 1858 (Scarpino 1985). However, the river was navigable to St. Paul only during high water stages. Federally authorized modifications of the system were undertaken as early as 1824 to aid movement of people and commodities. Snags, sandbars, and rocks were removed to eliminate rapids; sloughs were closed and backwaters cut off to confine most of the flow to the main channel (Fremling and Claflin 1984). Under the "4.5 foot" channel project initiated as part of the Rivers and Harbors Act in the late 1800s, wing dams and revetments were constructed, additional side channels and chutes were closed, and some areas were dredged. The "6 foot" channel project, authorized in 1907, funneled as much water as possible into the navigation channel through the construction of closing dams and the placement of hundreds of rock and brush wing dams or dikes that extended out from

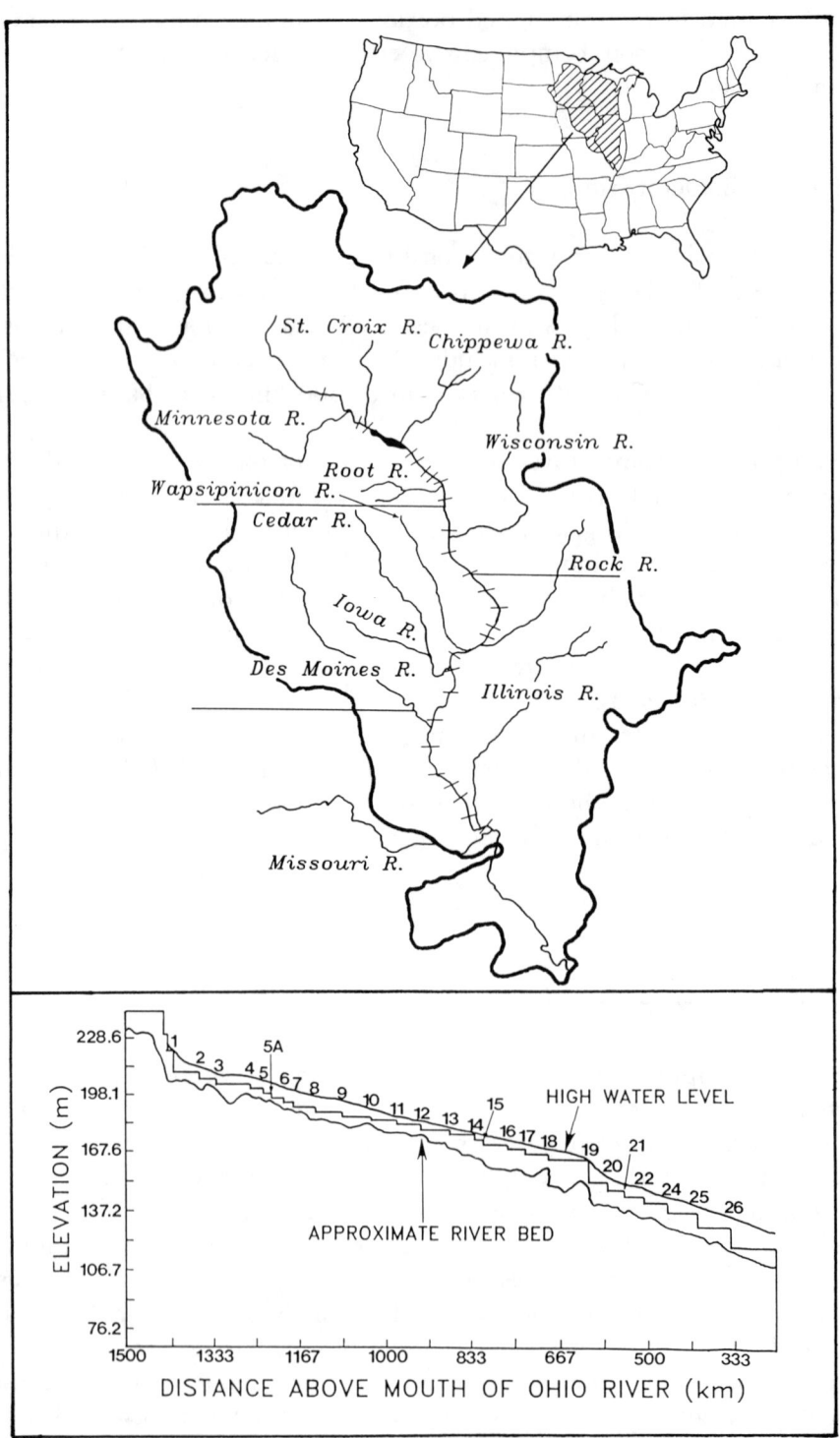

FIGURE 1. Upper Mississippi River and its drainage system showing navigation locks and dams. The side-view schematic illustrates the stairstep pattern, gives the elevations above mean sea level, and lists the numbering of locks and dams in the system.

shore. A "9 foot" navigation channel, authorized in 1927, resulted in the construction of 27 locks and dams and the creation of 26 associated "pools" or impoundments.

Physical Effects of Navigation Modifications

Early modifications to the upper Mississippi River had little effect on flood stages or on location of the main channel. However, the wing dams slowed the flow and enabled sediment to deposit in the dike fields. This reduced the surface area of the river and increased the size of many of the islands (Chen and Simons 1986). River width also decreased. For example, near Clarksville, Missouri (now Pool 24), the width of the river was reduced 11% between 1891 and 1927 (Olson and Meyer 1976).

The principal modern day features of the upper Mississippi River result from construction of the "9 foot" channel project by the Corps of Engineers between 1930 and 1950. The 27 navigation locks and dams produced a terraced pool system that stretches from Minneapolis, Minnesota to Alton, Illinois (Figure 1). Each pool is named for the number of its impounding dam. Locations along the river are generally referred to by their river distance (here in kilometers) above the confluence with the Ohio River.

Each navigational pool of the upper Mississippi River contains a variety of aquatic habitats: 1) main channel—the dredged navigation corridor; 2) main channel border—the zone between the channel and the main river bank, which includes wing dikes; 3) side channels—all departures from the main channel that contain flows during normal river stage; and 4) backwaters—including sloughs, river lakes, and ponds (Rasmussem 1979). The 26 pools differ in total area and in proportion of area contained in each specific habitat. As one proceeds downstream, the area in side channels and lentic backwaters decreases and that in main channel border habitats increases (Figure 2).

The low-head dams caused significant changes in surface area features and flow stage patterns of the river. The impounded waters permanently inundated marshlands and bottomland hardwoods in the middle and lower areas of most pools. Braided channels more typical of the pre-impoundment river persisted in the upper portions of each pool. The extent to which pool areas were increased and bottomlands were inundated depended on the topography of the local floodplain and, thus, varied among pools. For example, the surface area of the lower portion of Pool 4 increased 118%, whereas that of Pool 24 increased only 6% (Chen and Simons 1986).

Although the dams were not designed to impound more water than required to maintain the "9 foot" navigation channel, the low-head dam system provided more stable levels of river stage than had previously existed. Before the locks and dams were constructed, stage readings varied by more than 2 m between seasonal highs and lows at Winona, Minnesota; now they vary only about 1 m (Chen and Simons 1986). Although significant annual variations in discharge occur on the river, they are not reflected by major variations in river stage (Figure 3).

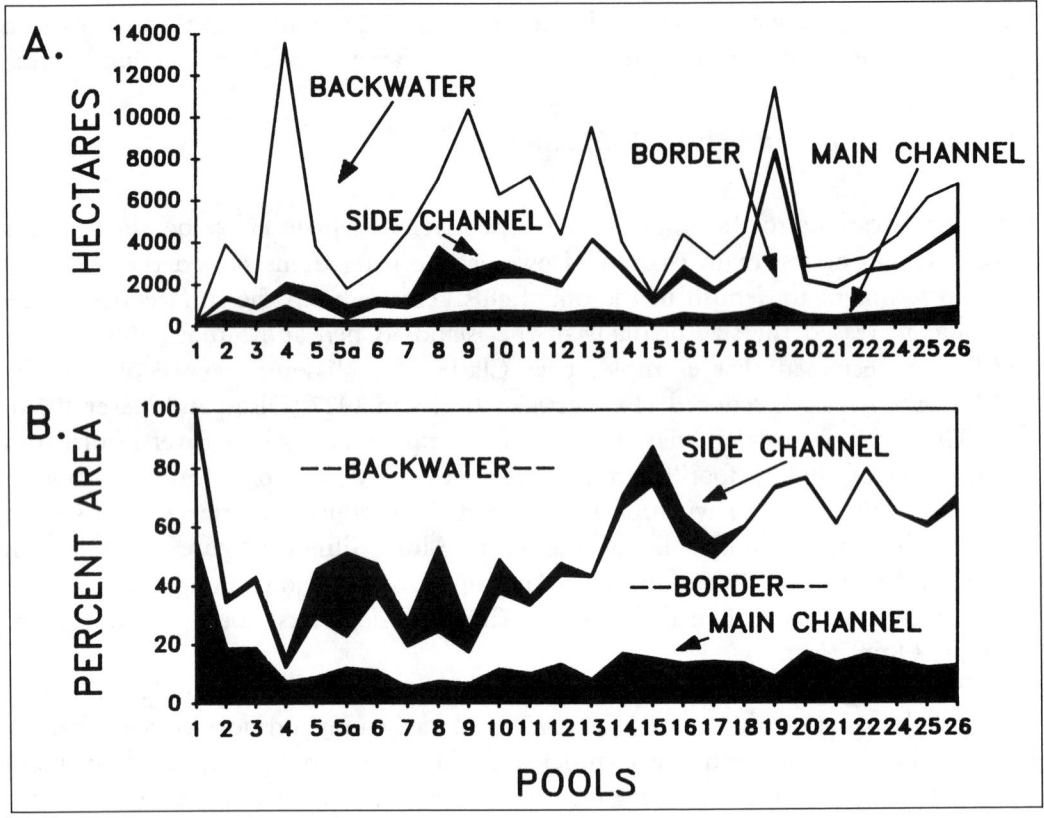

FIGURE 2. Area in hectares (A) and percentage of total water surface area (B) of four major aquatic habitat types in pools of the upper Mississippi River.

Water Quality Issues

General

Most waters of the upper Mississippi River and its tributaries are unable to fully meet surface-water quality criteria for designated uses (e.g., domestic water supply, fish consumption, and recreation) under the federal Clean Water Act. Each state establishes designated uses for water bodies within their jurisdiction that may vary significantly among systems. These uses often represent historic or existing patterns. Use classifications for Minnesota rivers, for example, include aquatic fish and wildlife, domestic water supply, recreation, agriculture, industrial, and navigation (MPCA 1988). The percent of monitored stream habitat (by length) in each state that partly or completely fails to support various designated uses is 73% in Minnesota (MPCA 1988), 42% in Wisconsin (WDNR 1988a), 99% in Iowa (IDNR 1988), and 55% in Illinois (IEPA 1988).

Because each state develops its own water quality criteria and because the upper Mississippi River is an interstate boundary, two different water-quality classifications may prevail in the same stretch of river. Although Minnesota declares that water quality in its section of the Mississippi River is completely unsuitable to provide fish

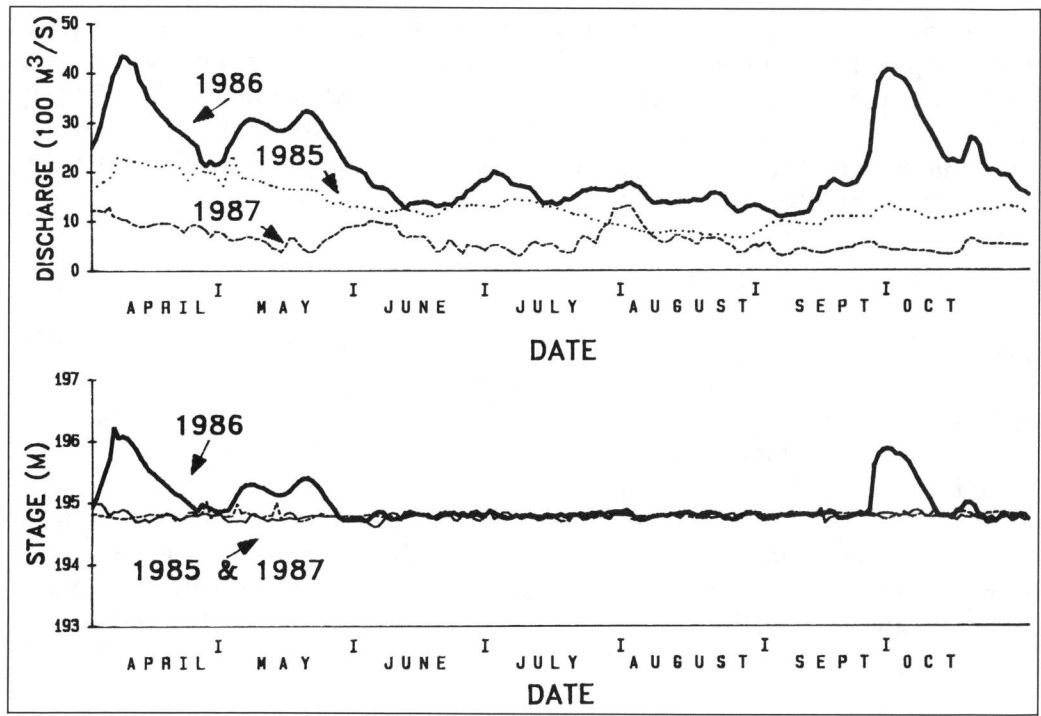

FIGURE 3. Discharge (100 m³/s) and stage (meters above mean sea level) patterns at Lock and Dam 7 near La Crosse, Wisconsin, during three years—1985 (similar to the 10 year average for 1978–1987), 1986 (highest average water discharge for the 10 year period), and 1987 (lowest average discharge for the 10 year period).

for consumption, Wisconsin declares that the same area is at least partly suitable for this purpose. Iowa classifies the river along its entire length as only partly supporting designated uses because of industrial and municipal discharges, agricultural nonpoint pollution, urban runoff, and navigation. Illinois, on the other hand, classifies these same waters as partly to fully meeting fishable or swimmable goals. Areas below the confluence of the Illinois River (Pool 26) have been degraded in water quality to where they partly or completely fail to support designated uses. These states consistently cite nonpoint sources of pollution for reducing water quality (e.g., Minnesota 73%, Wisconsin 98%, and Illinois 75%).

Point sources of pollution on the upper Mississippi River occur primarily at Minneapolis-St. Paul, Minnesota (2 million population), in the metropolitan area of Rock Island, Moline, Davenport, and Bettendorf (360,000 population) near Lock and Dam 15, and at a number of smaller communities (<80,000 population) between the two. The Minneapolis-St. Paul area has more than 1,300 potential industrial sources of trace element contamination (Boyer 1984). St. Louis, Missouri (located on the river below Pool 26), and Chicago, Illinois (on the headwaters of the Illinois River) play important roles in determining water quality in unpooled, middle reaches of the river.

The watershed of the upper Mississippi River drains into an aquatic zone that is only about 2% of the total area. Movements of materials from upland areas into the river pose the greatest long-term threat to fish habitats and the resulting effects on fish populations. Although little sediment enters the upper reaches of the river, 210,000 kg/hectare can be added each year in the lower reaches. This results in significant losses of fisheries habitat (GREAT I 1980a; UMRBC 1980). To further complicate the problem, toxic contaminants that are transported with fine sediments become more available to stream biota (Wiener et al. 1984a).

Spatial Water Quality Patterns

An evaluation of water quality characteristics along the upper Mississippi River reveals patterns of nonpoint and point-source pollution (Figure 4). Levels of un-ionized ammonia nitrogen peak at River Location 1329 in Pool 2 below St. Paul, implicating a major sewage treatment plant (River Location 1345) as the likely source. Values for ammonia then drop noticeably until another increase at La Crosse, Wisconsin (River Location 1123). Total unfilterable residues (total suspended solids) show major inputs from three sources—Minneapolis-St. Paul (River Location 1382 to 1352), the Minnesota River that passes through the southern portion of the metropolitan area and enters the upper Mississippi River at about River Location 1358, and the St. Croix River that enters the main stem above River Location 1283 (Figure 4). Reduced

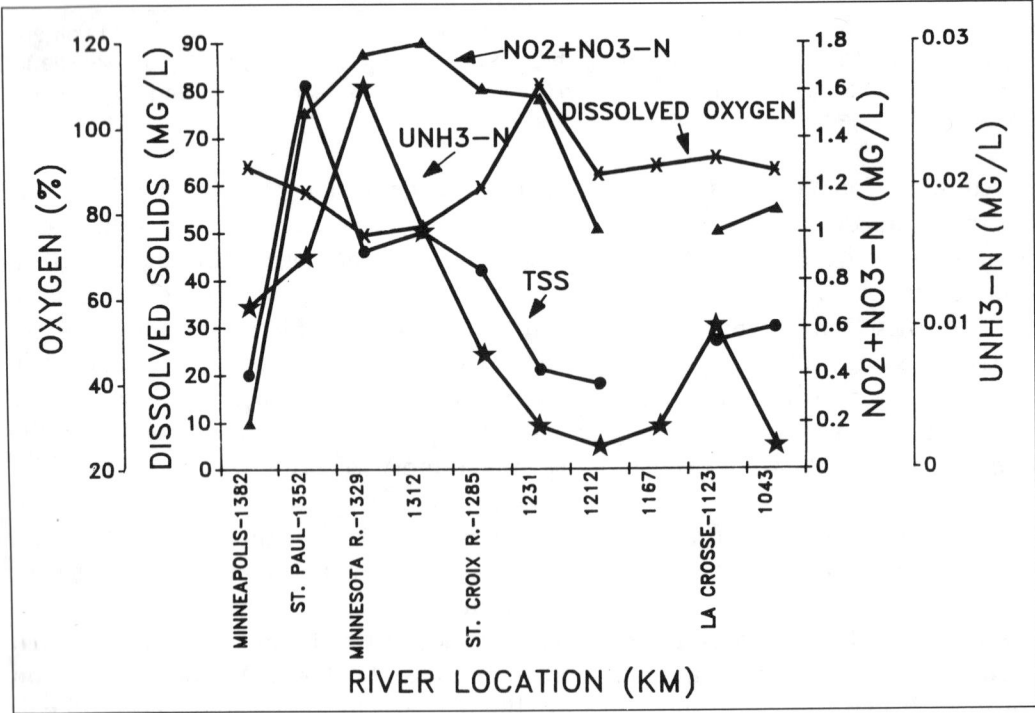

FIGURE 4. Representative changes in four water quality characteristics at 10 stations along 322 km of the upper Mississippi River from Minneapolis. The location of each station (River Location in kilometers above the confluence of the Ohio River) and the major point-sources of pollution (e.g., cities or tributaries) at or upstream from a station that might influence its water quality are listed on the x-axis. (Modified from Sullivan 1989.)

levels of suspended solids at River Location 1263 and downstream indicate that most suspended solids are deposited in Lake Pepin within Pool 4 (Nielsen et al. 1984). Total phosphate phosphorus is apparently supplied by both point and nonpoint inputs, whereas most levels of nitrite plus nitrate are derived from nonpoint agricultural runoff (Sullivan 1989).

Evaluations of sediments and organisms from the upper Mississippi River indicate that the greatest concentrations of polychlorinated biphenyls (PCBs) and trace metals are near Minneapolis-St. Paul, while the organic pesticide chlordane is of concern in the lower pooled and unpooled portions of the river (Wiener et al. 1984b; Sullivan 1988; MDC 1989). Levels of PCBs in sediments drop markedly between Pools 2 and 3 (Figure 5). A similar, but less rapid, longitudinal decrease in PCB levels occurs in fish. Both Minnesota and Wisconsin have issued fish consumption advisories for all sections of the upper Mississippi River within their boundaries based on PCB levels in certain species and sizes of fish (MDH 1987; WDNR 1988b). PCB concentrations in fish fall well below action levels in the lower river, but various fish consumption advisories exist based on chlordane. For example, in 1988, chlordane concentrations in several species of fish exceeded action levels in 4 of 5 lower pools sampled (Pools 20 to 26) and in much of the unpooled portions of the river. Either nonconsumption or limited consumption has been recommended for fish of specified

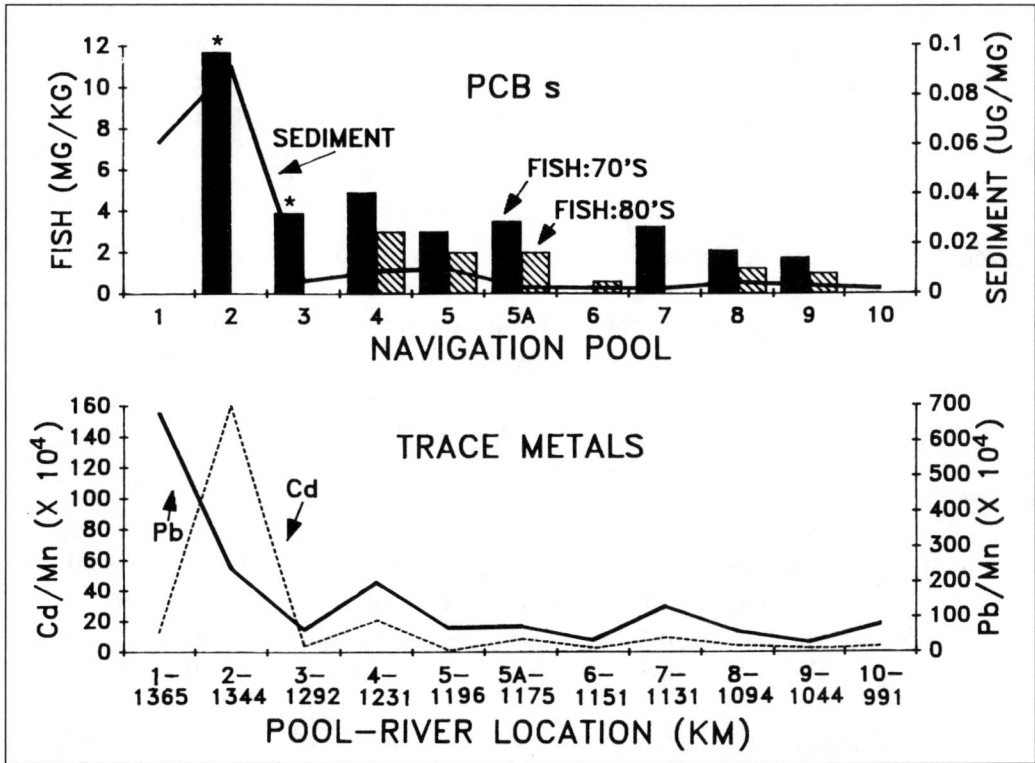

FIGURE 5. Contaminant concentrations recorded in the upper pools of the upper Mississippi River. Polychlorinated biphenyl (PCB) data in sediment and fish are from Sullivan (1988) and Hora (1984, with asterisk) with modifications. Trace metal data are from Wiener et al. (1984b).

species throughout much of the upper Mississippi River bordering Missouri (MDC 1989). Trace metals in sediments (Figure 5) and in fish are generally highest in the Minneapolis/St. Paul area. Concentrations of some trace metals, such as cadmium in common carp (*Cyprinus carpio*), are relatively low. Others, such as mercury and lead, are relatively high, compared with those in other areas of the nation (Wiener et al. 1984b). However, a lack of adequate information makes it difficult to determine the effect of observed levels of contaminants on the biota.

Sediment loads vary significantly throughout the river system. Sediment yield is low in the upper Mississippi River basin, and most yields fall in an Erodibility Index isocline of 1, equivalent to an average annual sediment loss of < 175 tonne/ km^2. However, some areas in the southern end of the basin are subject to major erosion, giving rise to an Erodibility Index of 12 (Figure 6). The long-term annual sediment production increases by 45 times between Pool 4 and Pool 22 (from 1.4 to 63.4 tonne/km^2, respectively). These sediments are transported as suspended load or as bedload. Variability in the suspended load, which consists of fine-sized particulates, is correlated with watershed characteristics in each subregion along the upper Mississippi River. Pools that receive drainage from agricultural, loess-covered glaciated areas, such as Pools 8, 11, 14, 18, 19, and 20 (Root, Turkey, Wapsipinicon, Iowa, Skunk and Des Moines river drainages), also receive significant inputs of suspended sediments. Annual yields of suspended sediments in these rivers range from 56 to 350 tonne/km^2, and ratios of bedload to suspended sediment load range from 1:6 to 1:26 (Nielsen et al. 1984). Although suspended sediments can affect habitats far downstream, bedload sediments only affect habitat near their point of entry. Dredging requirements for pools that receive sediments from tributaries are 3 to 10 times greater than for adjacent pools that have no sediment-bearing tributaries (GREAT II 1980b).

Temporal Water Quality Patterns

Dawson et al. (1984) showed that the hydrologic regime was the single most important factor affecting temporal patterns in the water chemistry of Pools 7 and 8 on the upper Mississippi River, but they could not detect significant changes in quality over a 10-year (1972–1981) study. However, Sullivan (1989), who examined temporal trends over a broad stretch of the river (Pools 1–9), reported that water quality in the Minneapolis-St. Paul and adjacent areas improved between 1977 and 1987. Reductions in ammonia nitrogen and in 5-d biological oxygen demand reflected successful abatement of point-source pollution in the metropolitan area. Average annual patterns in dissolved oxygen and in biological oxygen demand during summer varied significantly over the last 30 to 50 years. However, available data suggest that water quality, in general, has improved in the 1980s over that in the 1960s and 1970s (Figure 7).

General improvements are evident in PCB levels in fish from the upper pools along the upper Mississippi River (Hora 1984; MPCA 1985; Sullivan 1988). Health consumption advisories still exist for fish along much of the river but PCB levels have declined consistently since the late 1970s (Figure 5). Contamination was reduced

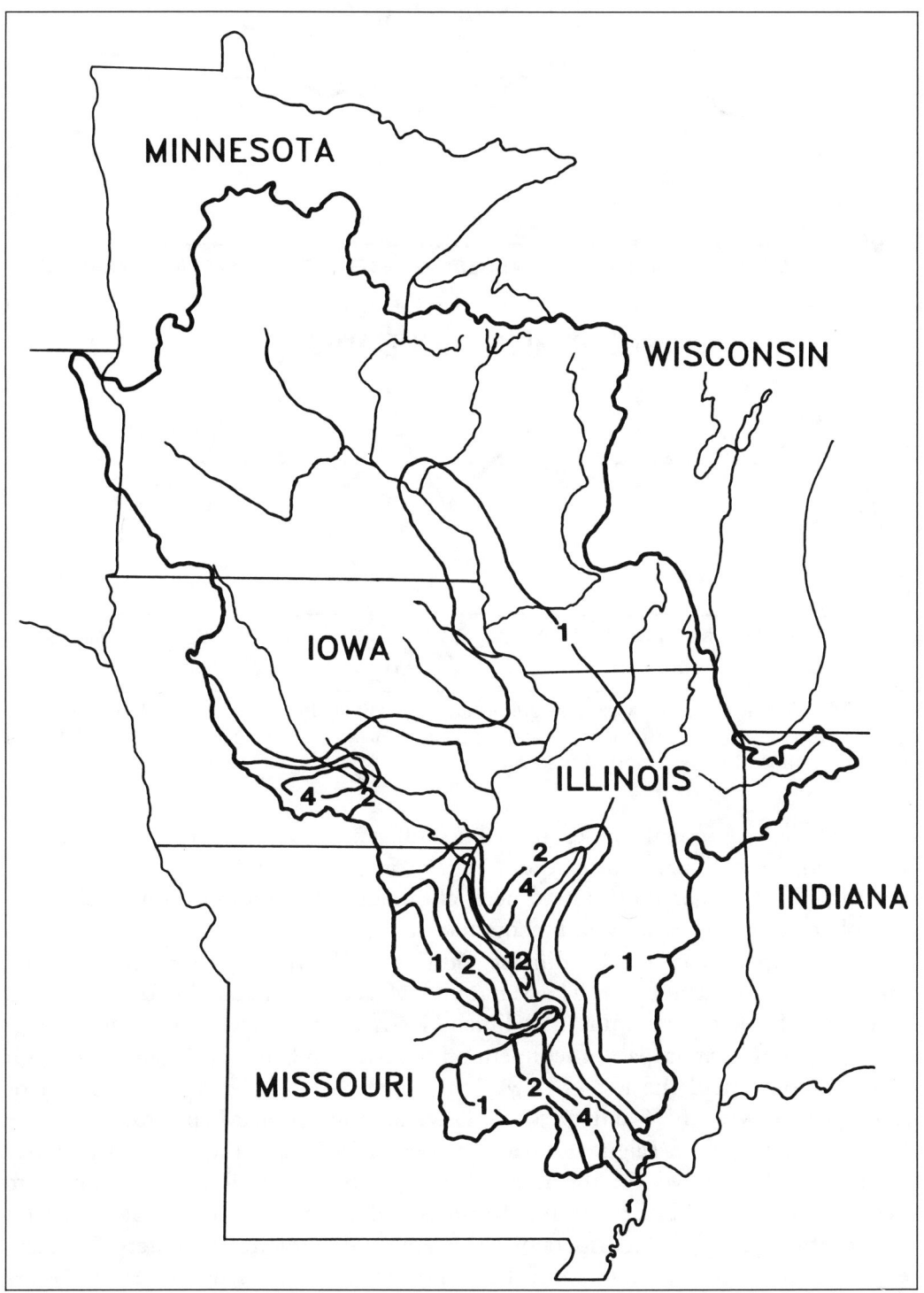

FIGURE 6. Erosion patterns in the upper Mississippi River Basin. An erodibility Index of 1 T is equivalent to an annual soil loss of about ≤ 175 tonne/km².

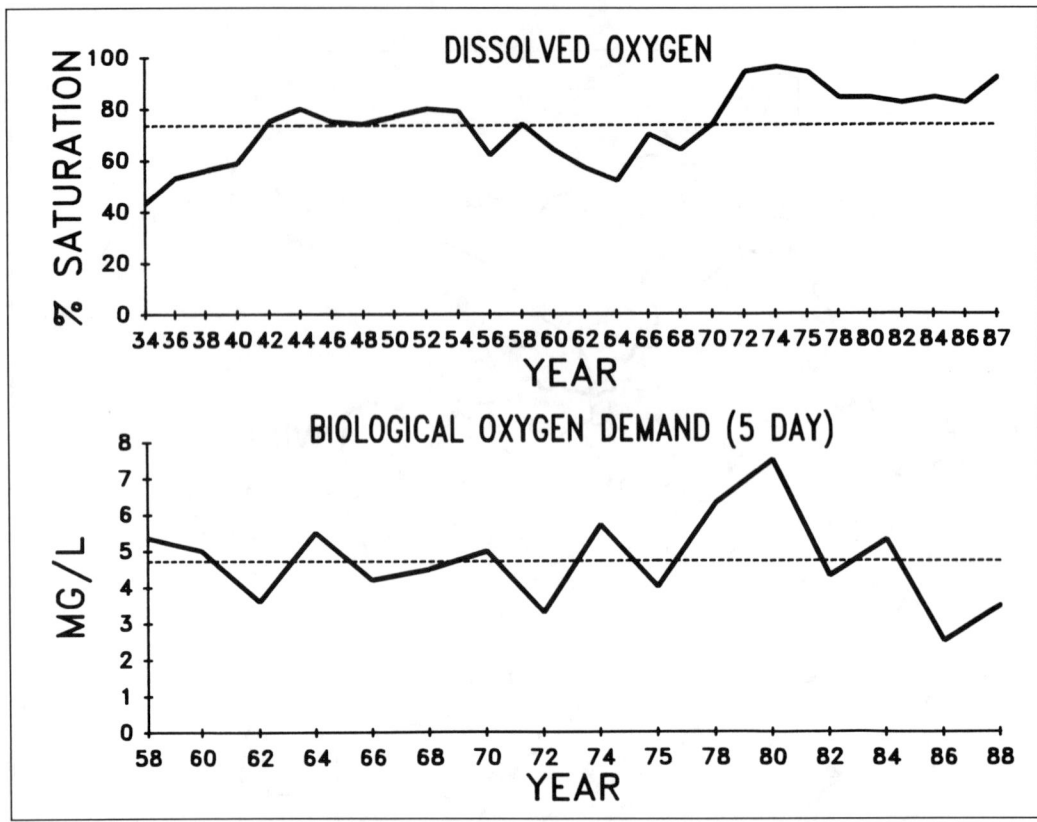

FIGURE 7. Average dissolved oxygen during summer (1934–1987) and 5-d biological oxygen demand (1958–1988) in the Minneapolis-St. Paul area (River Mile 815.3). Dashed line indicates historical averages for the period of measure.

40 to 90% among several species in Pool 4 (Sullivan 1988). However, concentrations of PCBs in the sediment still exceed 1 mg/kg in areas near Minneapolis-St. Paul, even though the major sources have been eliminated. Bioaccumulation by fish will probably continue as existing sediments release PCBs.

The temporal pattern of sediment addition to the river is positively correlated with flow volume (Figure 8). The variability of annual stream discharge patterns makes it difficult to determine if sediment loads have changed over time. Nevertheless, the long-term impact of sediment on the river has been significant. Although sedimentation caused the loss of only 5 to 9% of the total open water habitat of many pools between 1956 and 1975, the losses occurred primarily in productive side channels and backwaters (Figure 9). An estimated 22 to 49% of the existing backwater habitats will be lost within the next 50 years if present sedimentation rates are allowed to continue (GREAT II 1980b). Projected increases in barge traffic would significantly increase the transport of suspended sediment to backwaters, intensify sedimentation problems, and cause further negative impacts on fishery resources (UMRBC 1982).

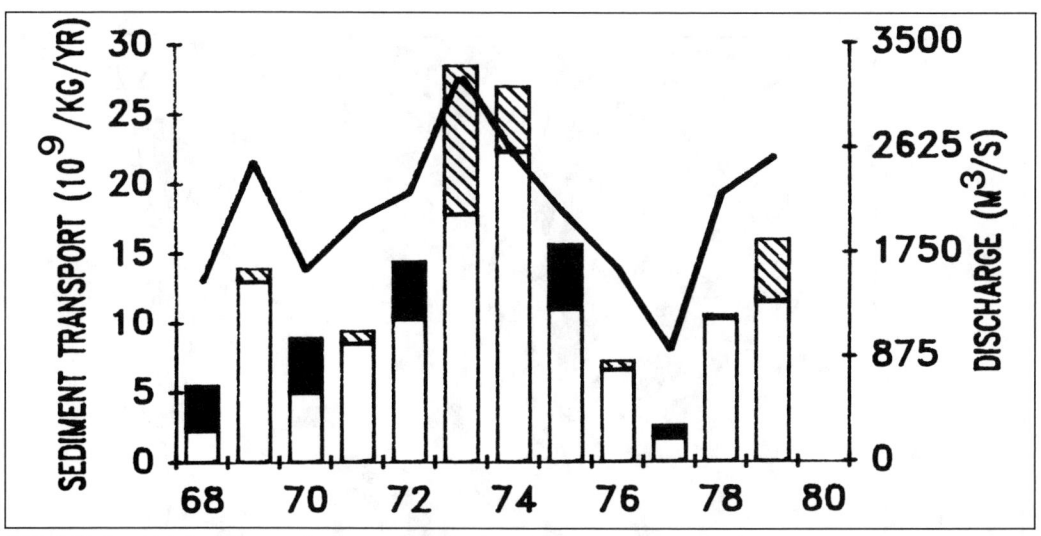

FIGURE 8. Total sediment transport at Pool 19, including amounts of net deposition (solid bar) and net scour (slashed bar) from 1968 to 1979 in relation to average river discharge (solid line). (Modified from Bhowmik and Adams 1986.)

Trends in Fish Populations

It is difficult to demonstrate the effects of toxic contaminants and sediments on the upper Mississippi River ecosystem from available data. Losses of backwater habitat are expected to change community diversity and biomass of many of the 86 fish species in the river, of which 50% require lentic habitats during part of their life history. However, few distinct patterns can be documented because a long-term, comparable data base is lacking. There is no distinct pattern of species loss or relative decrease in species diversity that corresponds to known losses of habitat or increases in navigational activity (Table 1). Relative abundance provides a qualitative measurement that reflects only major changes in the fish community.

Commercial fishing statistics provide some consistent information, but catches often reflect sources of variability rather than actual decreases in biomass. Kline and Golden (1979), who evaluated 35 years of commercial harvest data, found an overall decline in total harvest of fish; however, they suggested that commercial harvest trends might also vary due to annual changes in hydrologic and climatic conditions, making interpretation of trends difficult. Recent declines in harvest are within the limits of previous fluctuations (Figure 10A). Lubinski et al. (1986) felt that PCB advisories reduced market demand and decreased the harvest of common carp. Despite these problems, commercial fishery statistics have documented a decline in harvest from other rivers where water-quality has been degraded more than in the upper Mississippi River. For example, data for the Illinois River, where water quality has declined significantly, clearly show a decline in the fishery (Sparks 1984).

Pitlo (1987) indicated that few useful, quantitative data on fish standing crop exist for the upper Mississippi River system. The large variability in standing crop

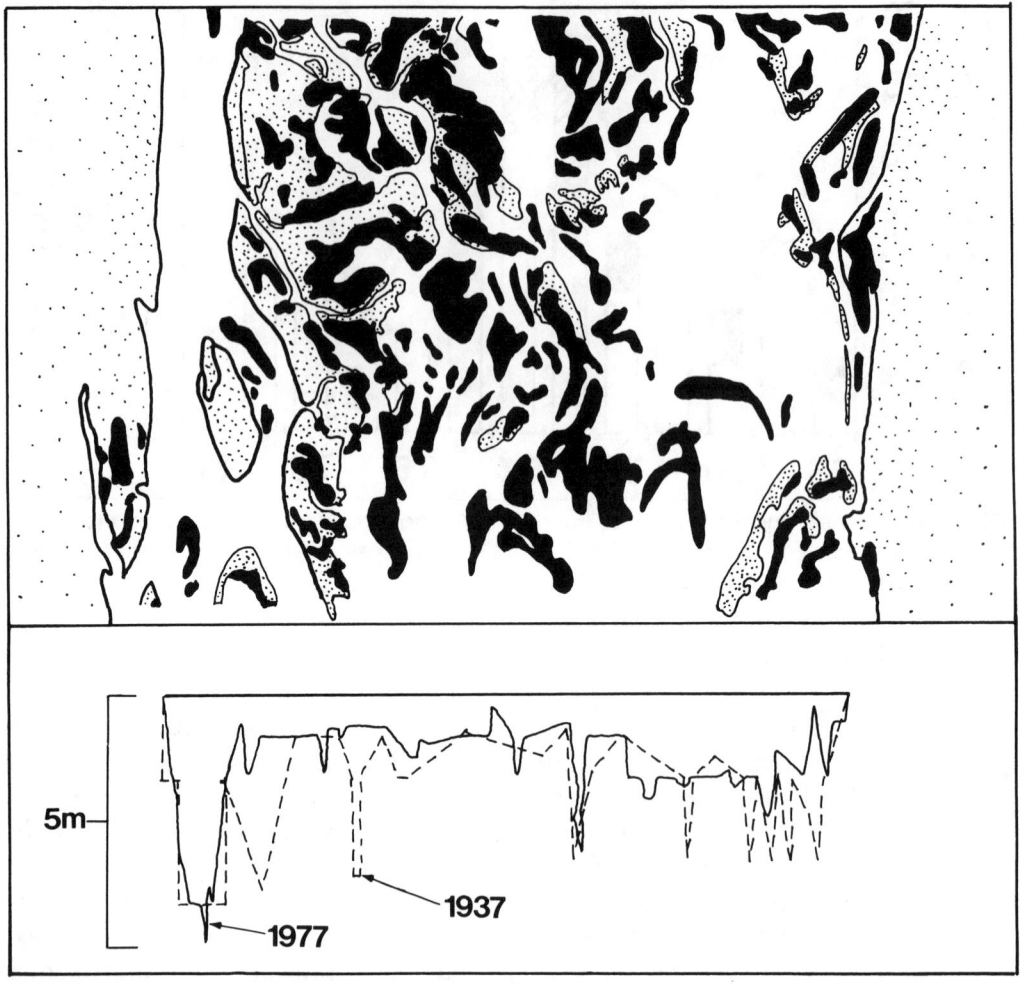

FIGURE 9. Losses of water surface areas (black areas) due to sediment deposition in a portion of a Navigation Pool 8 (near River Mile 688) of the upper Mississippi River from 1939 to 1973 (modified from GREAT I 1980b) and depth profile changes along a transect through the area between 1937 and 1977 (from McHenry et al. 1978).

estimates within even a single pool make it difficult to compare data from upstream and downstream pools. Only two studies provide data for a temporal estimate of changes in standing crop, and these are only for one site. For Pool 5a, a comparison of 1983 data with data collected in four consecutive earlier years (1948-1951) suggested that the biomass of top level predators decreased, whereas the total biomass and the biomass of rough and forage fish increased (Figure 10B). However, limited replications and high variability in the data make it difficult to determine if the present fish community is healthy, recovering, or declining.

Existing methods of evaluation suggest that the fish populations in the upper Mississippi River are adjusting, overall, to the current levels of human and environmental stress. However, incremental, pervasive changes in the fish populations may

TABLE 1. Relative abundance and number of species of fish (86 possible) in the 26 navigation pools of the upper Mississippi River. (Modified from Van Vooren 1983.) Navigation rates (tows/year) for 1988 and ratios of backwater to main channel surface area are included.

Pool	Navigation rate (tows/year)	Backwater to channel ratio	Present (current or historical)[a]	Number of species[b]					Species Index[c]
				A	C	O	U	R	
1	1118	0.0	44/3	5	24	11	3	1	1.9
2	1525	1.9	60/2	5	26	23	4	2	2.4
3	1451	1.4	70/6	6	27	27	6	4	2.7
4	1435	7.1	69/11	6	28	23	8	4	2.7
5	1528	2.4	75/7	6	29	25	9	6	2.8
5A	1441	3.4	70/6	6	29	23	8	4	2.7
6	1499	1.7	70/10	6	30	21	7	6	2.7
7	1511	4.1	72/7	6	30	21	8	7	2.7
8	1490	4.2	69/15	6	29	23	7	4	2.7
9	1410	7.2	73/12	6	29	21	11	6	2.8
10	1871	1.6	72/8	6	29	17	12	8	2.7
11	1780	2.0	74/10	6	30	17	13	8	2.7
12	2157	1.3	70/0	6	27	17	11	8	2.5
13	2272	1.3	72/3	7	26	16	16	8	2.6
14	3009	0.5	64/1	6	23	17	15	14	2.5
15	3405	0.3	60/5	6	23	17	9	5	2.3
16	3209	0.9	69/3	6	24	16	17	6	2.5
17	3026	1.0	65/4	6	23	19	11	6	2.4
18	3077	0.7	71/3	6	25	16	16	8	2.5
19	3105	0.4	66/5	6	24	20	12	4	2.5
20	3237	0.3	69/3	7	23	21	12	6	2.6
21	3353	0.6	70/3	7	22	25	12	4	2.6
22	3385	0.3	70/3	7	23	23	10	7	2.6
24	3513	0.5	68/1	7	22	25	10	4	2.6
25	3508	0.7	70/2	7	22	27	11	3	2.7
26	7521	0.5	73/2	7	22	25	13	6	2.7

[a] Historical—Records for these additional species of fish are available, but no collection has been documented in the last 10 years.

[b] A = Abundantly taken in all surveys; C = Commonly taken in most samples and can make up a large portion of some; O = Occasionally collected, not generally distributed, but local concentrations may occur; U = Uncommon, does not usually appear in samples, populations are small but the species does not appear to be near extirpation; and R = Considered rare, and some species may be near extirpation.

[c] The index is the average abundance score for the possible 86 species for that pool. Relative abundances were assigned values A = 5, C = 4, O = 3, U = 2, R = 1, and Absent = 0. Maximum score is 5, if all species were present and abundant in the pool.

occur undetected because of the limited data bases and the complexities of the river system.

Water Quality Control Efforts

More than 17 federal agencies and 30 state departments or divisions have responsibility and authority for maintaining water quality in the upper Mississippi River. Although

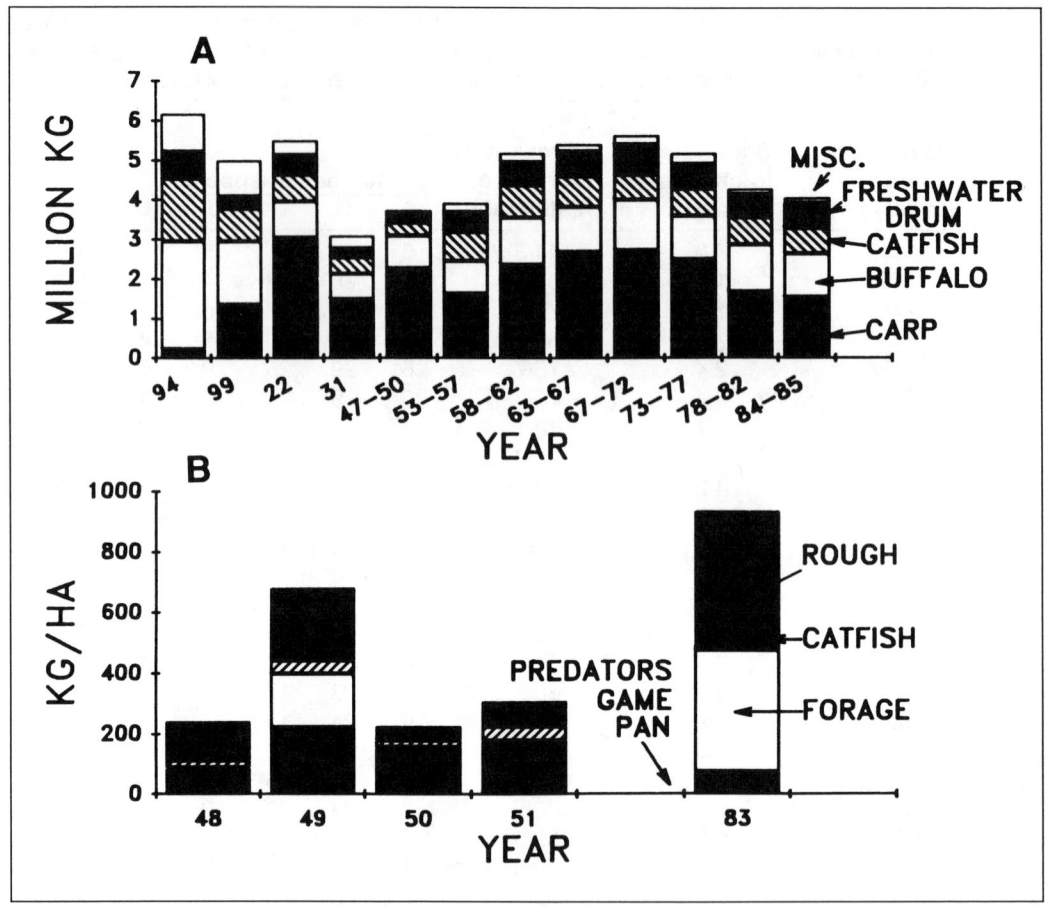

FIGURE 10. Historic changes in (A) total commercial fish catch and catch of specific groups from 1894 through 1985, and (B) biomass estimates from Pool 5 between 1948–1951 and 1983. (Data from Carlander 1954; Kline and Golden 1979; Pitlo 1987; and unpublished state records.)

no single group or commission is legally responsible for developing water-quality standards for the river, the various states and federal agencies interact through advisory commissions. For example, the appointed members of the Minnesota-Wisconsin Boundary Area Commission develop coordinated water-quality positions for their respective governors. The Upper Mississippi River Basin Association, established by presidential executive order in 1972, carries out the mandates of the 1965 Water Resources Planning Act, and prepares and updates a coordinated federal, state, and local development plan for water and related land resources. Numerous advisory groups (e.g., the Upper Mississippi River Conservation Committee) bring together state fishery, wildlife, and water-quality representatives at a technical level.

Although each state sets its own water-quality criteria independent of its neighboring states, the federal Clean Water Act of 1927 (with amendments) provided some uniform direction to the water-quality maintenance efforts of states that share the waters of the upper Mississippi River basin. Compliance with the wastewater treatment requirements of the 1977 amendments to the Clean Water Act initiated construction on more than 800 municipal wastewater treatment projects in the upper Mississippi River basin in 1979–1980 (UMRBC 1980). Completion of many of these projects significantly reduced point-source problems. For example, 53 stream segments in Wisconsin were considered degraded by point-source pollution in 1972, but only 14 still suffered from similar levels of degradation in 1988.

The issue of nonpoint pollution was addressed through initiatives under the Water Quality Act of 1987, and under conservation provisions of the Food Security Act of 1985 (Farm Act). The 1987 Water Quality Act required states to prepare an assessment report and management program for nonpoint-source pollution within 18 months. Their efforts to date were described in state water quality reports to Congress (IDNR 1988; IEPA 1988; MPCA 1988; WDNR 1988a). The Conservation Title of the Farm Act includes four programs with different objectives: the Conservation Reserve Program, which withdraws highly erodible lands from crop production for 10 years; the Sodbuster Program, which discourages conversion of erodible lands to cropland; the Swampbuster Program, which discourages conversion of wetlands to cropland; and the Conservation Compliance Program, which requires farmers using highly erodible lands to design and implement an approved conservation plan by 1990 to remain eligible for federal subsidies. Recent amendments to the Conservation Reserve Program allow inclusion of some croplands along rivers and cropped wetlands, previously not considered highly erodible, in the program.

Future Prospects

The upper Mississippi River of today continues to be a system of great diversity and productivity, but it is a multiuse system with a difficult future. Although point-source pollution problems have been addressed during the last decade through legislation, nonpoint-source pollution continues to be a pervasive and critical issue for the river. New federal legislation and associated state initiates for the control of nonpoint-source pollution could help rejuvenate the upper Mississippi River ecosystem. However, the success of these initiatives depends on whether future federal farm policies are consistent with goals and requirements of the Clean Water Act, whether nonpoint-source programs are focused and complement each other, and whether funding is consistent and stable.

The lack of a historic, long-term, statistically valid framework for trend analyses limits our ability to precisely evaluate biological responses to the new nonpoint-source program. However, several new programs on the upper Mississippi River offer hope for developing a better information base. When Congress approved construction of the additional navigation lock at Lock and Dam 26 in 1986 (Public Law 99–662), it accepted the "Comprehensive Master Plan for the Management of the Upper Mississippi River System." This plan authorized environmental programs that would improve habitat for fish and wildlife; monitor and analyze the river's physical, chemical, and biological features; and expand recreational opportunities. These programs, known collectively as the Upper Mississippi River System Environmental Management Plan, should provide a consistent framework by which to monitor trends in fish populations and water quality. Monitoring these trends is critical for maintaining the upper Mississippi River as a true multiuse resource.

Acknowledgments

I thank John F. Sullivan, Wisconsin Department of Natural Resources; Terri O'Dea, Metropolitan Sewage District of Minneapolis/St. Paul; and Michael Dewey and Michael Schueller, National Fisheries Research Center, La Crosse, Wisconsin for their assistance in this work.

References

Bhowmik, N. G., and J. R. Adams. 1986. The hydrologic environment of Pool 19 of the Mississippi River. Hydrobiologia 136:21-30.

Boyer, H. A. 1984. Trace elements in the water, sediments, and fish of the upper Mississippi River, Twin Cities metropolitan area. Pages 195-230 in J. G. Wiener, R. V. Anderson, and D. R. McConville, editors. Contaminants in the Upper Mississippi River. Proceedings of the 15th Annual Meeting of the Mississippi River Research Consortium. Butterworth Publishers, Boston.

Carlander, H. B. 1954. History of fish and fishing in the upper Mississippi River. Upper Mississippi River Conservation Committee, Rock Island, Illinois.

Chen, Y. H., and D. B. Simons. 1986. Hydrology, hydraulics and geomorphology of the upper Mississippi River system. Hydrobiologia 146:5-20.

Dawson, V. K., G. A. Jackson, and C. E. Korschgen. 1984. Water chemistry at selected sites on Pools 7 and 8 of the upper Mississippi River: A ten-year survey. Pages 270-198 in J. G. Wiener, R. V. Anderson, and D. R. McConville, editors. Contaminants in the upper Mississippi River. Proceedings of the 15th Annual Meeting of the Mississippi River Research Consortium. Butterworth Publishers, Boston.

Fremling, C. R., and T. O. Claflin. 1984. Ecological history of the upper Mississippi River. Pages 5-24 in J. G. Wiener, R. V. Anderson, and D. R. McConville, editors. Contaminants in the upper Mississippi River. Proceedings of the 15th Annual Meeting of the Mississippi River Research Consortium. Butterworth Publishers, Boston.

GREAT I. (Great River Environmental Action Team). 1980a. A study of the upper Mississippi River. Volume 1-Main report. GREAT I, Twin Cities, Minnesota.

GREAT I. (Great River Environmental Action Team). 1980b. A study of the upper Mississippi River, Volume 4 Part F. Water quality, Part G. Sediment & erosion. GREAT I, Twin Cities, Minnesota.

GREAT II (Great River Environmental Action Team). 1980a. Upper Mississippi River (Gutenberg, Iowa to Saverton, Missouri). Main report. GREAT II, Rock Island, Illinois.

GREAT II (Great River Environmental Action Team) 1980b. Upper Mississippi River (Guttenberg, Iowa to Saverton, Missouri). Side channel work group appendix. GREAT II, Rock Island, Illinois.

Pitlo, J., Jr. 1987. Standing stock of fishes in the upper Mississippi River. Upper Mississippi River Conservation Committee, Rock Island, Illinois.

Rasmussen, J. L. 1979. Description of the upper Mississippi River. Pages 3–20 *in* J. L. Rasmussen, editor. A compendium of fishery information on the upper Mississippi River. Upper Mississippi River Conservation Committee, Rock Island, Illinois.

Scarpino, P. V. 1985. Great river: An environmental history of the upper Mississippi 1890–1950. University of Missouri Press, Columbia.

Sparks, R. E. 1984. The role of contaminants in the decline of the Illinois River: Implications for the upper Mississippi. Pages 25–66 *in* J. L. Wiener, R. V. Anderson, and D. R. McConville, editors. Contaminants in the upper Mississippi River. Proceedings of the 15th Annual Meeting of the Mississippi River Research Consortium. Butterworth Publishers, Boston.

Sullivan, J. F. 1988. A review of the PCB contaminant problem in the upper Mississippi River system. Mississippi River Work Unit. Wisconsin Department of Natural Resources, La Crosse.

Sullivan, J. F. 1989. Water quality characteristics and trends in the main channel of the upper Mississippi River: A review of Wisconsin's 10-year ambient monitoring program. Western Boundary Rivers Unit Report. Wisconsin Department of Natural Resources, La Crosse.

UMRBC (Upper Mississippi River Basin Commission). 1980. Water resources management plan. Volume 1: Upper Mississippi River. UMRBC, St. Paul, Minnesota.

UMRBC (Upper Mississippi River Basin Commission). 1982. Comprehensive master plan for the management of the upper Mississippi River system. UMRBC, St. Paul, Minnesota.

UMRCBS (Upper Mississippi River Basin Comprehensive Basin Study). 1972. Upper Mississippi River comprehensive basin study. Volume 1—Main report. UMRCBS Coordinating Committee, St. Paul, Minnesota.

UMRNW & FR (Upper Mississippi River National Wildlife and Fish Refuge). 1987. Environmental impact statement. Refuge master plan. U. S. Fish and Wildlife Service, Minneapolis.

Van Vooren, A. 1983. Distribution and relative abundance of upper Mississippi River fishes. Upper Mississippi River Conservation Committee, Rock Island, Illinois.

WDNR (Wisconsin Department of Natural Resources). 1988a. Wisconsin Water Quality Report to Congress 1988. WDNR, Madison.

WDNR (Wisconsin Department of Natural Resources). 1988b. Fish Advisory. WDNR, Madison.

Wiener, J. G., R. V. Anderson, and D. R. McConville, editors. 1984a. Contaminants in the upper Mississippi River. Proceedings of the 15th Annual Meeting of the Mississippi River Research Consortium. Butterworth Publishers, Boston.

Wiener, J. G., G. A. Jackson, T. W. May, and B. P. Cole. 1984b. Longitudinal distribution of trace elements (As, Cd, Cr, Hg, Pb, and Se) in fishes and sediments in the upper Mississippi River. Pages 139–170 *in* J. L. Wiener, R. V. Anderson, and D. R. McConville, editors. Contaminants in the upper Mississippi River. Proceedings of the 15th Annual Meeting of the Mississippi River Research Consortium. Butterworth Publishers, Boston.

GREAT II (Great River Environmental Action Team). 1981. Executive summary. GREAT II, Rock Island, Illinois.

GREAT III (Great River Environmental Action Team). 1982. Great river resource management study—14028, Mississippi River (Saverton, Missouri to Cairo, Illinois). GREAT III, St. Louis District, St. Louis, Missouri.

Hora, M. E. 1984. Polychlorinated biphenyls (PCBs) in common carp (*Cyprinus carpio*) of the upper Mississippi River. Pages 231–240 *in* J. G. Wiener, R. V. Anderson, D. R. McConville, editors. Contaminants in the upper Mississippi River. Proceedings of the 15th Annual Meeting of the Mississippi River Research Consortium. Butterworth Publishers, Boston.

IDNR (Iowa Department of Natural Resources). 1988. Water Quality in Iowa during 1986 and 1987. IDNR, Des Moines, Iowa.

IEPA (Illinois Environmental Protection Agency). 1988. Illinois Water Quality Report, 1986 and 1987. IEPA, Springfield, Illinois.

Kline, D. R., and J. L. Golden. 1979. Analysis of the upper Mississippi River commercial fishery. Pages 82–139 *in* J. L. Rasmussen, editor. A compendium of fishery information on the upper Mississippi River. Upper Mississippi River Conservation Committee, Rock Island, Illinois.

Lubinski, K. S., A. Van Vooren, G. Farabeen, J. Janecek, and S. D. Jackson. 1986. Common carp in the upper Mississippi River. Hydrobiologia 136:141–154.

McHenry, R. J., J. C. Ritchie, and C. M. Cooper. 1978. An assessment of the sediment accumulation in Pool 8 of the upper Mississippi River. Report to Great River Environmental Team (GREAT I). Sedimentation Laboratory, United States Department of Agriculture, Oxford, Mississippi.

MDC (Missouri Department of Conservation). 1989. Fish contaminant study of the Missouri and Mississippi rivers. MDC, Jefferson City, Missouri.

MDH (Minnesota Department of Health). 1987. Minnesota Fish Consumption Advisory. MDH, Minneapolis, Minnesota.

MPCA (Minnesota Pollution Control Agency). 1985. Polychlorinated biphenyls (PCBs) in common carp (*Cyprinus carpio*) in the upper Mississippi River (1975-1982). MPCA, Roseville, Minnesota.

MPCA (Minnesota Pollution Control Agency). 1988. Minnesota Water Quality, Water Years 1986–1987. MPCA, Roseville, Minnesota.

Nielsen, D. N., R. G. Rada, and M. M. Smart. 1984. Sediments of the upper Mississippi River: Their sources, distribution, and characteristics. Pages 67–98 *in* J. G. Wiener, R. V. Anderson, and D. R. McConville, editors. Contaminants in the upper Mississippi River. Proceedings of the 15th Annual Meeting of the Mississippi River Research Consortium. Butterworth Publishers, Boston.

Olson, K. N., and M. P. Meyer. 1976. Vegetation, land and water surface changes in the upper navigable portion of the Mississippi River basin over the period 1939-1973. Report to the St. Paul District Army Corps of Engineers. Remote Sensing Laboratory, University of Minnesota, St. Paul.

La Grande Rivière (Northern Québec)

DOMINIQUE ROY
Société d'énergie de la Baie James
800, de Maisonneuve est Montréal, H2L 4M8, Canada

ABSTRACT. *The La Grande Rivière is the largest river flowing into the eastern side of James Bay. After the partial diversion of the Eastmain and Caniapiscau rivers into La Grande Rivière, the drainage area expanded to 177,685 km² and the long-term average discharge reached 3,396 m³.s⁻¹. The region drained by this river is part of the Canadian Shield: the bedrock consists mainly of metamorphic and igneous Precambrian rock and the loose material is mainly of glacial origin except near James Bay where deep deposits of marine clay are found. The La Grande Rivière is located in the cold continental subarctic with short, mild summers and long, harsh winters. The frost-free period averages 75-d and the mean annual temperature is around −3°C. Until the early 1970s, the use of natural resources was limited to subsistence fishing and hunting by Cree Indians, the fur trade, and a few outfitting facilities. Since 1972, the La Grande Hydroelectric Project leads the economy of the region. Before the creation of the reservoirs, the water of La Grande Rivière was clear, almost colorless, slightly acid, and low in dissolved salts and nutrients. A few years after the two diversions and the flooding of the reservoirs, the quality of the water is unchanged. The paucity of nutritive elements, the low organic contribution by land vegetation, and the rigors of a northern climate keep aquatic production at the level of oligotrophic lakes. The only exception was during the first years after the reservoirs filled when features were more mesotrophic. Less than 24 species of fish are found in waters of the La Grande Rivière and usually only 5 or less species form 80% of the catch by gill nets. After reservoirs were filled, numbers of northern pike* Esox lucius, *lake whitefish* Coregonus clupeaformis *and, occasionally, cisco* Coregonus artedii *increased rapidly.*

The La Grande Rivière is the largest river flowing into the eastern side of James Bay, Northern Québec (Figure 1). Until recently, the use of natural resources in the area was limited to subsistence fishing and hunting, the fur trade, and a few outfitting facilities. With the responsibility given to the Direction générale du Nouveau-Québec (Ministère des Ressources naturelles du Québec) and the start of exploration for hydroelectric development in early 1960, a detailed description of the main rivers began. The flows and levels of these rivers were measured regularly, but data on their water quality remained sketchy (Grisel and Bobée 1974).

The first phase of the La Grande Hydroelectric Complex started in 1972. It called for the creation of five large reservoirs (Figure 2). The total flooded area reached 11,350 km² at maximum water levels; 85% of this area was previously land (Table 1). Powerhouses were built next to the La Grande 2, La Grande 3, and La Grande 4 reservoirs. The present installed capacity of these powerhouses for hydroelectric power is 10,282 MW. An additional powerhouse currently under construction at La Grande 2 will bring the total capacity to 12,282 MW, 60% at the La Grande 2 site alone. This electrical energy is mainly for domestic and industrial uses in the province of Québec and, on a smaller scale, for adjacent Canadian provinces and the United States.

Diversions from other rivers were necessary to obtain enough water for maximum generating capacity. Most (90%) of the water in the Eastmain River basin was diverted to the La Grande 2 Reservoir, and part (40%) of the water in the Caiapiscau River basin was diverted to La Grande 4. The Opinaca and Caniapiscau reservoirs transfer

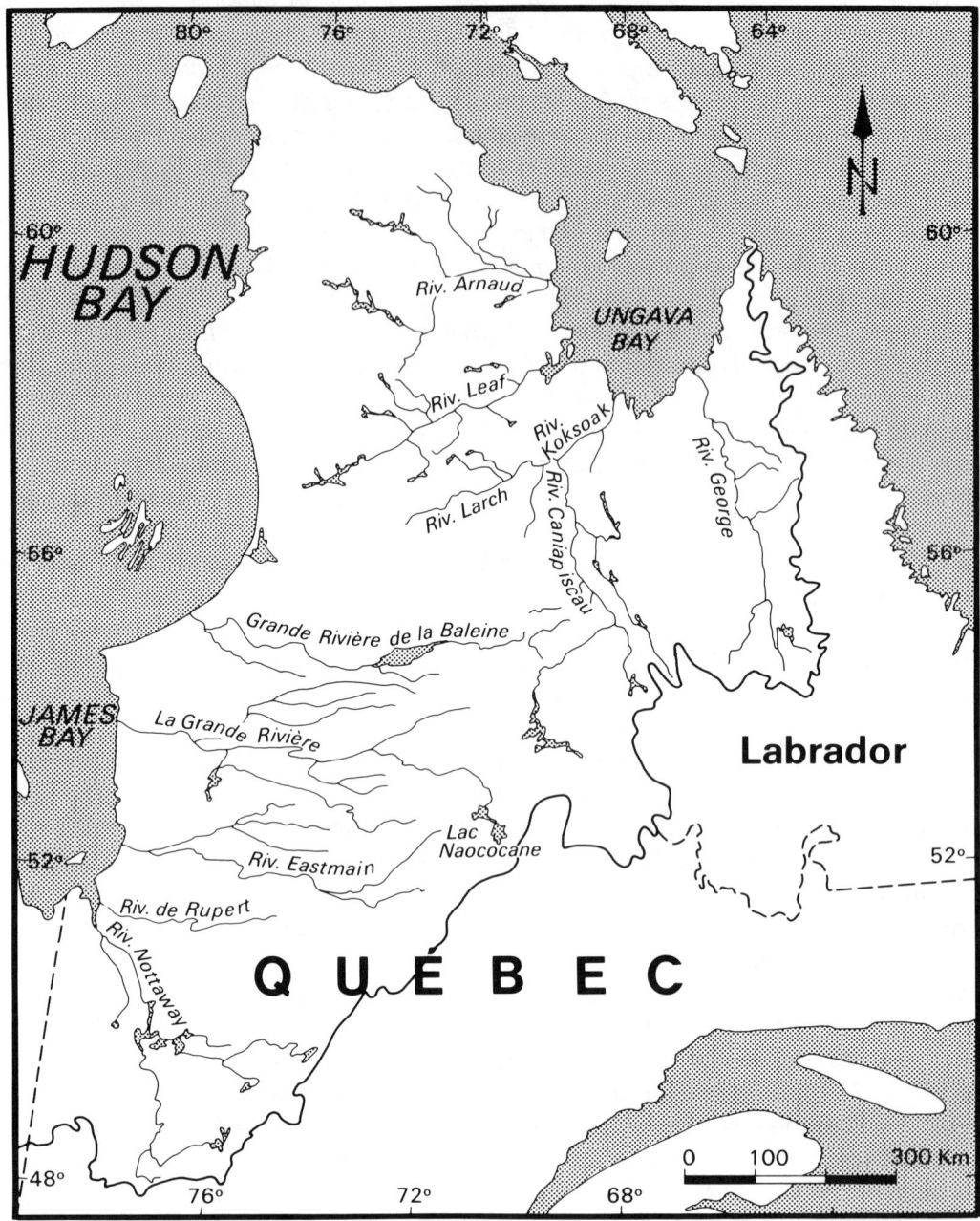

FIGURE 1. Main rivers of northern Québec.

these waters to the La Grande Rivière basin and, in the case of the Caniapiscau Reservoir especially, they provide additional storage for the winter season and for low dry years.

Several other sites may still be developed to optimize the La Grande Complex. Additional powerhouses are planned: La Grande 1, Laforge 1, Brisay, Laforge 2, Eastmain 1. A further 2,000 km² or so of land will be flooded, and another 3,516

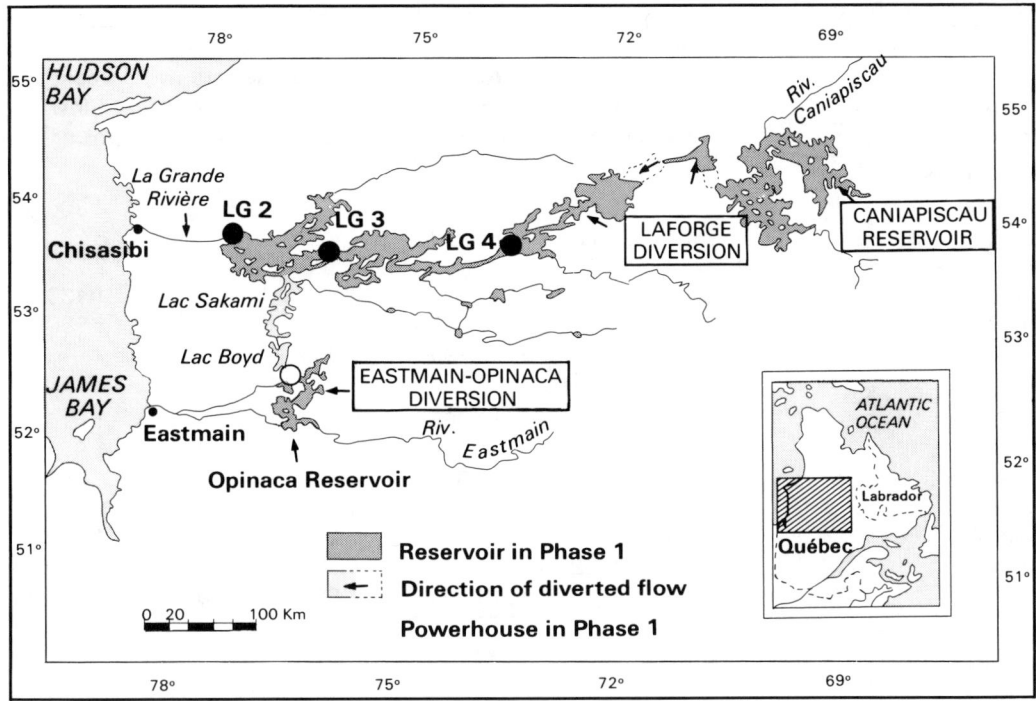

FIGURE 2. La Grande Hydroelectric Complex.

MW of generating capacity will be added. When these projects are completed, the La Grande Rivière will be, except in a few places, a continuous series of impoundments from the Caniapiscau Reservoir down to the estuary.

Water flow has been reduced drastically at the cutoff points to the Caniapiscau, Eastmain, and Opinaca rivers, the last being a tributary of the Eastmain River. The only overflows to the first two river systems result from occasional spilling of water from the Caniapiscau and Opinaca reservoirs. Otherwise, their water now comes solely from residual basins.

The Ministère des Richesses naturelles, in cooperation with the Société d'énergie de la Baie James (SEBJ), provided information on water quality in the La Grande Rivière, Eastmain, Nottaway, Broadback, and de Rupert rivers after 1974 (Grisel and Bobée 1974; Bobée et al. 1976). The SEBJ, for its part, completed similar investigations on the rivers and lakes likely to be affected by the hydroelectric developments, and added a biological component to its research. Without citing all reports stemming from these efforts, I should mention that more than a thousand lakes were studied at least once over a period of 11 years (Magnin 1977). Some of the lakes, along with a number of river sections, were included in the monitoring network and were visited repeatedly. For example, water quality samples were taken some 100 times at the control station for the La Grande 2 region, the site of La Grande 2, and the river's estuary region. Water quality determinations were made at another 24 stations in the ecological monitoring network at least 65 times (Boucher and Roy 1985; Roy 1985; Schetagne and Roy 1985).

TABLE 1. Main characteristics of the La Grande hydroelectric complex.

La Grande Project	Water level range (m above sea level)	Active storage ($10^6 m^3$)	Reservoir area (km^2)	Electrical generating capacity (MW)	Design Flow ($m^3 \cdot s^{-1}$)
Phase I (constructed)					
La Grande 2	175.3 - 167.6	19,365	2,835	5,328	4,300
La Grande 3	256.0 - 243.8	25,200	2,420	2,304	3,262
La Grande 4	377.0 - 366.0	7,160	765	2,650	2,520
Caniapiscau	535.5 - 522.6	39,070	4,275	—	1,130
Opinaca	215.8 - 211.8	3,395	1,040	—	1,980
Total	— —	94,190	11,335	10,282	—
Phase II (planned)					
La Grande 2A	175.3 - 167.6	19,365[a]	2,835[a]	2,000	1,610
La Grande 1	32.0 - 30.5	94	64	1,368	5,950
Eastmain 1	285.0 - 274.0	5,525	680	550	889
Laforge 1	439.0 - 431.0	4,890	1,002	820	1,544
Laforge 2	481.0 - 479.5	390	260	318	1,200
Brisay	535.5 - 522.2	39,070[b]	4,275[b]	460	1,130
Total	— —	10,899	2,006	5,616	—

[a]La Grande 2 Reservoir, already included in phase I.
[b]Caniapiscau Reservoir, already included in phase I.

The follow-up effort continues, and has been the responsibility of Hydro-Québec since 1986. Although the number of reports made by SEBJ and by its consultants (universities, private companies) is high, few papers have been published in scientific journals. I will mention only the most important summary reports; the reader will find references in them to more specific studies and comparisons with similar environments.

Morphology

The region drained by the La Grande Rivière is part of the Canadian Shield in Northern Québec, and it lies almost entirely within the James physiographic province. The area is relatively flat. It may be divided into three main morphological units, from west to east: a coastal plain, a rolling plateau, and a mountainous zone (Figure 3). Rivers flow generally from east to west.

The coastal plain runs along the edge of James Bay and consists of a strip of poorly drained lowlands some 150 km wide. The plain is covered with a layer of clay and sand, and crossed by a network of small and medium sized rivers with few lakes. There are a large number of depressions occupied by bogs and marshes, which decrease in size and number to the east.

The plateau stretches out behind the coastal plain, and has a relief that becomes more marked toward the east, reaching an elevation of 400 m about 500 km inland. The terrain continues to rise slowly beyond that point. Although more rugged than the coastal plain, the relief remains fairly unpronounced. In fact, the topography

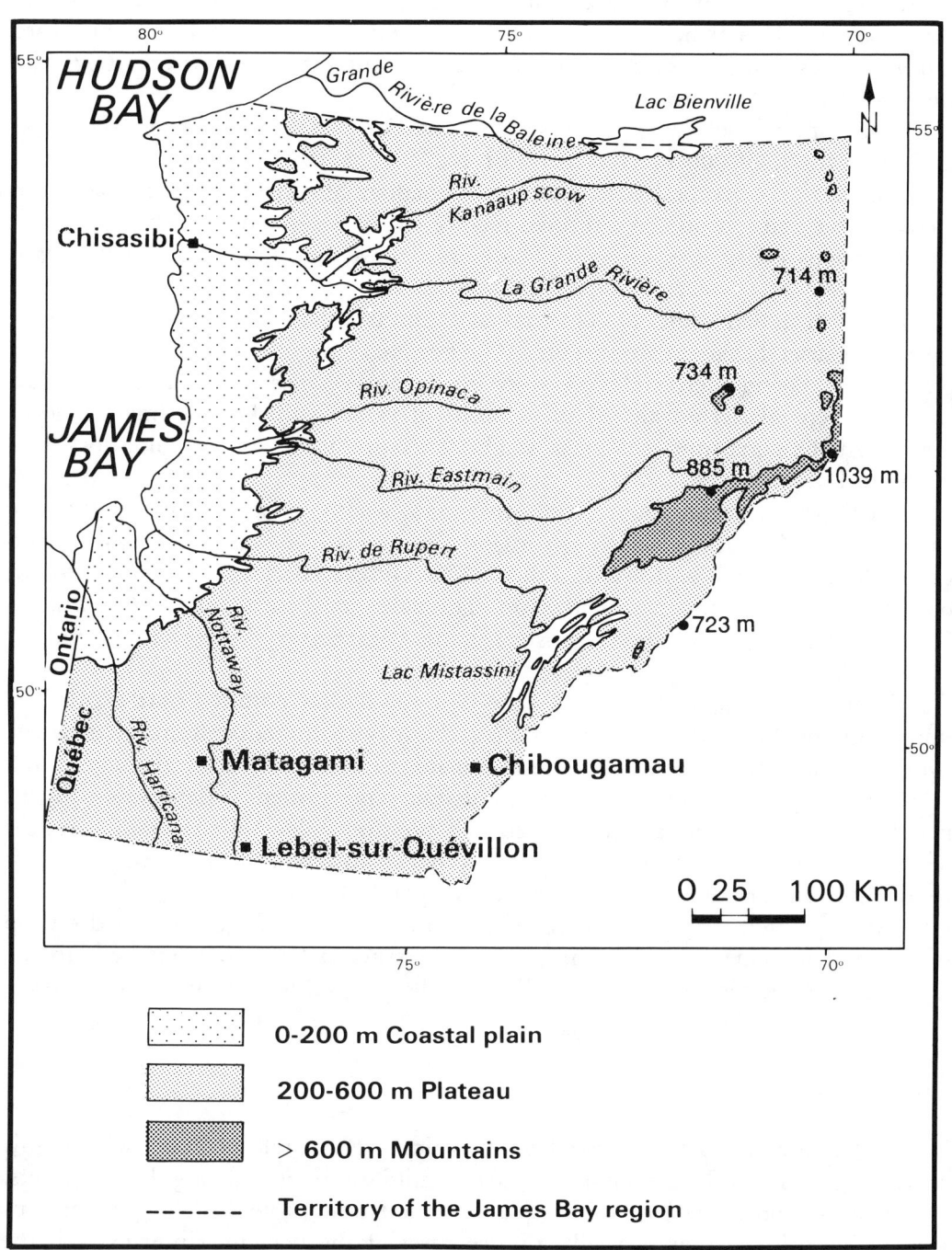

FIGURE 3. Topography of James Bay Territory.

consists of rolling hills that range in height between 15 and 60 m above the base relief. The bedrock outcrops frequently at the peaks, while morainic deposits of variable thickness cover the low areas. In contrast with the coastal plain, the plateau has many scattered lakes. Large rivers flow across the plateau and drain it. In the project region, these are the Eastmain and La Grande Rivière, which empty into James Bay; in the northeast part is the Caniapiscau River, which flows northward and discharges into Ungava Bay. The plateau also contains numerous peat bogs that are smaller than those near the bay.

The slope of the plateau increases near the Otish Mountains, where the peaks reach an altitude of 900 to 1100 m (SEBJ 1987).

The region's bedrock is part of the Canadian Shield. It consists mainly of metamorphic and igneous Precambrian rock (2.4×10^9 years old), which was shaped by the various Pleistocene ice ages. The loose deposits are mainly of glacial origin and, in some way, are associated with the last ice movements of the Pleistocene epoch, which ended some 10,000 years ago (Shilts 1986). The Nouveau-Québec-Labrador glacier began in northcentral Québec, near the eastern boundary of the project region, and then extended westward as far as James Bay and Hudson Bay. As the ancient glacier advanced, till deposits of variable thickness formed beneath the ice. At the same time, the great weight of ice produced considerable subsidence in the rock dome.

Deglaciation began about 10,000 years ago, leading to a division between the Labrador and Keewatin glaciers. An inland freshwater lake was created as the two glacial lobes receded northward, called Lake Ojibway-Barlow, with its northern boundary located at the Eastmain River. This lake acted as a huge sedimentation basin that left major lacustrine deposits (Figure 4) made up of alternating bands of silt and clay.

Then, around 8,000 years ago, Lake Ojibway-Barlow emptied northward. The upper Hudson Bay and James Bay drainage basin was immediately invaded by the Tyrrell Sea to a distance more than 200 km inland and reached an altitude of 180 to 270 m. Deposits of silty clays and fine deltaic sand formed in depressions in this marine environment. As a result of reduced pressure following the gradual retreat of the ice, the bedrock in the region slowly began to rebound. The isostatic uplift continues even now at a rate of 0.5 to 1.5 m per 100 years.

As the Tyrrell Sea disappeared, deltaic sands and beach gravels settled on top of clay and moraine deposits. The present rivers gradually carved out beds in the clay and moraine sediments, eventually creating the profile of today's riverbanks.

Climate

The La Grande Complex is located in the cold, continental subarctic, with highly contrasting seasons. This region features short, mild summers and long, harsh winters.

Hudson and James bays do not play the moderating role usually performed by seas during cold weather. Actually, the ice cover of the bays in winter extends the continental surface, allowing polar conditions to spread. In summer, the presence

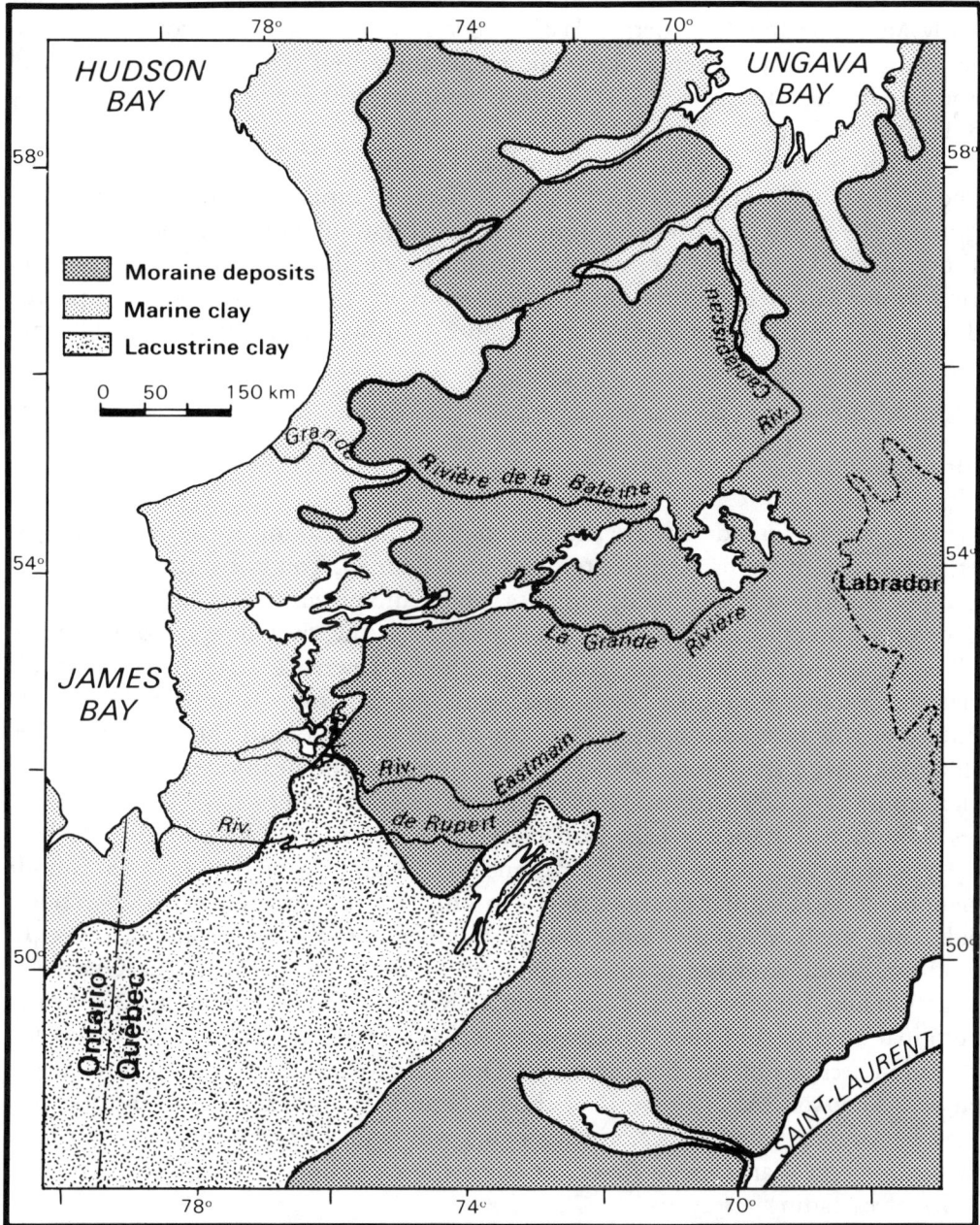

FIGURE 4. Main surface deposits in northern Québec.

of cold currents and ice in the surrounding seas slow the warming of the coastal climate. In addition, the relatively flat relief of the region makes it easy for masses of arctic air to flow unimpeded. The Arctic air provides the intense cold that prevails for many months and helps prolong the frost period until late in the spring.

Temperatures in the region show significant interseasonal fluctuations and highly pronounced extremes. The latitude and altitude cause the temperature to drop from southwest to northeast. Inland, ice forms generally in mid-October and melts away in early May. The frost-free period lasts 75-d on average, from the fourth week in June to the first week in September. The mean annual air temperature at the head of the La Grande Rivière is $-3.8°C$. Mean temperatures are $-22.9°C$ for January and $13.6°C$ for July.

Precipitation decreases from southeast to northwest. The wettest area is located in the highest zone, where the La Grande Rivière, Canipiscau, and Eastmain rivers originate. The mean value for total annual precipitation (rain and snow) over the territory of the La Grande Complex is 765 mm, one third of which falls as snow during winter. On average, precipitation between April 1 and September 30 is pluvial and totals 477 mm. Precipitation from October 1 to March 31 amounts to 288 mm (water equivalent).

Hydrology

Originally, the La Grande Rivière drained 98,820 km^2, making it the fourth largest river in Québec in terms of discharge and area drained. Following the diversion of the upper basins of the Eastmain, Opinaca, and Caniapiscau rivers, 78,865 km^2 were added for an overall drainage area of 177,685 km^2. The river now ranks second in Québec for area drained and discharge, after the St. Lawrence River. The La Grande Rivière has its source in Lake Naococane, some 850 km from James Bay. The main tributary of this lake adds another 150 km to the river's length, bringing the total to 1000 km. After the Caniapiscau River was diverted, the headwaters of La Grande Rivière must flow downstream 1200 km before reaching James Bay.

Only the tributaries of Lake Naococane that arise in the Otish Mountains have steep gradients—12 m.km^{-1} over a distance of 50 km. After that, the La Grande Rivière has a fairly regular gradient of 0.7 m.km^{-1} caused by a series of flat sections, rapids, and small waterfalls. Since the La Grande Complex was constructed, close to 400 km of the river has been incorporated into the La Grande 2, La Grande 3, and La Grande 4 reservoirs. The Caniapiscau Reservoir inundated about 200 km of its original stream, and the Opinaca reservoir flooded and joined together 40 km and 50 km sections of the Eastmain and Opinaca rivers before flowing into the La Grande 2 Reservoir. After Phase II of the La Grande Complex is built, the addition of the La Grande I and Vincelotte reservoirs will transform the entire watercourse flowing from the Caniapiscau Reservoir, with the exception of about 100 km, into a series of cascading reservoirs.

The basin of the La Grande Complex has an average specific discharge of 19.25 L.s^{-1}.km^{-2}, which corresponds to an annual net surface flux of 607 mm. Similar to the precipitation, the specific discharge gradually decreases from southeast to northwest. Maximum discharges occur in the plateau, where the complex's rivers have their sources, reaching 24 L.s^{-1}.km^{-2} in the upper Eastmain basin. Specific discharges drop along the coast, mounting to between 14.2 and 15.3 L.s^{-1}.km^{-2}.

La Grande Rivière's regime features a heavy spring flood, whose inflows from snowmelt generate about 40% of the annual discharge. This flood is followed by a summer period of low water that varies in degree, depending on year. The so-called fall flood, pluvial in origin, causes only a small increase of flow. The winter period of low water starts in November and lasts until the beginning of May. The hydrological regime of the La Grande, Eastmain, and Caniapiscau rivers is characterized by highly variable spring floods, whose peaks are the combined product of snowmelt and rain (Figure 5).

The long-term average discharge of the La Grande Rivière since 1959 is 1,760 $m^3.s^{-1}$. The long-term average discharge of the Eastmain River is 927 $m^3.s^{-1}$ while that of the Caniapiscau River is 1,804 $m^3.s^{-1}$. Since diversion of the Eastmain (845 $m^3.s^{-1}$) and the Canipiscau (790 $m^3.s^{-1}$) rivers took place, the long-term average discharge at the mouth of the La Grande Rivière has reached 3,395 $m^3.s^{-1}$, or almost twice the previous amount.

Water Quality and Resource Management

The basins of the La Grande Rivière and the Eastmain and Opinaca rivers are sparsely populated. In 1976, close to 3,000 people lived in three villages near the James Bay coast. Of this number, a little more than half were Cree hunters who trapped inland more than 150 km away from the coast. About 100 other Crees from the village of Mistassini, 400 km south of the Canipiscau Reservoir, must be added to this figure. During construction of the La Grande Complex from 1973 to 1986, the influx of people ranged between 1,000 and 18,000 with the maximum attained in 1978. These workers were housed in camps near job sites. In term of the territory's whole area, human occupation remained very low. Human density in 1978, at the peak of construction work, was about 0.12 inhabitants per km^2. At present, the only human activities in the La Grande Complex involve the initial construction of a new powerhouse, the hydroelectric exploitation of the rivers, and continued hunting, fishing, and trapping practices.

Before the reservoirs were created, the water in the La Grande Rivière was clear, almost colorless, slightly acid, and low in dissolved salts and nutrients (Table 2). The water upstream generally had a lower content of dissolved salts than the water downstream (Bobée et al. 1976; Schetagne and Roy 1985). Surface runoff over insoluble base materials (granite, gneiss) and percolation into coarse moraines contributed little in way of dissolved substances. The low chemical content reflected a poverty of the land where open forest and peat bogs abound (taiga) rather than dense forests and marshes. Only in the limited coastal plain did chemical enrichment of water increase—because of contact with marine clays.

The positive gradient of water enrichment in rivers is also noted in regional lakes. Inland, at the sources of the three main rivers, lake water has essentially the same water quality properties as water in the upstream portion of the La Grande Rivière, reflecting the impoverishment of the same land drained. Upstream areas, covered with scores of lakes, provides numerous contacts between land and water. Because the rock and loose deposits are chemically identical in nature, the lakes have more or less the same water qualities as the rivers.

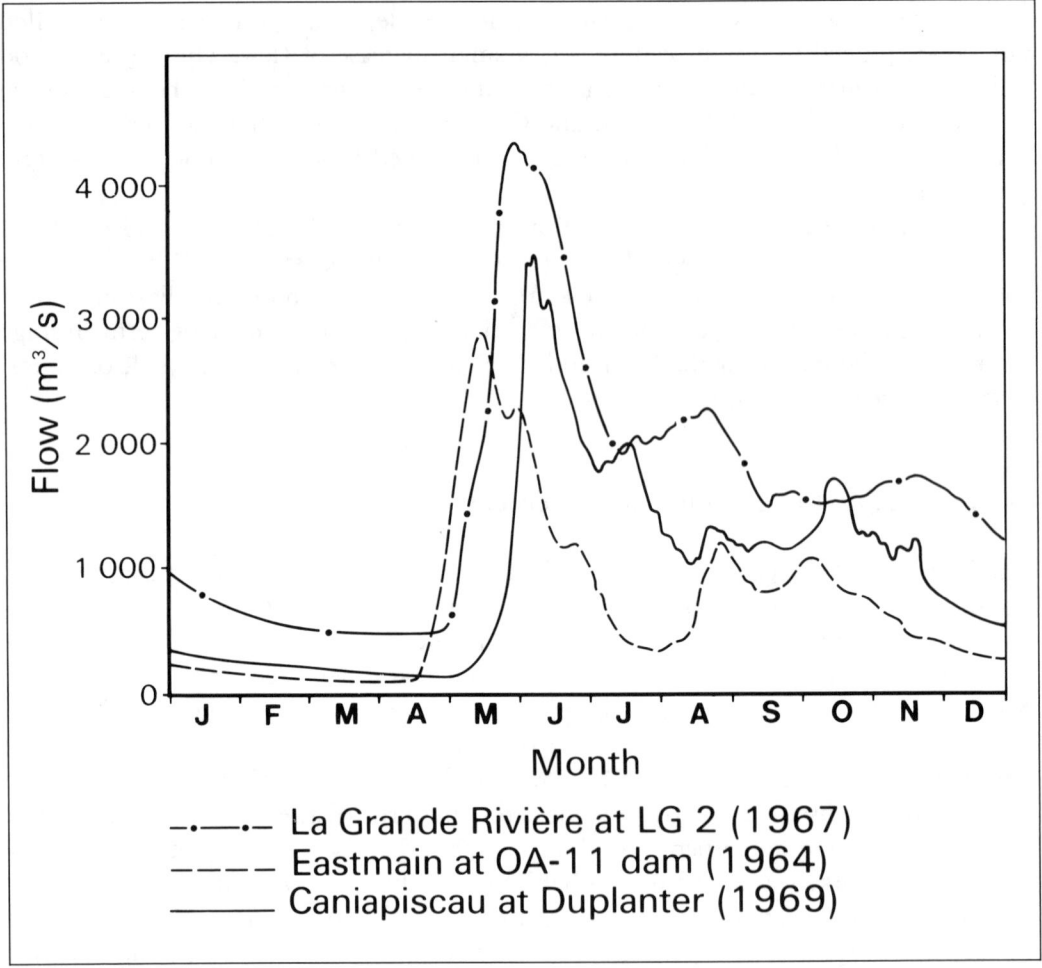

FIGURE 5. Daily discharge of Eastmain and Caniapiscau rivers at their diversion sites and La Grande Rivière at La Grande site.

Starting 500 km inland from the coast of James Bay, surface deposits aquire more fine elements. The fine elements may initially stem from the leaching of moraines by glacier water, greater grinding of rocks further away from the glacier center and, especially within 150 km of the coast, from deposits of marine clay. At this point, lake water becomes enriched with dissolved substances and organic carbon, a reflection of the greater richness of the soils and vegetation. However, because inflows to the river are diluted by water from upstream, their enrichment closer to the coast lags behind enrichment of the lakes. For example, the conductivity of lake water, which is equal to that of river water at the source, doubles about 200 km from the coast, and more than triples near the coast.

The initial changes in aquatic environments occurred in late 1978 as the first reservoir was filled. Prior to that, the application of environmental protection regulations to domestic waste water (secondary treatment) and to industrial water prevented severe deterioration in water quality. The main physical changes that effected

TABLE 2. Means (standard deviations) of water quality variables from La Grande Rivière before the filling of the first reservoir of the La Grande Complex and from a lake 100 km from the James Bay.

Variable	La Grande Rivière Downstream		La Grande Rivière Upstream		Lake Detchevery	
Number of samples	58.00		28.00		97.00	
Color (Hazen units)	23.00	(4)	19.00	(16)	29.00	(5)
Turbidity (NTU)[a]	1.80	(1.2)	0.70	(0.3)	0.40	(0.2)
Maximum temperature (°C)	16.50		18.50		16.60	
Oxygen saturation (%)	99.00	(5)	94.00	(5)	93.00	(4)
Conductivity (us.cm^{-1})	16.00	(2)	11.00	(2)	36.00	(2)
pH (units)	6.40	(0.2)	6.20	(0.3)	7.00	(0.2)
Chlorides (mg.L^{-1})	0.60	(0.3)	0.30	(0.2)	1.80	(0.5)
Bicarbonates (mg.L^{-1})	4.80	(1.4)	2.90	(1.5)	10.90	(1.3)
Sulfates (mg.L^{-1})	2.40	(0.8)	1.40	(0.6)	3.40	(0.9)
Sodium (mg.L^{-1})	0.90	(0.2)	0.50	(0.1)	2.20	(0.2)
Potassium (mg.L^{-1})	0.40	(0.1)	0.30	(0.1)	0.30	(0.1)
Magnesium (mg.L^{-1})	0.30	(0.1)	0.30	(0.1)	0.60	(0.1)
Calcium (mg.L^{-1})	1.30	(0.3)	0.90	(0.3)	3.80	(0.3)
Iron (mg.L^{-1})	0.17	(0.11)	0.08	(0.06)	0.05	(0.04)
Nitrates & nitrites (mg.L^{-1})	<0.02		<0.02		0.04	(0.02)
Kjeldahl nitrogen (mg.L^{-1})	0.15	(0.06)	0.11	(0.05)	0.17	(0.06)
Total phosphorus (mg.m^{-3})	8.00	(4)	5.00	(1)	4.00	(0.1)
Total inorganic carbon	5.60	(1.1)	4.40	(1.0)	7.60	(1.9)
Silica (mg.L^{-1})	2.60	(0.6)	1.80	(0.6)	2.10	(0.3)
Chlorophyll a (mg.m^{-3})	1.28	(0.57)	1.09	(0.54)	1.08	(0.43)

[a]Nephelometric turbidity units

water quality in the La Grande Rivière system were the creation of the reservoirs, the diversion of water from one drainage basin to another, the increase of flow along the main watercourse, and the reduction of flows below cutoff points of diverted streams.

Three reservoirs were the subject of ecological follow-up studies since 1978: La Grande 2, Opinaca, and Caniapiscau. The same trend in water quality modifications appeared in all reservoirs. Slight variations are explained largely by the individual characteristics of the reservoirs, e.g., morphometry, water retention time, relative area of flooded lands, nature and density of submerged vegetation, and duration of impounding period (Roy et al. 1986).

Water in the surface layer (1–10 m) occupies between 40% and 60% of the volume of the three reservoirs. Few changes in water quality occurred in the surface layer of La Grande Complex reservoirs during the ice-free period. The greatest variations in the reservoirs appeared where spruce stands had been dense, the inflow was limited, and exchange of water with the main reservoir was restricted (Table 3). While we cannot provide statistical proof, the leaching of organic soil and submerged plants may have helped the release of organic matter in dissolved or particulate form; favored the formation of phosphorus, iron, and organic complexes; and, despite the apparent enrichment, reduced phytoplankton production. Dilution by replenishment of water from the river and addition of decomposition products quickly masked the

TABLE 3. Surface (0–10 m) water quality variables in La Grande 2 Reservoir near the outlet and in a protected bay, before and after the filling in 1979. Values are means (standard deviations).

Variable	Outlet			Protected bay		
	1978	1981	1984	1978	1981	1984
Number of samples	5.00	14.00	11.00	16.00	14.00	11.00
Color (Hazen units)	23.00 (11.0)	26.00 (2.0)	25.00 (3.0)	60.00 (16.0)	50.00 (4.0)	43.00 (6.0)
Turbidity (NTU)[a]	2.10 (1.4)	0.80 (0.2)	0.60 (0.1)	1.20 (0.4)	1.00 (0.3)	1.10 (0.4)
Maximum temperature (°C)	16.00	16.60	13.70	16.60	16.20	16.50
Oxygen saturation (%)	102.00 (4.0)	82.00 (13.0)	86.00 (8.0)	87.00 (9.0)	61.00 (20.0)	81.00 (11.0)
Conductivity (uS.cm^{-1})	15.10 (2.0)	16.10 (0.7)	13.60 (0.9)	30.60 (3.0)	23.40 (2.2)	16.20 (0.6)
pH (units)	6.50 (0.1)	6.30 (0.2)	6.40 (0.2)	6.80 (0.2)	6.40 (0.2)	6.40 (0.2)
Chlorides (mg.L^{-1})	0.40 (0.2)	0.60 (0.1)	0.30 (0.1)	0.90 (0.7)	1.00 (0.3)	0.70 (0.2)
Bicarbonates (mg.L^{-1})	4.20 (0.7)	4.00 (0.3)	3.30 (0.1)	10.30 (0.9)	7.60 (0.9)	4.20 (0.2)
Sulfates (mg.L^{-1})	2.80 (0.5)	1.80 (1.0)	1.60 (0.2)	3.00 (0.8)	2.10 (0.7)	1.60 (0.2)
Sodium (mg.L^{-1})	0.80 (0.2)	0.90 (0.2)	0.70 (0.1)	1.90 (0.2)	1.10 (0.2)	0.90 (0.1)
Potassium (mg.L^{-1})	0.30 (0.1)	0.50 (0.1)	0.40 (0.1)	0.40 (0.1)	0.80 (0.1)	0.40 (0.1)
Magnesium (mg.L^{-1})	0.40 (0.1)	0.40 (0.1)	0.30 (0.1)	0.70 (0.1)	0.60 (0.1)	0.40 (0.1)
Calcium (mg.L^{-1})	1.30 (0.2)	1.30 (0.1)	1.20 (0.1)	3.70 (0.4)	2.30 (0.1)	1.70 (0.2)
Iron (mg.L^{-1})	0.21 (0.09)	0.17 (0.10)	0.15 (0.05)	0.32 (0.04)	0.38 (0.16)	0.31 (0.13)
Nitrates & nitrites (mg.L^{-1})	<0.10	<0.02	0.04 (0.02)	<0.10	<0.02	0.02 (0.01)
Kjeldahl nitrogen (mg.L^{-1})	0.17 (0.06)	0.13 (0.05)	0.15 (0.03)	0.29 (0.06)	0.27 (0.06)	0.26 (0.03)
Total phosphorus (mg.L^{-3})	8.00 (4.0)	12.00 (3.0)	11.00 (2.0)	10.00 (2.0)	20.00 (4.0)	15.00 (1.0)
Total inorganic carbon (mg.L^{-1})	1.20 (0.2)	1.90 (0.6)	1.50 (0.5)	2.30 (0.6)	3.10 (1.2)	1.70 (0.6)
Total organic carbon (mg.L^{-1})	5.90 (0.5)	5.50 (1.2)	5.10 (0.1)	11.60 (2.0)	9.40 (1.5)	8.30 (0.3)
Silica (mg.L^{-1})	2.30 (0.5)	1.80 (0.6)	1.10 (0.6)	3.50 (0.6)	1.50 (0.6)	0.60 (0.6)
Chlorophyll a (mg.m^{-3})	1.27 (0.62)	1.77 (1.69)	2.00 (1.43)	1.88 (0.93)	2.57 (1.53)	3.85 (2.01)

[a] Nephelometric turbidity units

effects of leaching decomposition (Schetagne and Roy 1985). The decomposition of organic matter and mixing affected water quality the most during the short term (0–5 years) but mixing alone remained important during the medium and long terms.

Decomposition phenomenon in the La Grande reservoirs peaked the second or third year after impounding. Decomposition effects were evident in the concentration of dissolved oxygen (-20%), organic carbon (-15%), total phosphorus ($+100\%$), potassium ($+100\%$), and the level of conductivity ($+10\%$). Phytoplankton responded to higher nutrient concentrations by an increase in chlorophyll a ($+100\%$), but only after the first year. This latency period was probably explained by the delayed appearance of phosphorus compounds easily assimilated by algae. The same phenomena were observed on the Nelson-Churchill complex (Jackson and Hecky 1980). Photosynthetic activity in the new reservoirs peaked the third year. In following years, spurts of primary production took place in the spring, but they were rapidly checked after mid-summer by a lack of silica.

At the end of winter, in sheltered reservoir areas, water acidity (pH 5.6 to 6.0) and concentrations of dissolved oxygen (2.2 to 3.3 ppm), total phosphorus (26 to 49 ppb), bicarbonates (9.0 to 13.7 ppm), and total inorganic carbon (6.0 to 8.6 ppm) in the surface water were affected by anoxic conditions near the bottom. However, no changes could be detected in surface waters located above great depths or where the water still flowed.

Finally, the high rate of water replenishment and the mixing of residual water with inflows (spring and fall turnover, wind-generated wave action, fluctuation in discharge volume) made the quality of reservoir water more like the quality of the original river. This return process is usually slower when reservoirs are built in series, but it was already detectable in La Grande 2 Reservoir after five years although this impoundment received water from four reservoirs.

Immediately below a reservoir, water quality was similar to that of the source impoundment. After agitation and stirring in the first downstream rapids, the water regained the characteristics of an unmodified river (Table 4).

Each time a new reservoir in the La Grande Complex was impounded, flow of river water was totally cut off. For nearly eight months after closure of its outlet dam, the river section below the La Grande 2 Reservoir received inflow only from the residual drainage basin, or an amount equal to 3% of the initial flow at the river mouth. One year later, flow below the La Grande 2 Reservoir was re-established and regulated to meet electrical requirements. The filling of the other two reservoirs had less impact on downstream flow because they were immediately above another reservoir.

The main causes of fluctuations in water quality of the La Grande Rivière during cutoff in 1979 and, since 1980, cutoff of the Eastmain and Opinica rivers, were the nature of inflows from the residual basin, erosion of the banks, and increase in the time of contact with sediments. The tributaries of sections located below the cutoff points of these rivers drained deposits of marine origin partly covered by bogs. These rivers, therefore, introduced neutral water that was buffered; richer in mineral, organic matter, and nutritive elements; and undiluted by more pristine water from

TABLE 4. Means (standard deviations) of water quality variables observed at the mouth of La Grande Rivière before, during, and after the flooding of La Grande 2 Reservoir.

Variable	1974–1978	1979	1982	1984
Number of samples	58.00	13.00	13.00	11.00
Color (Hazen units)	23.00 (4.0)	58.00 (26.0)	27.00 (5.0)	24.00 (2.0)
Turbidity (NTU)[a]	1.80 (1.2)	11.70 (7.9)	2.00 (1.6)	1.50 (1.1)
Maximum temperature (C)	16.50	17.50	10.00	12.00
Oxygen saturation (%)	99.00 (5.0)	96.00 (7.0)	95.00 (3.0)	96.00 (4.0)
Conductivity ($uS.cm^{-1}$)	16.00 (2.0)	50.00 (37.0)	16.00 (1.0)	14.00 (1.0)
pH (units)	6.40 (0.2)	6.90 (0.3)	6.30 (1.0)	6.40 (0.1)
Chlorides ($mg.L^{-1}$)	0.60 (0.3)	5.00 (3.9)	0.60 (0.1)	0.40 (0.1)
Bicarbonates ($mg.L^{-1}$)	4.80 (1.4)	8.70 (4.1)	4.30 (0.4)	3.40 (0.1)
Sulfates ($mg.L^{-1}$)	2.40 (0.8)	5.70 (1.5)	1.50 (0.7)	1.60 (0.2)
Sodium ($mg.L^{-1}$)	0.90 (0.2)	3.90 (1.9)	0.90 (0.2)	0.80 (0.2)
Potassium ($mg.L^{-1}$)	0.40 (0.1)	1.00 (0.2)	0.60 (0.1)	0.40 (0.1)
Magnesium ($mg.L^{-1}$)	0.30 (0.1)	1.20 (0.2)	0.50 (0.1)	0.30 (0.1)
Calcium ($mg.L^{-1}$)	1.30 (0.3)	3.00 (0.8)	1.50 (0.3)	1.20 (0.3)
Iron ($mg.L^{-1}$)	0.17 (0.11)	1.32 (0.54)	0.20 (0.10)	0.20 (0.10)
Nitrates & nitrites ($mg.L^{-1}$)	<0.02	0.06 (0.06)	0.04 (0.02)	0.03 (0.01)
Kjeldahl nitrogen ($mg.L^{-1}$)	0.15 (0.06)	0.20 (0.05)	0.16 (0.02)	0.16 (0.02)
Total phosphorus ($mg.m^{-3}$)	8.00 (4.0)	24.00 (13.0)	15.00 (6.0)	12.00 (3.0)
Total inorganic carbon ($mg.L^{-1}$)	0.90 (0.4)	1.70 (1.2)	1.90 (0.3)	1.40 (0.2)
Total organic carbon ($mg.L^{-1}$)	5.60 (1.1)	11.40 (3.3)	5.30 (0.6)	5.30 (0.6)
Silica ($mg.L^{-1}$)	2.60 (0.6)	3.10 (0.7)	1.30 (0.5)	1.10 (0.3)
Chlorophyll a ($mg.m^{-3}$)	1.28 (0.57)	1.39 (0.37)	1.40 (1.15)	1.74 (0.86)

[a] Nephelometric turbidity units

glacial deposits. The erosion of clay or silty clay banks augmented the concentration of suspended matter and, indirectly, the degree of mineralization. Finally, water retention time increased in the three rivers, further favoring solubilization and mineralization (Table 4). Reduced turbulence and longer water retention restored concentrations of dissolved-oxygen to the level encountered in lakes, or slightly below the maximum saturation level. Unlike the reservoir, the cutoff of rivers brought few additional organic materials and did not lead to marked deficiencies of dissolved oxygen (Messier and Roy 1987a).

Since 1980, the water downstream from La Grande 2 Reservoir has retained its original quality despite an increase in discharge volume and its passage through the reservoir. The physical component most affected by impoundment release was water temperature (Table 4). A reduction of 5°C in maximum temperature and a delay of 2 or 3 weeks in the seasonal temperature curve were due to the transfer of the top 40 m layer of water from the La Grande 2 Reservoir.

Properties of the Natural Ecosystem

The lack of nutritive elements in runoff and ground waters, the small organic contribution by land vegetation, and the rigors of a northern climate keep aquatic production in La Grande Complex at a level typical of oligotrophic lakes. Only 24 species of fish are present in La Grande Rivière and Eastmain watersheds and 17 species are present in the upper Caniapiscau basin (Table 5). The low rate of fish replacement in northern Québec and the vulnerability of fish stocks to intensive harvest has been emphasized previously (Power and LeJeune 1976). The low potential of annual fish yield in the lakes, estimated at 2.3 kg.ha^{-1}.yr^{-1} (Schlesinger and Regier 1982), is not initially apparent. The first fishing activities bring in many large, old fish. However, the large specimens are the products of 10-year or even 40-year growth, and their removal quickly lowers the catch weight per unit effort. The low rate of annual yield in natural lakes is then shown by the slow recovery of decimated fish stocks and by how long the fish take to regain their previous sizes. Low yield is more apparent for such predominant predatory species as walleye *Stizostedion vitreum* and lake trout *Salvelinus namaycush* than for more adaptable species whose numbers may fluctuate rapidly, e.g., lake cisco *Coregonus artedii*, lake whitefish *Coregonus clupeaformis*, and northern pike *Esox lucius*.

Flow velocities, seasonal fluctuations in water flow and level, and the formation of frazil ice in winter constrained populations of fish in the main sections of the La

TABLE 5. Relative importance of fish species from Eastmain, La Grande Rivière, and Caniapiscau watersheds (number = % of catch, + = present, and − = absent).

Name	Eastmain	La Grande Rivière	Caniapiscau
Lake sturgeon *Acipenser fulvescens*	3	+	−
Landlocked salmon *Salmo salar*	−	−	2
Brook trout *Salvelinus fontinalis*	+	+	2
Lake trout *Salvelinus namaycush*	2	3	19
Cisco *Coregonus artedii*	10	7	−
Lake whitefish *Coregonus clupeaformis*	19	14	13
Round whitefish *Prosopium cylindraceum*	+	+	9
Northern pike *Esox lucius*	13	9	2
Longnose sucker *Catostomus catostomus*	4	14	37
White sucker *Catostomus commersoni*	13	13	5
Burbot *Lota lota*	+	4	1
Walleye *Stizostedion vitreum*	33	33	−
Other species present:	2	3	10
Lake chub *Couesius plumbeus*	+	+	+
Emerald shiner *Notropis atherinoides*	+	−	−
Spottail shiner *Notropis hudsonius*	+	+	−
Longnose dace *Rhinichthys cataractae*	−	−	+
Creek club *Semotilus atromaculatus*	+	+	−
Fallfish *Semotilus corporalis*	−	+	+
Pearl dace *Semotilus margarita*	+	−	−
Brook stickleback *Culaea inconstans*	+	+	−
Threespine stickleback *Gasteroteus aculeatus*	+	+	+
Ninespine stickleback *Pungitius pungitius*	+	+	+
Trout−perch *Percopsis omiscomaycus*	+	+	−
Yellow perch *Perca flavescens*	+	+	−
Logperch *Percina caprodes*	−	+	−
Mottled sculpin *Cottus bairdi*	+	+	+
Slimy sculpin *Cottus cognatus*	+	+	+

Grande Rivière. Plankton production was almost nil, and benthic production was barely enough for the needs of fish that were 3 to 4 times less plentiful than in adjacent lakes. Comparable numbers of fish remained in small tributaries, even though the physical constraints were less acute.

Three studies on the harvest of Cree Indians near La Grande Rivière and Eastmain Rivers are available. The last one covers more than 5 years and describes the catch for each community and each year. The mean annual harvest for the Chisasibi band, at the mouth of La Grande Rivière, is presented in Table 6. The catch of walleye and northern pike was low compared to catches of the whitefish and brook trout, showing that the Chisasibi band fishes mainly in rivers and lakes near James Bay and along the coast, north and south of the mouth of La Grande Rivière. Almost no brook trout and lake trout are caught in La Grande Rivière watershed except at a distance more than 400 km from the mouth, and few hunters move so far upstream (JBNQNHRC 1982).

After high levels of mercury were found in northern pike and walleye in inland waters in 1976 (Penn 1978), the number of fish harvested decreased, although the main catch of the Cree Indians came from anadromous stocks whose levels of mercury were low. Another cause of reduced harvest was related to the abundance of small game. Fishing effort was low when the hare population was high, and hare and ptarmigan populations both peaked between 1978 and 1982. Finally, during the filling of La Grande 2 reservoir in 1979, fishing effort was reduced near the village to protect anadromous stocks (JBNQNHRC 1982).

Changes from Hydroelectric Development

I will not list all environmental losses and gains caused by the flooding of land in the creation of reservoirs in the La Grande Complex, nor will I deal specifically with impacts on users of aquatic and land resources. I will discuss only the quality of aquatic environments in reservoirs.

TABLE 6. Mean numbers of fish harvested by the Fort George (Chisasibi band) community between 1974 and 1980.

Species	Mean	(Standard deviation)	Percent
Whitefishes[a]	48,807	(19,844)	48
Burbot	3,355	(4,619)	3
Brook trout	21,615	(6,553)	21
Lake trout	5,310	(1,882)	5
Arctic char	512	(261)	1
Northern pike	4,942	(1,839)	5
Sucker[b]	15,009	(5,792)	14
Lake sturgeon	615	(264)	1
Walleye	1,936	(392)	2
Total	102,101		100

[a]Includes the cisco, the lake whitefish, and the round whitefish.
[b]Includes the longnose sucker and the white sucker.

For the first 2 or 3 years after the start of inundation, nutrients gradually were enhanced because of leaching and, especially, the decomposition of organic matter in submerged soils and vegetation. In comparison with natural lakes in northern Québec, enrichment resulted in an increase of 50% in phytoplankton, a doubling of zooplankton biomass and, after 5 years, the establishment of equivalent, if not denser fish populations than in non-impounded lakes.

In all cases, relative compositions of fish communities in the La Grande Complex were modified in favor of pioneer-type species, to the detriment of species more commonly found in the area's natural lakes. Northern pike, lake whitefish, lake cisco, and burbot (*Lota lota*) moved rapidly into the new impoundments and reproduced successfully (Figure 6). On the other hand, walleye, suckers (*Catostomus catostomus* and *C. commersoni*), and lake trout did not benefit as much, since relatively few young specimens of these species appeared in stomach contents of predator fish or were caught in nets and seines after the impounding. Apparently, suitable spawning sites and rearing conditions for these species were lacking, at least during the initial years of the reservoir. However, growth rates and condition indexes were better after impoundment in these two groups of fish. Lake whitefish, longnose sucker (*C. catostomus*), and walleye showed an additional growth in length of 25%, while northern pike added 13%. In young individuals, walleye and longnose sucker maintained their original growth rates for length, while northern pike gained 10% and lake whitefish gained 50%. On the basis of these observations, I conclude that the environment in the new reservoirs changed only slightly, going temporarily from oligotrophic to mesotrophic.

Temporary eutrophication was accompanied by release of mercury, which accumulated in aquatic organisms, particularly in fish. Although not detectable in the water, this metal built up in predatory fish in natural lakes of the region to a critical toxic level for humans. Five years after the reservoirs were filled, the concentration of mercury in fish reached 4 to 5 times the previous levels in all species (Figure 7 and 8). Non-predatory fishes (lake cisco, lake whitefish, suckers) acquired mercury at levels close to or above the Canadian marketing standard of 0.5 ppm in individuals of average size. All predatory species showed concentrations far above the standard, often reaching more than 2.5 ppm mercury in large specimens (Messier and Roy 1987b). Monitoring of mercury levels in fish since 1978 showed that the highest accumulation rate was reached in non-predatory species after 6 years and, in predatory species, it will be reached after 8 years.

In cutoff rivers, zooplankton benefitted from the sections that became lacustrine, sometimes reaching numbers and biomasses characteristic of natural lakes in northern Québec. Benthic organisms maintained densities equivalent to those in the original river. After La Grande 2 reservoir began to release its flow, the abundance of plankton and benthic organisms, although relatively high for a river, remained low in density and biomass. The downstream passage of a large number of young ciscos, and occasionally other species, increased densities and biomasses of fish below the reservoir as far as James Bay (Boucher et al. 1984). The survivors raised the number of anadromous fish that enter James Bay each summer to feed but must overwinter in the lower stretch of La Grande Rivière. The downward movement of small fish from

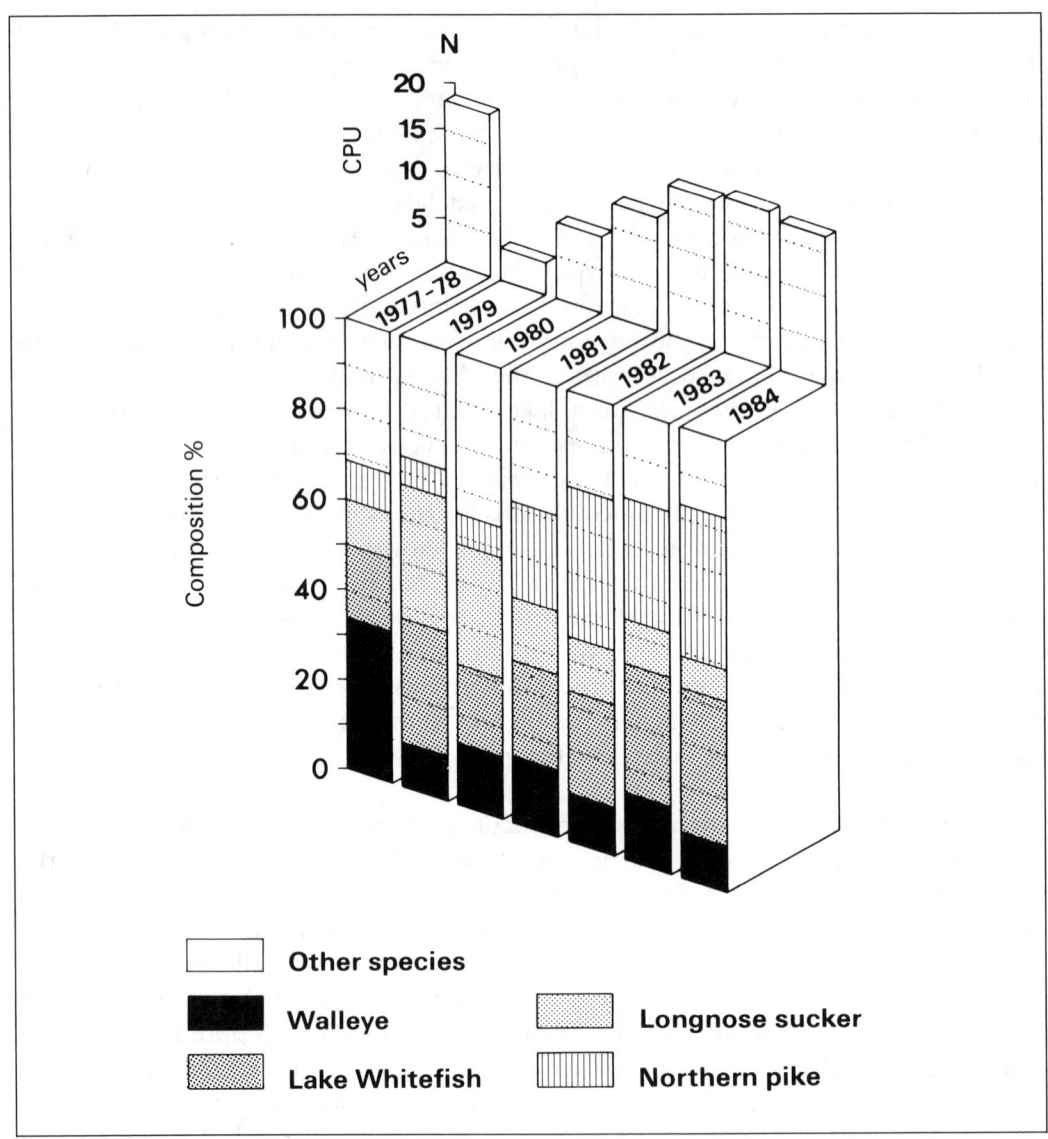

FIGURE 6. Catch per net per day (CPU) and percent of catch before, during, and after the filling of La Grande 2 reservoir (1979).

La Grande 3 to La Grande 2 reservoirs and from Opinaca Reservoir to Lake Boyd is also significant, but it is minor at the outlet of the other reservoirs (Roy and Messier 1987).

Fish entrained from reservoirs carried a certain amount of mercury, which contributed to the contamination of regular predators (northern pike, walleye, lake trout) or occasional predators (lake whitefish, lake cisco, suckers) downstream. The highest concentrations of mercury were encountered near the powerhouses or flow control structures in all species except northern pike and walleye, which showed the same values as in the reservoir upstream. Mercury remained high in fish for more than 100 km downstream from the reservoirs. The spread of mercury was apparently

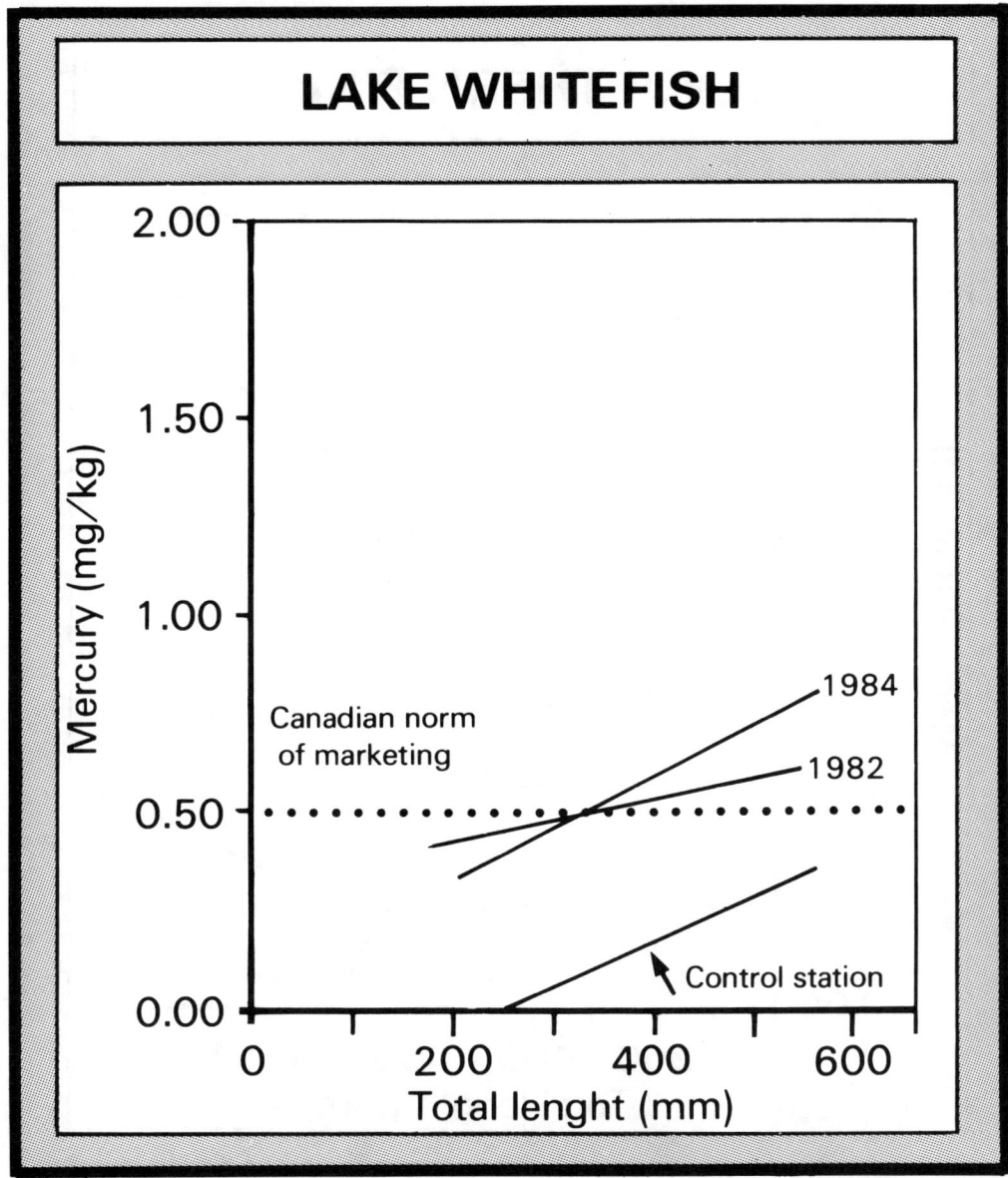

FIGURE 7. Total mercury concentrations in lake whitefish from La Grande 2 reservoir.

caused more by the downward passage of contaminated organisms from reservoirs than by the metal's presence in dissolved form. However, no increase in mercury was detected in Opinaca and Eastmain rivers, where the flow had been reduced.

Five years of monitoring on Eastmain River showed that the reproduction and growth of fish were the same after cutoff as in small rivers and adjacent lakes. Two exceptions were the lake whitefish and lake sturgeon for which condition factors decreased. These two species are not common in turbid water and a decrease in population might be expected (Messier and Roy 1987a). On Caniapiscau River, no

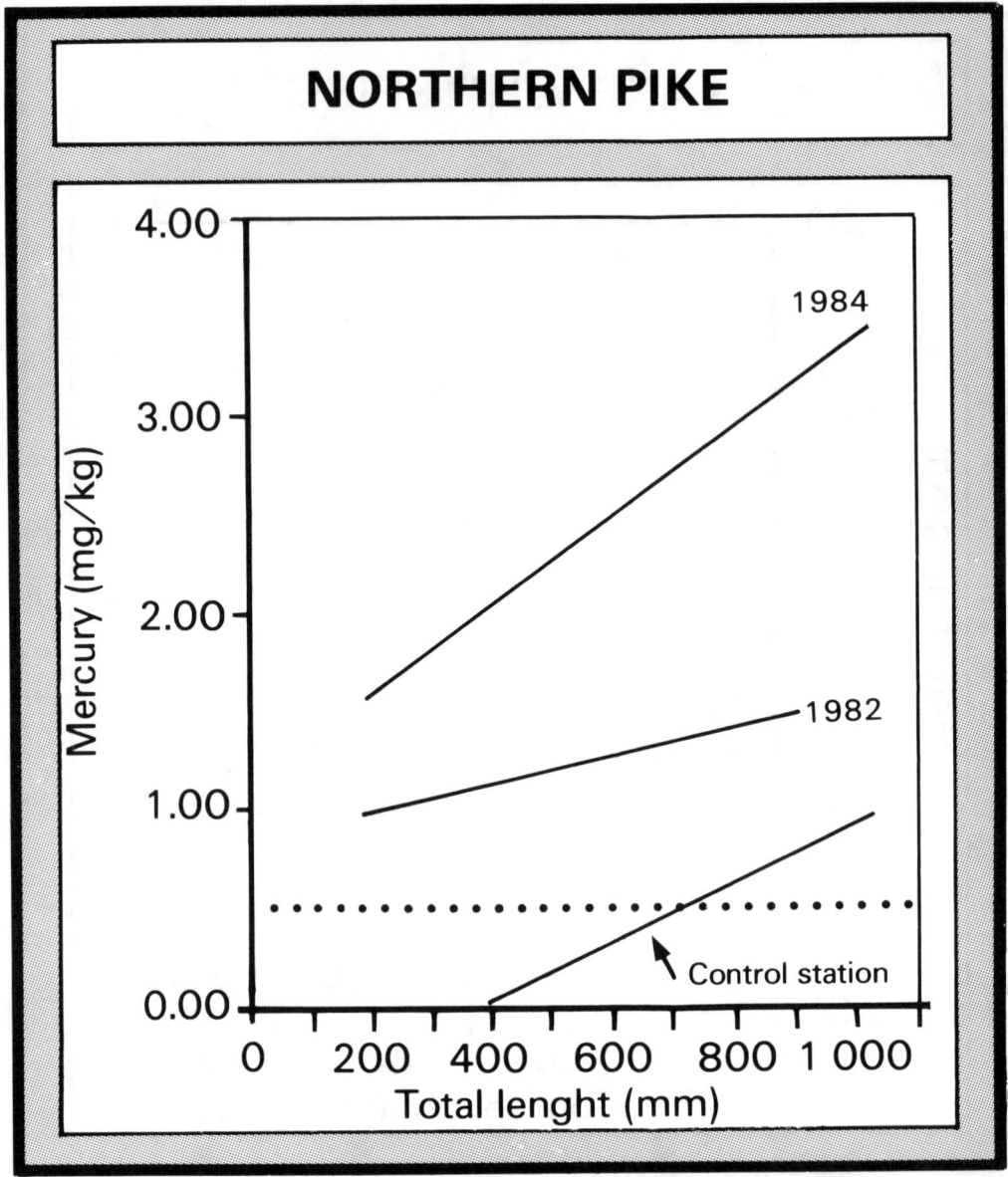

FIGURE 8. Total mercury concentrations in northern pike from La Grande 2 reservoir; the horizontal broken line indicates the Canadian legal limit for commercial fish products (0.5 mg/L).

major modifications were predicted or identified except for the first 100 km downstream from the dam. No migratory populations of fish were present to encounter the new obstacle and no fishway was needed. These sections of the rivers were not usually exploited by aboriginal hunters.

A few hundred fish were caught yearly in pre-reservoir areas before they were filled. More fish are now available but fishing effort is kept low because of the high concentrations of mercury in some species. Mercury does not affect the total harvest of Chisasibi band significantly because their catch is as low as before. With time, as the mercury levels decrease (Messier et al. 1985), these water bodies will become more exploited because of the high production of fish and good accessibility by road.

With the operation of La Grande 2 powerhouse, the annual harvest by the Chisasibi band from the river and James Bay seems as high as before. But winter fishing near the mouth of the river is more difficult because ice conditions are bad (Berkes 1983).

Control of Water Quality

The Ministère de l'Environment du Québec is responsible for maintaining the quality of river and lake water in the province. In other environments (estuaries, seas), this responsibility falls to the federal department counterpart, Environment Canada and the Department of Fisheries and Oceans. Through their respective charters, the Société de développement de la Baie James (SDBJ), the Société d'énergie de la Baie James, and Hydro-Québec must apply, in their own areas of activity, the laws of both levels of government. These three organizations may even issue specific regulations, although the regulations do not have the force of laws. Laws and regulations concerning the James Bay Territory and those generally decreed for the province must, therefore, both be consulted.

Laws and regulations have been applied more strictly in James Bay Territory than in the province overall, particularly those concerning drinking water, waste and industrial water, the disposal of solid waste, and construction activities in general. Application is through the joint efforts of the government agencies involved and the employees hired by SDBJ, SEBJ, and Hydro-Québec for this purpose. Mercury contamination is followed by a special committee that has the responsibility of monitoring mercury levels in fish and humans.

Future Projects

Apart from wildlife harvest, the only resources that can be utilized in the drainage basin of the La Grande Complex are hydroelectric energy and minerals. There are few mining deposits and, in view of the low market for ferrous and nonferrous metals as well as high regional operating costs, it is unlikely that mines will be developed in the near future. Electrical generation will, therefore, remain the driving force of the economy for a long time to come.

Acknowledgments

Data and support for this publication was provided by the Société d'énergie de la Baie James. Danielle Messier and Marcel Laperle provided constructive criticism of the manuscript. I am grateful to all the biologists and technicians for the collection and analysis of biological information during the last fifteen years.

References

Berkes, F. 1983. Fish and wildlife harvesting by the people of Chisasibi. Report to the Société d'énergie de la Baie James, Montréal.

Bobée, B., D. Cluis, A. Tessier, and R. Robitaille. 1976. Analyse des données de qualité de l'eau 1975–1975 du réseau de la Baie James. Institut national de la recherche scientifique-eau, rapport scientifique no 66, Québec.

Boucher, R., and D. Roy. 1985. Réseau de surveillance écologique du Complexe La Grande 1978–1984: poissons. Société d'énergie de la Baie James, Montréal.

Boucher, R., P. Wickham, and D. Roy. 1984. Modifications observées sur La Grande Rivière en aval du réservoir de LG 2, volume 2. Société d'énergie de la Baie James, Montréal.

Grisel, H., and B. Bobée. 1974. Analyse des données de qualité de l'eau du réseau rivière, Baie James 1968–1973. Ministère des Richesses naturelles, Québec.

Jackson, T. A., and R. E. Hecky. 1980. Depression of primary productivity by humic matter in lake and reservoir waters of the boreal forest zone. Canadian Journal of Fisheries and Aquatic Sciences 37:2300–2317.

JBNQHNRC (James Bay and Northern Native Harvesting Research Committee). 1982. The wealth of the land; wildlife harvests by the James Bay Cree, 1972–73 to 1978–79. JBNQHNRC, Québec.

Magnin, E. 1977. Ecologie des eaux douces du Territoire de la Baie James. Société d'énergie de la Baie James, Montréal.

Messier, D., D. Roy, and R. Lemire. 1985. Réseau de surveillance écologique du Complexe La Grande 1978–1984: évolution du mercure dans la chair des poissons. Société d'énergie de la Baie James, Montréal.

Messier, D., and D. Roy. 1987a. Effets de la coupure des rivières Eastmain-Opinaca et Caniapiscau en aval des ouvrages de dérivation. Pages 376–384 *in* Proceedings of the symposium on interbasin transfer of water: impacts and research needs for Canada. Saskatoon, Saskatchewan.

Messier, D., and D. Roy. 1987b. Concentration en mercure chez les poissons au complexe hydroélectrique de La Grande Rivière (Québec). Naturaliste Canada 114: 357–368.

Penn, A. F. 1978. The distribution of mercury, selenium and certain heavy metals in major fish species from Northern Québec. A report to the screening program for mercury in fish: Mistassini and Waswanipi regions, northwestern Québec, summer 1976. Report to Fisheries and Environment Canada, Ottawa.

Power, G., and R. Le Jeune. 1976. Le potentiel de pêche du Nouveau-Québec. Cahiers de Géographie de Québec 20:409–428.

Roy, D. 1985. Réseau de surveillance écologique du Complexe La Grande 1978–1984: zooplancton. Société d'énergie de la Baie James, Montréal.

Roy, D., J. Boudreault, R. Boucher, R. Schetagne, and N. Thérien. 1986. Réseau de surveillance écologique du Complexe La Grande 1978–1984: synthèse des observations. Société d'énergie de la Baie James, Montréal.

Roy, D., and D. Messier. 1987. Répercussions du transfert des eaux des rivières Eastmain-Opinaca et Caniapiscau dans La Grande Rivière (Québec). Pages 169–183 *in* Proceedings of the symposium on interbasin transfer of water: impacts and research needs for Canada. Saskatoon, Saskatchewan.

Schetagne, R., and D. Roy. 1985. Réseau de surveillance écologique du Complexe La Grande 1978–1984: physico-chimie et pigments chlorophylliens. Société d'énergie de la Baie James, Montréal.

Schlesinger, D. A., and H. A. Regier. 1982. Climatic and morphoedaphic indices of fish yields from natural lakes. Transactions of the American Fisheries Society 111:141–150.

SEBJ (Société d'énergie de la Baie James). 1987. Le complexe hydroélectrique de La Grande Rivière. Réalisation de la première phase. SEBJ, Montreal.

Shilts, W. W. 1986. Glaciation of the Hudson Bay region. Pages 55–78 *in* I.P. Martini (editor), Canadian inland seas. Elsevier Oceanography Series, Elsevier Press, New York.

Historical Changes in Water Quality and Fishes of the Ohio River

WILLIAM D. PEARSON
Water Resources Laboratory, University of Louisville
Louisville, Kentucky 40292, USA

ABSTRACT. *The Ohio River has been altered extensively by humans since 1800. The effects of siltation after clearing of forests in the 19th Century, and the construction of navigation dams between 1900 and 1927, has affected the entire river. Pollution became severe in the upper third of the Ohio River in the 1940s. Dissolved oxygen (DO) concentrations became very low during the summer below the three metropolitan areas by 1940. However, DO levels have increased over the past 40 years and are now sufficient to support aquatic life. Counts of coliform bacteria and other pollution indicators have been greatly reduced in the upper third of the river, particularly in the 161 km below Pittsburgh. Between 1819 and 1988, 159 species of fish were reported from the Ohio River, of which 14 were introduced. Only 13 species present before 1970 have not been found since. Of the 13 missing species, the lake sturgeon* Acipenser fulvescen *was the only economically important component. However, several other unique fishes declined in abundance and distribution after 1900 (i.e., shovelnose sturgeon* Scaphirhynchus platorynchus, *paddlefish* Polyodon spathula, *and blue sucker* Cycleptus elongatus). *Between 1957 and 1985, densities of fish increased in the upper 161 km of the Ohio River where water quality improved the most. Fish populations remained relatively stable in the lower two thirds of the river between 1957 and 1985. Many historical changes in fish abundances were associated with the effects of siltation and canalization on substrates. Much progress has been made in pollution abatement in the Ohio River system, but continuing vigilance and monitoring will be required.*

The first human inhabitants along the Ohio River arrived 6,000 to 13,000 years ago. The early arrivals appeared after the last glaciers retreated, and they occupied fields in the Ohio River valley near the mouths of major tributaries that the European settlers would later claim. The Indians depended heavily on shellfish and fish from the river, judging from the immense heaps of shells and fishing artifacts found in their middens.

The first European to record his visit to the Ohio River was LaSalle in 1669. By 1700 the river, which the French called "La Belle Rivière"— the beautiful river— was well-known to Europeans. The English name "Ohio" was derived from either the Indian word "oyo," meaning "beautiful," or "ohiopeekhanne," meaning "the white foaming river."

When the French were defeated in the French and Indian Wars, the Ohio River valley passed into the uncontested hands of the British. Their American colonists then began to drift slowly into the valley. Most the early (1790–1820) settlers traveled overland to Pittsburgh and then down the Ohio River on flatboats. As many as 3,000 flatboats may have descended the Ohio River each year between 1810 and 1820.

The fishes of the Ohio River have always been important to humans for food, for making a living, for recreational angling, as natural curiosities, and eventually as indicators of overall water quality. When Europeans began to arrive, fish were reportedly large, abundant, and easily captured at traditional fishing sites used by the aboriginal people (e.g., the Falls of the Ohio at Louisville, Kentucky). As the human population expanded, the character of the river was altered by their activities in the watershed. Initially, the clearing of the vast hardwood forests and the introduction of agriculture were predominant influences.

The Ohio River basin is one of the great coal-producing regions of the world. The realized and potential impacts of exploiting coal fields along the river are enormous. The Ohio River basin today supports a large human population of nearly 25 million including, on the banks of the river, the metropolitan areas of Pittsburgh, Cincinnati, and Louisville. These areas have many large industries, including steel and other heavy manufacturing, chemical plants, refineries, distilleries, meat-packing plants, and electric generating facilities. Plans are now underway to develop plants that convert coal to synthetic fuels in the next decade. The U. S. Army Corps of Engineers maintains a 2.7 m deep channel in the Ohio River for the passage of about 135 million metric tonnes of waterborne cargo each year through offstream reservoir releases, dredging, and operation of 20 locks and dams.

The impacts of all human activities on water quality, aquatic habitats, and aquatic life in the Ohio River have drawn notice and concerned action.

My objectives in this report are to: 1) summarize the effects of human activities on the Ohio River, 2) discuss the effects of these changes on aquatic biota, and 3) consider the future of water quality and fish.

Morphology and Hydrology

The Ohio River forms at Pittsburgh, Pennsylvania, where the Allegheny and Monongahela rivers join, and then flows 1,578 km southwesterly to join the Mississippi River at Cairo, Illinois. The Ohio River and its predecessors have channeled water to the Mississippi River basin for over 200 million years. The ancient Teays River, which headed in the Piedmont region of North Carolina and flowed north to Huntington, West Virginia, was one early forerunner of the Ohio River. The Teays River flowed northwest across Ohio River from Huntington, then turned westerly across Indiana and Illinois to join the Mississippi River 320 km north of the Ohio River's present union (Janssen 1952). The Pleistocene glacial invasion, which began 1 million years ago and ended just 10,000 years ago with the retreat of the Wisconsin glacier, worked enormous changes on the drainage pattern of the Ohio River basin. The present course of the Ohio River arises from reversed flow of the Allegheny, Monongahela, and several smaller river systems following glacial blockage. The Ohio River valley now, in most places, lies just 15 to 120 km south of the southernmost extent of the ice sheets. The Ohio River basin drains part of 14 states and covers about 528,360 km^2.

The Ohio River is about 366 m wide and its mean annual disccharge is 906 m^3/s upstream near its source at Pittsburgh. The river increases very little in size over its uppermost 240 km, but the contributions of the Muskingum, Little Kanawha, and Hanawha rivers widen the main stem to about 460 m and increases its discharge to 2210 m^3/s at Huntington, West Virginia (River Kilometer 496 = RKm 496, measured downstream from Pittsburgh). The Ohio River gradually increases in width to about 550 m and in mean annual discharge to 3796 m³/s between Huntington and Evansville, Indiana (RKm 1274). Major tributaries in this section are the Big Sandy, Scioto, Great Miami, and Salt rivers (Figure 1). Its largest tributaries, the Wabash, Cumberland, and Tennessee rivers, join the Ohio River below Evansville. The Ohio River is almost 1097 m wide at its mouth, with a mean annual discharge of 7307 m³/s, larger than the Mississippi River at that point. The flood plain of the Ohio River widens from about 1.3 km at Pittsburgh to 11.0 km at the mouth. The Ohio River falls only 136 m during its course, and the gradient gradually decreases with distance from the source.

Discharge Patterns

Flows in the Ohio River are usually highest in March during most years, and lowest in September and October (Figure 2). Peak flows in recent years have been 8 to 11 times greater than recorded low flows. Maximum flows ranged from 11,668 m³/s in the upper 161 km of the river to 50,409 m³/s near the mouth during a 1937 flood, the greatest flood on record (USACE 1969).

Floods occurred regularly along the Ohio River during winter and late spring even before the watershed was cleared. Flood stage rises of 18 m were often reported before 1800. Discharges during the great 1937 flood reached a stage of 24 m at Cincinnati, nearly 8.5 m above flood stage. Damage to the surrounding countryside was extensive.

To reduce flood damages, the Corps of Engineers and other agencies constructed 77 reservoirs on tributaries that allow control of flows from about one third of the

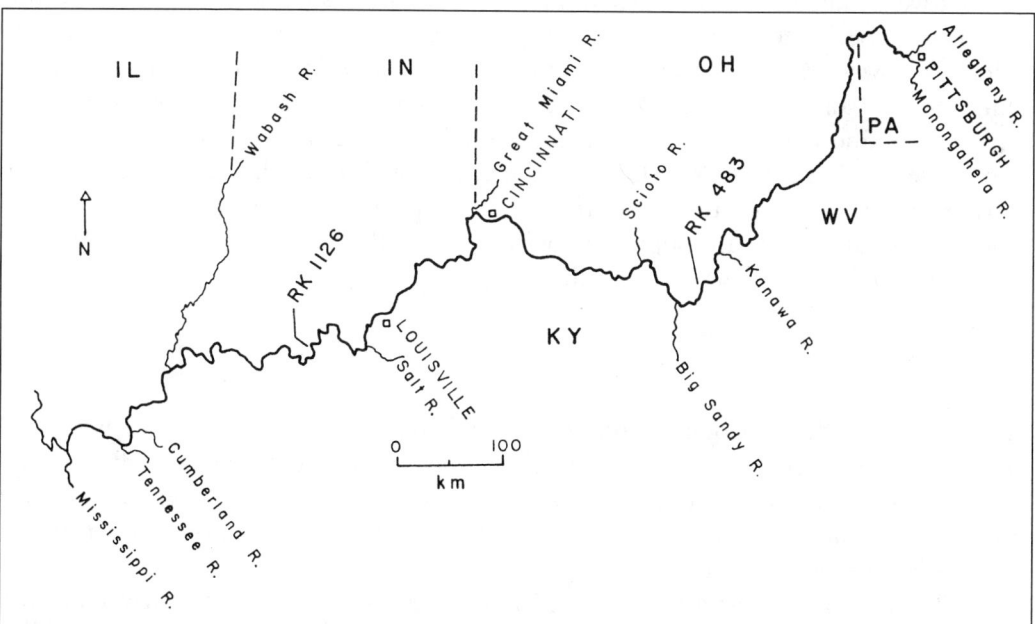

FIGURE 1. The Ohio River and its major tributaries. The numbers are River Kilometers (RKm) below the river's source at Pittsburgh, Pennsylvania as conventionally shown on navigation charts by the U. S. Army Corps of Engineers.

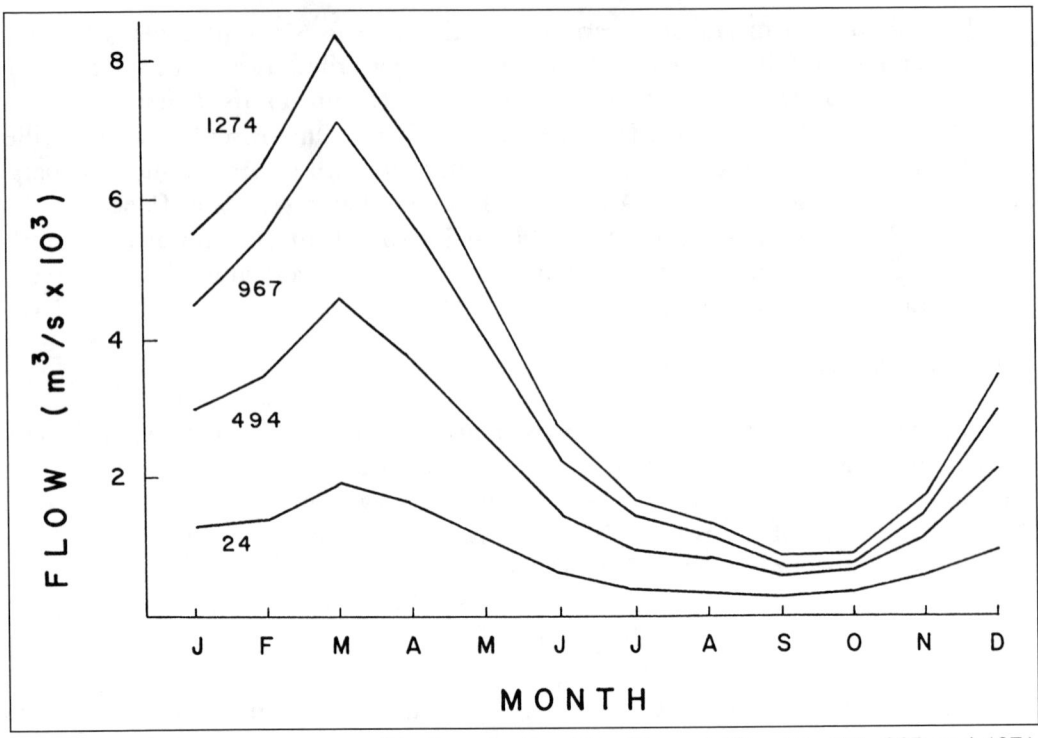

FIGURE 2. Long-term mean monthly flows for the Ohio River at RKm 24, 494, 967 and 1274. (Adopted from ORSANCO 1978b.)

Ohio River drainage-basin (USACE 1978). This vast system of tributary reservoirs has three major effects on the main-stem Ohio River: 1) flood waters are stored and released slowly, reducing the maximum seasonal flow rates and lengthening the duration of bank-full stages; 2) low flows during drought are nearly twice those before impoundment; and 3) the water released to the main stem from the reservoirs is often less turbid than before. Overall, the effects of flow regulation might eventually have greater significance than pollution in altering the character of the nation's large rivers (Wolman 1971), although this point was hardly mentioned in a recent paper on water quality trends (Smith et al. 1987).

Navigation

The first steamboat to operate on the Ohio River was the "New Orleans," which was actually launched on the Monongahela River at Pittsburgh in 1811. By 1835, at least 684 steamboats were operating on the Ohio and Mississippi river systems. The first "Rivers and Harbors Act" of 1824 authorized funds to remove snags from the Ohio River to aid navigation.

The Federal Government continued snag removal on the Ohio River between 1837 and 1966 and also constructed 47 back-channel dams and 111 training dikes. Some dredging and rock-removal operations also began during this period. Between 1834 and 1975, dams passable by gated locks were installed on the smaller Green,

Monongahela, and Muskingum rivers, and rivermen sought similar constructions on the Ohio River.

The first cross-channel dam on the Ohio River was constructed at Davis Island, 8 km below Pittsburgh in 1885. Between 1909 and 1910, the government decided to maintain a 2.7 m channel throughout the Ohio River, and 12 additional dams were constructed between Louisville and Pittsburgh. A new movable dam (#41) was completed at Louisville in 1927, the first with a permanent "high-lift" feature. All previous dams on the Ohio River had a navigable section over which tows and fish could pass during high flows. The new dam provided an 11.2 m lift at lock 41, making it impossible for fish to pass upstream at Louisville by any means other than locking through with tows, negotiating the small gated sections, or topping the dam at flood stages. There were 50 dams on the Ohio River by 1929.

Barge traffic on the Ohio River increased between 1937 and 1950, as did the average size of both tows and towboats. Therefore, the Corps of Engineers began a modernization program in 1955 by replacing the older, low, movable wicket dams with fewer high, non-navigable, gated, high-lift dams. At present there are 20 lock and dam structures on the Ohio River, with the last of the small wicket-type dams (#52 and #53) in the lower river tentatively scheduled for replacement.

The new, non-navigable dams above Louisville have lifts of 3.0 to 10.6 m and, below Louisville, lifts of 4.9–7.6 m, or nearly twice the lift of older structures. New structures above RKm 1158 are non-navigable, gated dams with fixed weirs, while those below RKm 1158 are gated with navigable weirs. The primary difference is that the lower dams have a fixed weir section over which barges can be towed when the locks are closed during high water. The gated structures above RKm 1158 have tainter, roller, or vertical lift gates that release water near the bottom rather than over the top. Bottom releases flush oxygen-deficient water and silt from the pools above the dams, but also make upstream movement of fish difficult if not impossible at low flow.

The amount of freight traffic on the Ohio River has increased from about 6.6 million tonnes in 1915 to 13.6 million tonnes in 1929, 77.2 million tonnes in 1962, and 145 million tonnes in 1986. The distribution of traffic density on the main stem has also changed. In 1935, traffic in the Pittsburgh area was about twice that in the Huntington area and four times that below Cincinnati. In the ensuing years, traffic has shifted its density to the middle and lower river reaches.

In recent years about 80 bars occasionally require dredging by the Corps of Engineers or their contractors. About 20 of these bars must be dredged each year, and an average of 2.4 million m^3 of material is moved annually (USACE 1979).

The navigation dam system had several effects on the riverine environment. The Ohio River has been converted from its original free-flowing state, in which mid-channel depths during dry seasons were less than 1 m at many locations, to a canalized form in which channel depths always exceed 2.7 m. The most important effect of canalization, coupled with the effects of land-use practices and erosion in the Ohio River basin, has been the inundation and siltation of the extensive gravel and rubble substrates typical of the original, glacial-formed Ohio.

Water Quality

Early Condition of the Ohio River

The quality of water in the Ohio River probably declined between 1820 and 1940 in proportion to the number of people living in the basin and their combined technological capabilities. As Wolman (1971) pointed out, we do not have reliable, long-term records of water quality for the major rivers of North America. Most of the continuous water quality and hydrologic records extend over just the last 65–70 years.

Accounts of early travelers in Kentucky, Indiana, and Ohio repeatedly refer to troublesome marshes, mires, and soft boggy areas. Newcomers also mentioned the yellowish appearance of early settlers, the fevers and "vapors" contracted in low-lying areas near the river, and the clouds of mosquitos that made life miserable. The natural vegetation and filtering action of extensive swampy areas must have caused the Ohio River to seem very clear. Indeed, almost without exception, early reports describe the Ohio's water as clear except during floods.

Floods have been a regular, annual feature of the Ohio River. Seven floods with crests exceeding 18 m were reported between 1762 and 1792. The powers of such discharges are not to be denied, and there must have been a great deal of erosion along river banks despite the natural riparian vegetation. The banks of the Ohio River in 1818 were described as "...all alluvial and of a deep and rich soil, seldom quite sandy or muddy. There are in many bottoms a second and a third bank, all very steep and from ten to forty feet high. The first bank is almost everywhere overflowed to high waters, the second never....Many banks sink or are washed away in inundations, when the channel sets against them" (Rafinesque 1820).

Early floods occurred mainly in the spring and fall, although minor "freshes" could be expected after any rainfall. Flooding was eagerly awaited in the summer by travelers wishing to ride their flatboats over the bars and rapids. Water levels sometimes fell so low in summer that the river could be forded in many places between Cincinnati and Pittsburgh, at the Falls, and near Shawneetown and Cairo.

Rafinesque (1820) claimed there were about 130 islands in the Ohio River when he journeyed down it in 1818. Cramer (1818) lists the number of islands at over 100 and mentions that, although beautiful, the islands usually had many shoals, sandbars, and troublesome currents, which added to the hazard of navigation.

No temperatures were recorded for the Ohio River before 1818. We surmise that the greater shading of the river and its tributaries by trees, and the more abundant spring flows would have resulted in slightly lower summertime water temperatures than occur today.

Trends in Water Quality

The few observations recorded in the 19th century indicate that the overall quality of water in the Ohio River basin declined through the 1940s. Mean turbidity, total dissolved solids, chlorides, nitrates, and sulfates increased up to the 1940s. Dissolved oxygen values declined during the same period, particularly after 1900 and below major metropolitan areas. Acid mine drainage at the head of the Ohio River basin lowered pH values to less than 4.0 in the upper 160 km of the river before 1950. Mean monthly counts of total coliform bacteria often exceeded 20,000/100 mL in the 1940s.

The nadir in poor water quality probably occurred during the droughts of 1930, 1931, and 1934 when dilution effects were minimal (Cleary 1967), domestic sewage treatment facilities were rare in the Ohio River basin, and cases of water-borne disease were common (Ohio River Committee 1944).

Water quality variables include those important in determining fish distribution and reproductive success (i.e., turbidity, dissolved oxygen, temperature, dissolved solids, and pH); heavy metals and other contaminants of fish flesh (i.e., arsenic, mercury, lead, PCBs, and pesticides); and bacterial indicators of domestic sewage. Trends in the concentrations of many of these variables are summarized in annual reports from the Ohio River Valley Water Sanitation Commission, known as ORSANCO (ORSANCO 1949a–1987a).

Turbidity. Early reports described water in the Ohio River as "clear and salubrious." With the clearing of forests, the beginnings of intensive agriculture, and the draining of swampy areas, the waters of the Ohio became more turbid and silt deposits accumulated on the substrate.

Fishes of the main-stem Ohio River were accustomed to, and adapted for, occasional exposure to high turbidities that always accompanied the annual spring "freshes." Fish that preferred smaller forest brooks were not so adapted. The emphasis placed on the adverse effects of turbidity and siltation on fishes of the region by Pflieger (1971), Gammon (1977), Smith (1979), and Trautman (1981) applies more to fishes of small streams rather than to those inhabiting the main-stem Ohio River. Exceptions would include fishes that require clear, silt-free riffles in smaller streams for spawning (i.e., the lake sturgeon *Acipenser fulvescens* and some lampreys), and those that require silt-free expanses of coarse sand and/or gravel in the Ohio River channel (i.e., gravel chub *Erimystax x-punctatus*, crystal darter *Ammocrypta asperella*, channel darter *Percina copelandi*, and river darter *P. shumardi*).

Between 1952 and 1955, mean turbidity values of 2 to 576 JTU were reported monthly at 43 monitoring stations along the Ohio River (ORSANCO 1957a). In 1962–63, the reported range was 2 to 1301 JTU. A significant, although slight, decrease in turbidity occurred at all monitoring stations between 1953 and 1975 (ORSANCO 1977b). Similar slight decreases were reported between 1975 and 1985 (ORSANCO 1986b). The probable cause was the construction of numerous flood-control reservoirs on most of the major tributaries of the Ohio River in the last 45 years. These impoundments trap sediments and release clear water. The range of turbidity to which fish in the Ohio River are exposed in modern times is probably not significantly greater than in prehistoric times. In 1980, for example, mean daily suspended sediment concentrations at Louisville ranged from 8 to 510 mg/L and were directly related to river discharge (Figure 3). However, mean turbidities are higher now than in prehistoric times. These higher turbidities have influenced the

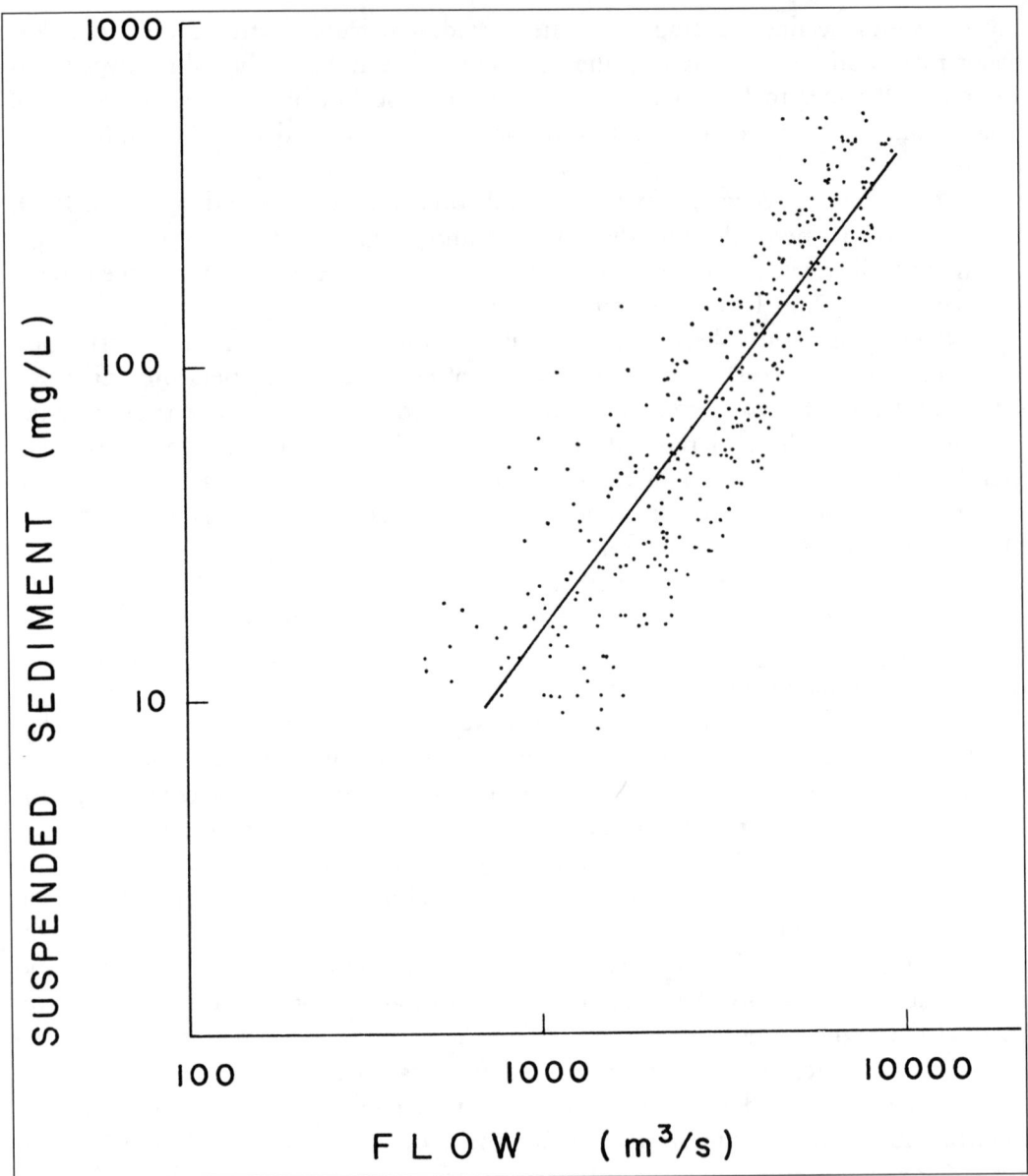

FIGURE 3. Log plot of mean daily suspended sediment concentrations in mean daily discharge, Ohio River at Louisville, 1980 water year (after Ruhl 1987).

fish community of the Ohio River more indirectly (i.e., by reducing primary production, favored food supplies, visibility of food organisms, and suitable spawning areas) than directly.

Dissolved Oxygen. Measurements of dissolved oxygen (DO) were insufficient to permit annual comparisons until the 1930s. Mean monthly DO values reported in 1939–1941 were usually above 6.5 mg/L in all seasons and in all reaches of the Ohio River except for the first 15–50 km below Pittsburgh, where mean monthly values dropped as low as 4.0 mg/L during low flows (late summer and fall), and individual readings at Emsworth Dam (RKm 10) were as low as 2.8 mg/L (Ohio River Committee 1944). Oxygen levels in the Ohio River after 1930 were lowest immediately below the three major population centers of Pittsburgh, Cincinnati, and Louisville, but concentrations usually returned to acceptable levels 30–60 km downstream. The exception may have been below Pittsburgh where industrial facilities line the river and oxygen levels were occasionally depressed significantly for 60–150 km below a source.

During low flows, particularly June through September, oxygen concentrations less than 4.0 mg/L have been reported occasionally at nearly all monitoring stations on the Ohio River. However, with the completion of secondary sewage treatment facilities at most major cities in the late 1970s, the number of oxygen records below 5.0 mg/L during low flows was reduced considerably. September mean values were 0.5 to 2.6 mg/L higher in 1983 than in 1963 at four locations along the river (Figure 4). Low oxygen events usually occur when sewage treatment facilities are temporarily shut down for repairs or maintenance, or in emergencies such as contamination of the Louisville treatment plant with hexachlorocylopentadiene in 1977 (ORSANCO 1980a).

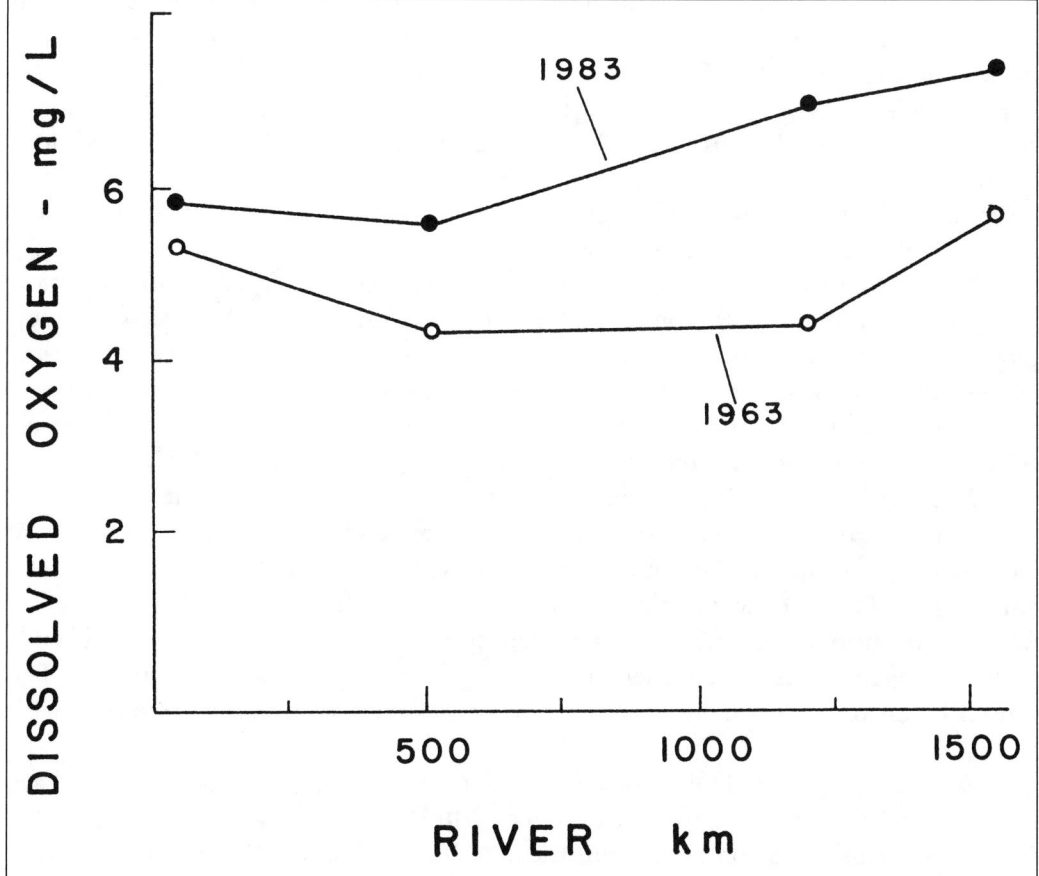

FIGURE 4. Mean monthly dissolved oxygen (DO) levels from four locations on the Ohio River in September of 1963 and 1983 (ORSANCO 1987a).

Although DO concentrations have improved (ORSANCO 1986b), fish are still subjected to chronically low levels below metropolitan areas on the main-stem Ohio River. Mean daily DO levels were less than 5.0 mg/L from 13 to 20% of the time at RKms 854, 967, and 1273 in June 1982. In August and September 1982, very low flows reduced mean daily DO concentrations to less than 5.0 mg/L for 20 to 76% of the time at RKms 788, 854, 967, and 1007 (ORSANCO Quality Monitor Data).

Temperature. No long-term, continuous temperature records are available before 1820. One may assume that, with the increase in turbidity associated with intensive agriculture, the mean annual temperature on the Ohio River may have increased slightly between 1820 and 1900, at which time keeping of good temperature records began. This slight increase may have been caused by absorption of sunlight by suspended particles, clearing of shade trees along tributaries and the main-stem Ohio River, and faster runoff of rainwater that reduced ground water contributions during low-flows.

No significant changes have occurred in the temperature of the Ohio River during the past 30 years (Wolman 1971; Butz et al. 1974; ORSANCO 1977b, 1986b). Mean daily temperatures are usually lowest in February, ranging from 6.9°C at RKm 25 to 8.8°C at RKm 1274 during the winters of 1964–74. Mean daily temperatures are usually highest in July-September, ranging from 30.3°C at RKm 25 to 31.1°C at RKm 1274 in 1964–74. Surface water temperatures often drop to 0°C in winter, but the river seldom freezes solidly from shore to shore below Louisville. Maximum summer water temperatures of 33.0°C (ORSANCO 1977b) have occurred in the lower half of the river, although higher records exist below heated effluent sources and in shallow backwaters warmed by insolation.

The Ohio River has 35 power generating facilities with a combined electrical capacity of over 30,000 MWe on its banks. Most of these facilities have flow-through cooling systems, each capable of raising the local river temperature 4°C at low-flow seasons. In the most severe low-flow situations, the maximum rise in river temperature due to the combined plant discharges is thought to be less than 0.5°C (Butz et al. 1974), and to present little danger to aquatic life (ORSANCO 1975b).

Dissolved Solids. Records for total dissolved solids (TDS) begin in the late 19th century. Apparently chloride, sulflate, nitrate, and total dissolved solids have increased significantly in the Ohio River main-stem since that time (Wolman 1971). Smith et al. (1987) conclude that nitrates and arsenic have increased in the Ohio River basin since 1974 from atmosphere deposition. The increases in dissolved solids probably began much earlier when the first forests were cleared for agriculture. The concentration of TDS usually increases downstream and is inversely correlated with flow.

Wolman (1971) cited ORSANCO data that indicate chloride concentrations have increased by a factor of 0.5 to 1.14 between 1910 and 1966 at various points on the Ohio River. This trend continued into the 1980's (Smith et al. 1987), probably from increased application of road salt in winter. The maintenance of TDS levels at most stations in the past 15 years, despite the increasing population in the Ohio River basin, had been attributed to a reduction of point source pollution and to a reduction of mine drainage in the upper reaches.

Hydrogen-ion Concentration (pH). The pH of the Ohio River was probably near neutrality (7.0) in prehistoric times with values ranging upwards to perhaps 8.5. When coal-mining became common in the Allegheny and Monongahela river basins in the last half of the 19th century, acid mine wastes depressed values in the upper Ohio River above RKm 277 to as low as pH 4.7 (Ohio River Committee 1944). In fact, records of less than pH 5.0 (often pH 4.0) became common in the upper 150 km. Below this point, mean values usually increased to pH 7.2 to 7.8 between Cincinnati (RKm 745) and Cairo (RKm 1578). During this period unusually high recordings of pH 9.0 to 10.0 appeared occasionally in all reaches of the river, usually during low flows when growth rates of algae were high.

The damaging effects of acid mine drainage on both the Monongahela and Allegheny rivers were recognized over 60 years ago when the Federal Government began sealing abandoned coal mines (Ohio River Committee 1944). During the 1960s and 1970s, additional progress was made in reducing acid pollution from both mine drainage and industrial discharges in the upper Ohio River basin. Annual reports of ORSANCO for 1974 to 1986 reflect the success of the efforts. In most of these years, compliance with the established standards for pH (6.0 min, 9.0 max) was achieved at all monitoring stations.

Coliform Bacteria. The direct effects of coliform bacteria on fishes are negligible but their concentrations indicate the extent that a river is contaminated by domestic sewage. Municipal sewage affects fishes by altering nutrient levels, BOD, DO, and suspended solids, and by adding many organic compounds.

Domestic sewage pollution has always been the most serious in the upper quarter of the Ohio River where the human population is greatest. More than 40 municipal discharges between Pittsburgh (RKm 0) and Wheeling (RKm 146) are major sources of coliform bacteria. No other comparable distance on the Ohio River has more than 14 municipal discharges per 161 km, while the lower section may have as few as seven discharges per 161 km.

When the Ohio River Committee (1944) reported on the condition of the Ohio River from 1939 to 1943, mean total coliform counts greater than 20,000/100 mL were listed 31 to 61% of the time at monitoring stations. When the ORSANCO compact was signed in 1948, only 1% of all domestic sewage was treated before dumped into the river.

By 1964, 97% of the people along the Ohio River were served by primary sewage treatment. As a result, concentrations of coliform bacteria in the upper 161 km of the river had fallen dramatically, for example, from a mean at Wheeling of 62,000/100 mL in 1952 to just 950/100 mL in 1964 (ORSANCO 1964a, 1965a). In the middle and lower reaches of the river, however, coliform bacteria remained at nearly the same levels as in 1953, or had declined slightly.

Today, 94% of the people in the Ohio River basin are served by secondary facilities and all communities with more than 10,000 people have secondary treatment facilities in place or under construction. With the installation of secondary sewage treatment facilities between 1960 and 1987, levels of coliform bacteria in the river were expected to decline markedly again. This decline did not occur. In the mid-1970s, ORSANCO began reporting only fecal coliform concentrations as a better indicator of sewage contamination. In 1976, a new drinking water standard of 2000 fecal coliform bacteria/100 mL was exceeded at just one station (RKm 100), and then by just two of the 12 mean monthly values recorded there (ORSANCO 1977b). The maximum mean monthly value at this station was 2510/100 mL; since fecal coliforms are usually 20% of total, this would represent a total coliform value of 12,550/100 mL.

The largest decrease in coliform bacteria in the mainstem Ohio River apparently occurred when primary sewage treatment facilities were constructed before the 1960s, and mean monthly total coliform counts dropped to less than 20,000/100 mL at most stations. The most dramatic decrease occurred in the upper 161 km of the Ohio River. Fecal coliform counts in this area have apparently continued to decrease between 1960 and 1986. However, most violations of the standards in recent years still occur in the upper 161 km, and primary recreation standards (<200/100 mL) are seldom met in this section.

In the lower two-thirds of the Ohio River, coliform counts appear to have declined only slightly, if at all, since the 1960s (ORSANCO 1977b, 1986b) despite the large-scale improvements in sewage treatment facilities. The relatively small declines have been attributed to sewer overflow, inadequate disinfection at treatment plants (particularly at Louisville and Charlestown in the late 1970s and early 1980s), poor secondary treatment plant operation and design (particularly at Cincinnati), and nonpoint source releases (ORSANCO 1986b, 1987b). Fecal coliform counts in the lower two-thirds of the Ohio River have, however, usually met the recreation standard (<200/100 mL) in the 1980s (ORSANCO 1984a, 1987a). Perhaps more important, universal secondary sewage treatment has reduced BOD loads and DOD (Smith et al. 1987).

Toxic Substances and Fish Contaminants. While changes in DO concentration, BOD, pH, and coliform counts over the past two decades have improved water quality in the Ohio River, several new sets of toxic substances have recently appeared or have been recognized as problems. Heavy metals (e.g., zinc, cooper, lead, nickel, mercury, and arsenic) frequently exceed established criteria since 1974, and levels of arsenic and lead have increased (Smith et al. 1987). Total phenolics and cyanide concentrations are frequently detected and actually exceed established criteria in many areas (ORSANCO 1987a).

Many organic chemicals, particularly halogenated compounds, are routinely transported and manufactured on and along the Ohio River. ORSANCO has established an Organics Detection System that measures 16 halogenated compounds at its 13 monitoring locations daily. The most frequently detected are chloroform, methylene chloride, tetrachloroethylene, and 1,1,1-trichloroethane, all of which are common at less than 1.0 μg/L (ORSANCO 1987b). Detection of some some compounds has been greatly reduced between 1982 and 1986. However, the 16 compounds measured represent only a fraction of all organic compounds that might be present.

Selected fish fillets have been examined for contaminants since 1976. Catfish and carp fillets are analyzed every other year. About 31 to 37% of all catfish fillets contain PCBs in excess of the FDA limits of 2.0 mg/kg. Chlordane exceeds FDA limits of 0.3 µg/kg in 22 to 25% of these fillets. Both mercury and DDE have also been detected in catfish and carp fillets at levels approaching, but not exceeding, FDA limits. In response to these findings, ORSANCO (1986b) has published advisory statements recommending that "... occasional consumption of fish taken from the Ohio River should not pose an unacceptable health risk, but that they should not comprise a major portion of anyone's diet." Such findings and recommendations compromise the recovery of many fish populations in the upper Ohio River.

Changes in Fish and Other Biota

Physical and chemical alteration of the Ohio River has greatly affected aquatic biota. Effects on plant life, plankton, and most invertebrates have not been documented as well as those on fishes. ORSANCO (1962b) and others have noticed that bottom fauna are scarce throughout much of the Ohio River. However, this apparent scarcity may be due, in part, to the difficulty inherent in sampling large rivers (Neff et al. 1981). In any event, there is little historical quantitative data on plants, plankton, and invertebrate populations useful for detecting long-term trends.

One exception is the mussel community. The great abundance and diversity of mussels in the Ohio River was reported in the 1800s by Rafinesque, Say, and Call. Today, mussels of the Ohio River have been reduced in both abundance and diversity by a combination of canalization, siltation, and overharvesting (Williams 1969; Taylor 1980; Neff et al. 1981; and Miller 1983). Many fishes have returned to the upper 161 km of the Ohio River in the last 20 years, but most mussels have not. The recent invasion of the Asian clam *Corbicula* has probably significantly impacted native mussels.

Fishery Investigations on the Ohio River

Accounts of fishes collected from the Ohio River before 1920 are not reliable for determining relative abundance. However, they show the disappearance of some species and severe reductions in other species that were once common.

The earliest general account of fishes in the Ohio River was provided by C. S. Rafinesque, who traveled down the river in the summer of 1818. He described 52 species of fish and commented on their relative abundance (Rafinesque 1820). Records on fishes of the Ohio River over the next century, between 1820 and 1920, are scarce.

Several investigators provided baseline data on Ohio River fish populations between 1920 and 1969, and interpreted their findings in light of human disturbances. Trautman (1957, 1981) discussed many Ohio River fishes. The ORSANCO (1962b) study of 1957–60 provided the first comprehensive review of fish populations in the Ohio River from its mouth to source. This work was based on rotenone samples from 123 lock chambers, supplemented by 125 collections made with other types of gear.

The number of fishery investigations on the Ohio River increased markedly after 1970, primarily under requirements of the National Environmental Policy Act of 1969 and the Clean Water Act of 1972. These acts, and their subsequent amendments, required environmental impact statements or other environmental reports for every sizeable project along the Ohio River. I examined 108 reports published between 1970 and 1988 that contained original data on Ohio River fishes; most are cited in Pearson and Krumholz 1984; valuable sources published since then include Geo-Marine 1984, 1986 and ESE 1987, 1988.

Abundance and Diversity Through Time

Estimates of the relative abundance of Ohio River fishes before 1900 are impressions of early workers rather than quantitative measurements. However, the following fish were probably reduced in abundance by 1950: lake sturgeon, shovelnose sturgeon *Scaphirynchus platorynchus*, muskellunge *Esox masquinongy*, bigeye chub *Hybopsis amblops*, and blue sucker *Cycleptus elongatus*. (All common names follow Robins et al. 1980.) Fish that apparently increased in relative abundance by 1950 included the introduced common carp *Cyprinus carpio* and the gizzard shad *Dorosoma cepedianum*. Important fish that apparently changed little in abundance were the freshwater drum *Aplodinotus grunniens*, channel catfish *Ictalurus punctatus*, and emerald shiner *Notropis atherinoides*.

Quantitative data after 1950 permit a more detailed description of changes in relative abundances of Ohio River fish (ORSANCO 1962b, 1987b; Preston 1969; Pearson and Krumholz 1984). During this 37 year period, three additional introduced species (goldfish *Carassius auratus*, white catfish *Ictalurus catus*, and banded killifish *Fundulus diaphanus*) became established. Also, the grass carp *Ctenopharyngodon idella* may now be established in the lower Ohio River.

Sampling of fish in lock chambers with rotenone has been carried out on the Ohio River since 1957. The technique has been described previously (Preston and White 1978; Pearson and Krumholz 1984). Most lock chambers have a surface area of 0.61 hectare. In a typical sampling, the downstream gate of the lock chamber is left open 2 to 12 hours. The lower gates are then closed. About 18 to 38 L of 5% rotenone solution are added to the lock chamber to achieve a concentration of 0.5–1.0 mg/L. Surfacing fish are then removed by dip netting over several hours.

I divided the densities of all fish species in rotenone samples from 280 lock chambers (Pearson and Krumholz 1984; ORSANCO 1987c) between 1957 and 1985 into four time periods: 1957–60, 1967–70, 1974–80, and 1981–85 (Figure 5). Densities of all fishes combined increased significantly after 1957–60 between RKm 0 to 161, RKm 644 to 965, and RKm 1287 to 1448. No significant decrease in densities occurred after that time at any location. In fact, biomass of all fish species combined doubled between 1957–60 (205 kg/hectare) and the four subsequent periods (Figure 6).

Abundance and Diversity Along the River

Many species of fish were more widely distributed along the length of the Ohio River in the early 1800s than today, according to records left by Rafinesque, Lesueur,

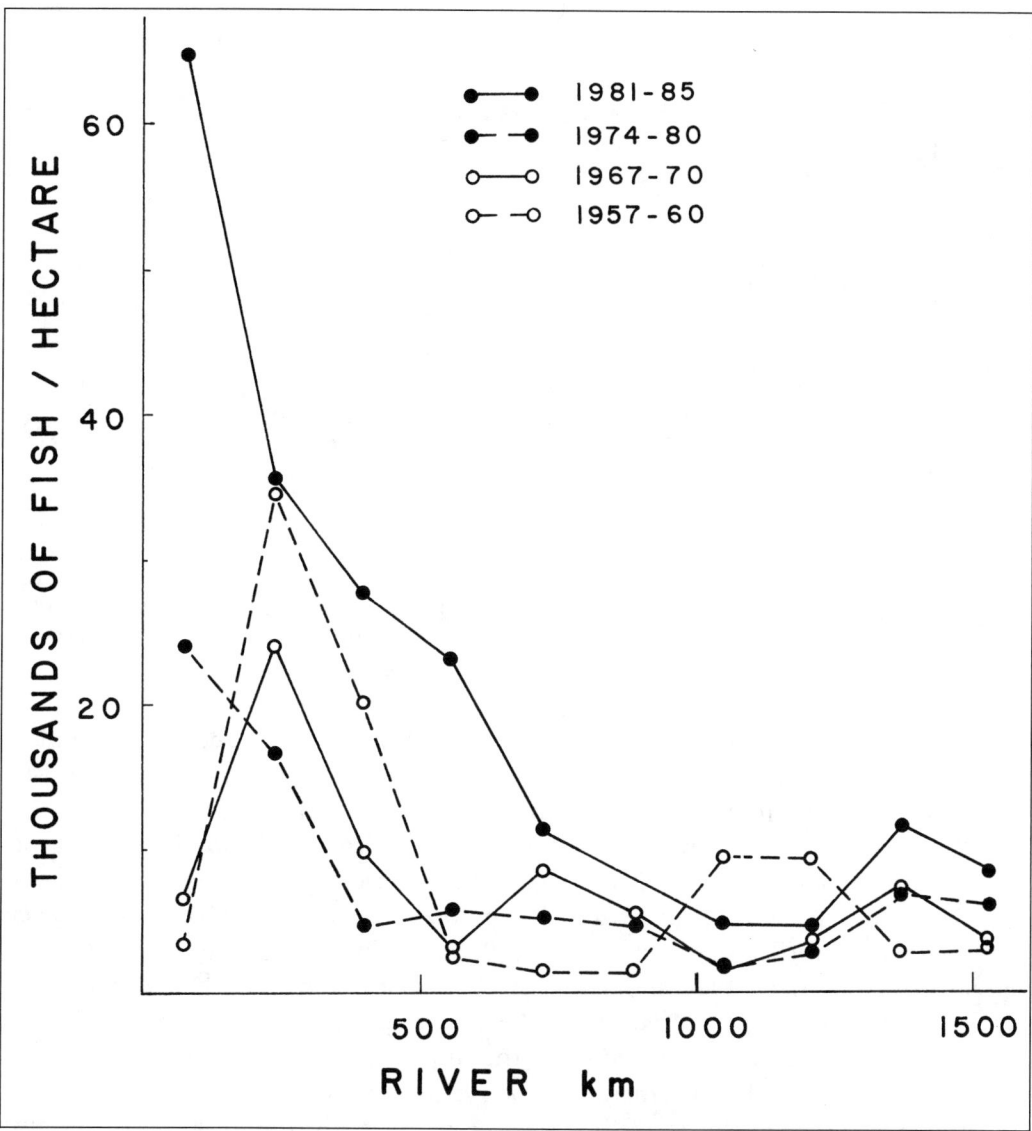

FIGURE 5. Density of all fish in lock chamber samples from 161-km sections of the Ohio River in the four periods: 1957–60 (open circle-dashed line), 1967–70 (open circle-solid line), 1974–80 (solid circle-dashed line), and 1981–85 (solid circle-solid line).

and others. For example, Lesueur described the blue sucker at Pittsburgh in 1817, and Rafinesque found it there in 1818. The blue sucker was never reported above RKm 483 between 1950 and 1980, but it has been found there several times since 1980. Other species of fish reported from the upper 280 km of the Ohio River by early workers (before 1900) were not reported there between 1920 and 1980, but have reappeared since 1980 (i.e., highfin carpsucker *Carpiodes velifer* and mooneye *Hiodon tergisus*) (ESE 1987, 1988). Several species present in the upper 280 km before 1920 have not been seen there since (i.e., least brook lamprey *Lampetra aepyptera*, shovelnose sturgeon, shortnose gar *Lepisosteus platostomus*, speckled chub

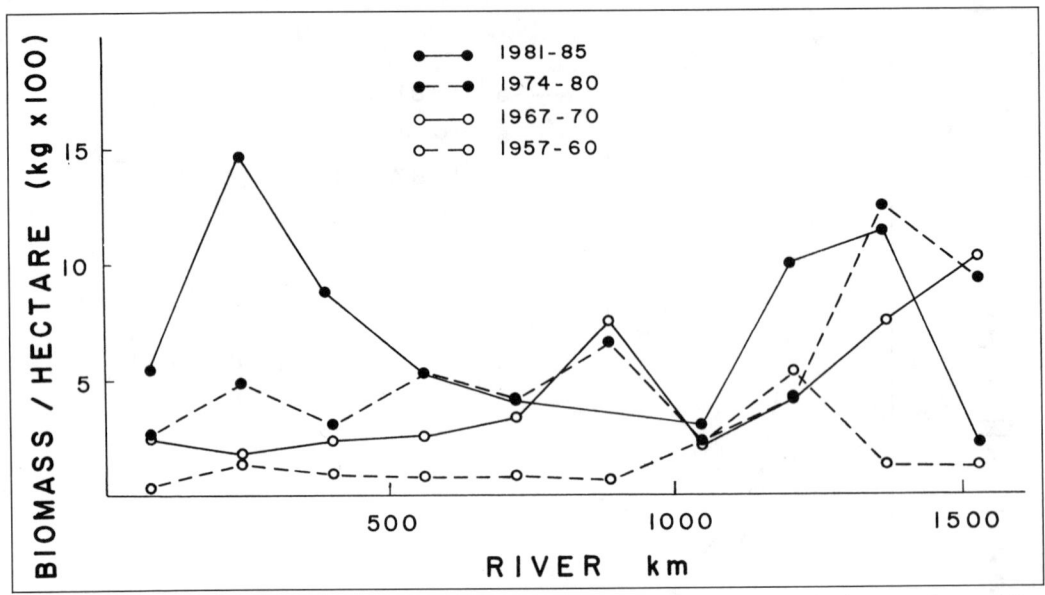

FIGURE 6. Biomass of all fish combined in lock chamber samples from 161-km sections of the Ohio River in the four periods: 1957–60 (open circle-dashed line), 1967–70 (open circle-solid line), 1974–80 (solid circle-dashed line), and 1981–85 (solid circle-solid line).

Hybopsis aestivalis, eastern sand darter *Ammocrypta pellucida*, and blackside darter *Percina maculata*.) Many species still found in the lower two-thirds of the Ohio River are most abundant in the last 320 km before it joins the Mississippi River (i.e., paddlefish, shortnose gar, spotted gar *Lepisosteus oculatus*, and shovelnose sturgeon).

A total of 154 species of fish (including introduced species) were reported from the Ohio River in 1984 (Pearson and Krumholz 1984), and five additional species were added later (Pearson and Pearson 1988). Reports before 1920 list only 111 species, those between 1920 and 1969 list 121 species, and those between 1970 and 1988 list 140 species. When data from these three periods are combined, 122, 132, and 119 fish species were reported from the upper, middle, and lower thirds of the Ohio River, respectively—a remarkably even distribution of species along the river's length. Eighty-nine of the 159 fish species occurred in all three sections of the river. Only 9, 9, and 16 species were found only in the upper, middle and lower thirds, respectively. Species diversity indices increased significantly in the upper 161 km of the river between 1957 and 1987 (Pearson and Krumholz 1984).

When recoveries from all lock chamber samples from 1957 to 1985 (n = 280) were sorted by location into 161 km sections and analyzed, the highest density of all fishes combined occurred in the upper third of the Ohio River (Figure 5). Densities in the middle and lower thirds were about half that of the upper third. The high densities in the upper third were due primarily to large numbers of emerald shiners and mimic shiners *Notropis volucellus* before 1980, and to the two shiner species and gizzard shad after 1980.

Densities of nearly all species of fish increased between 1957 and 1985 in the upper 161 km of the Ohio River. The most dramatic increases occurred after 1974.

Below RKm 161, the densities of emerald shiner, mimic shiner, silver chub *Hybopsis storeriana*, and several other species of minnows decreased after 1957, while densities of common carp, white crappie *Pomoxis annularis*, sauger *Stizostedion canadense*, white bass *Morone chrysops*, and freshwater drum increased (Table 1, Part A). Gizzard shad densities increased in samples above RKm 965, but decreased slightly below RKm 965. Densities of paddlefish, longnose gar, and shipjack herring *Alosa chrysochlorus*, increased slightly between 1957 and 1980 in the lower third of the river. Densities of bullheads (*Ictalurus melas*, *I. natalis*, and *I. nebulosus*) declined in most sections of the Ohio River between 1957 and 1985, and the species mix apparently shifted from black bullhead to brown bullhead in the upper third.

The biomass of all fishes combined, however, was usually greatest in the lower third of the Ohio River (Figure 6; Table 1, Part B). Biomass in the upper and middle thirds of the river was usually about half that in the lower third. Large catches of gizzard shad, common carp, and bigmouth buffalo *Ictiobus cyprinellus* contributed to the high biomass in the lower third. In the 1981 to 1985 samples, an abundance of gizzard shad and carp in the upper third of the Ohio River gave a biomass equal to or exceeding biomass in the lower river (Table 1, Part B).

TABLE 1. Density (number/hectare) and biomass (kg/hectare) of the 10 most abundant fishes in lock-chamber samples taken from 1957 to 1980 (243 samples) and from 1981 to 1985 (37 samples) in the upper, middle, and lower thirds of the Ohio River (ORSANCO 1987c).

	1957–1980				1957–1980		
	River kilometer[a]				River kilometer[a]		
Species	0–483 (upper)	483–1126 (middle)	1126–1578 (lower)	Species	0–483 (upper)	483–1126 (middle)	1126–1578 (lower)
Part A, Density							
Emerald shiner	11043.0	434.0	151.0	Emerald shiner	29021.0	235.0	236.0
Gizzard shad	1306.0	2308.0	3387.0	Gizzard shad	11553.0	5046.0	4528.0
Freshwater drum	97.0	388.0	764.0	Freshwater drum	1233.0	2474.0	1739.0
Mimic shiner	1272.0	40.0	16.0	Channel catfish	320.0	639.0	1390.0
Channel catfish	637.0	334.0	260.0	Common carp	421.0	72.0	126.0
Common carp	278.0	91.0	82.0	Mimic shiner	960.0	0.0	0.0
Bullheads (all)	303.0	3.0	2.0	Skipjack herring	8.0	461.0	419.0
Skipjack herring	41.0	91.0	63.0	Sand shiner	211.0	0.0	0.0
White crappie	34.0	70.0	80.0	Blue catfish	0.0	0.0	600.0
Threadfin shad	0.0	16.0	142.0	Threadfin shad	0.0	56.0	415.0
Part B, Biomass							
Gizzard shad	37.3	74.1	345.1	Common carp	543.8	158.9	275.7
Common carp	131.2	135.5	160.7	Gizzard shad	205.4	386.8	107.0
Freshwater drum	8.9	34.2	72.5	Freshwater drum	65.4	79.8	265.5
Channel catfish	32.0	42.3	20.7	Emerald shiner	45.9	0.6	0.4
Bigmouth buffalo	0.2	9.2	23.3	Channel catfish	28.4	32.2	42.0
Smallmouth buffalo	3.3	14.7	10.4	Smallmouth buffalo	14.0	25.9	13.9
Emerald shiner	15.9	1.2	0.6	Flathead catfish	6.2	17.7	10.9
Paddlefish	0.1	8.9	6.6	Skipjack herring	2.6	8.1	12.7
Flathead catfish	2.4	5.3	6.8	White bass	6.0	1.2	4.2
Bullheads (all)	13.4	0.1	0.2	Sauger	4.6	1.3	1.8

[a]Measured downstream from the river's source at Pittsburgh, Pennsylvania.

In terms of numbers, emerald shiners, gizzard shad, and freshwater drum were the three most abundant fishes in lock chamber samples from 1957 to 1985 (Table 1, Part A). Densities of emerald shiner were much greater in the upper third of the Ohio River than below. Densities of gizzard shad were more uniform throughout the river (although increased in the 1980s), while densities of freshwater drum were greatest in the middle and lower thirds. Mimic shiners, fourth in abundance, were concentrated in the upper third of the river. Channel catfish, fifth in abundance, were distributed uniformly along the river. The remaining five species on the 10 most-abundant list (Table 1, Part A) were: common carp, bullheads (all species combined), skipjack herring, white crappie, and threadfin shad *Dorosoma petenense*.

The fishes with the greatest biomass in lock chamber samples were gizzard shad and common carp (Table 1, Part B). These two species were distributed throughout the Ohio River, but stocks were greatest in the lower 322 km for most of the period. The biomass of freshwater drum and channel catfish ranked third and fourth, respectively, from 1957 to 1980 and they were uniformly distributed upstream and downstream. The bigmouth buffalo ranked fifth in biomass, primarily due to a few unusually heavy catches in the lower half of the river after 1974. The four species ranked sixth to tenth in biomass from 1957 to 1980 were smallmouth buffalo *Ictiobus bubalus*, emerald shiner, paddlefish, flathead catfish *Pylodictis olivaris*, and bullheads (all species combined). After 1980 bullheads, paddlefish, and bigmouth buffalo were replaced in the top ten in biomass by skipjack herring, white bass, and sauger.

Summary of Changes in Fish Populations

1. Ohio River fishes that spawned over clean gravel substrate declined in relative abundance (sturgeons, redhorses, darters, blue sucker, and paddlefish) while pelagic or semi-pelagic spawners and debris spawners increased (freshwater drum, gizzard shad, emerald shiner, and common carp).
2. Fish numbers, biomass, and diversity were reduced in the upper 161 km of the Ohio River by 1920–40. A few species (Alabama shad *Alosa alabamae*, lake sturgeon, burbot *Lota lota*, crystal darter, and gilt darter *Percina evides*) were probably extirpated by 1970.
3. Of the 159 fish species reported from the Ohio River, 14 were introduced but only common carp, goldfish, white catfish, and banded killifish have established reproducing populations.
4. Fishes with reduced numbers or distribution in the Ohio River include the lampreys, sturgeons, paddlefish, alligator gar *Lepisosteus spatula*, and blue sucker.
5. Many species that took refuge in the lower half of the Ohio River during the worst decades of pollution have begun to move back upstream since 1970–75. These include the paddlefish, blue sucker, mooneye, goldeye *Hiodon alosoides*, carpsuckers *Carpiodes* spp., silverjaw minnow *Ericymba buccata*, white bass, white crappie, and smallmouth buffalo (Pearson and Krumholz 1984; Duquesne Light Company 1984, 1986; Geo-Marine 1986; ESE 1987, 1988).
6. Densities, biomass, and diversity of the fish community have all increased in the upper 161 km of the Ohio River since 1970–75 as water quality conditions improved.

7. Increasing populations of gamefish throughout the Ohio River, along with reductions in taste and odor problems, have led to improved sport fishing. Fishing tournaments have been established, particularly in tailwaters and in areas where flooded creek outlets provide suitable off-channel habitat for centrarchids (Jackson 1986; Henley 1987).

Water Quality Control Efforts

The need to control pollution in the Ohio River basin was recognized as early as 1908. The dreadful conditions associated with the droughts of the 1930s caused the states bordering the Ohio River to form a regional commission to deal with the problem. These efforts lead to the establishment of the Ohio River Valley Water Sanitation Commission (ORSANCO) in 1948. The function of ORSANCO was to help the states formulate policies for pollution abatement, set standards and goals, facilite cooperation, and provide some overall research and monitoring. The Commission's standards were implemented through the National Pollutant Discharge Elimination System (NPDES), which was administered by the individual states. When the ORSANCO compact was signed by the eight member states (Illinois, Indiana, Kentucky, New York, Ohio, Pennsylvania, West Virginia, and Virginia), only 1% of the communities in the Ohio Valley had primary sewage treatment facilities. That figure increased to 97% by 1964. Similar successes were achieved in abatement of many industrial discharges as well.

The passage of the National Environmental Policy Act of 1969 and the Clean Water Act of 1972 had far-reaching effects on water quality and fish in the Ohio River and stimulated many investigations. The U. S. Environmental Protection Agency was called upon to review the various environmental impact statements and reports mandated by the new legislation. The Corps of Engineers, the Nuclear Regulatory Agency, many utility companies, and other organizations concerned with the Ohio River began preparing environmental reports. Some of these investigations extend long-term monitoring efforts. Notable examples are the lock chamber studies conducted biannually by the states with the cooperation of ORSANCO, the EPA, and others; the biological studies conducted at the Beaver Valley Power Plant by Duquesne Light Company; and the Ohio River Ecological Research Program on the upper Ohio River sponsored by a consortium of three power companies.

Federal expenditures to install secondary treatment facilities under the Construction Grants Program totaled $35 billion between 1972 and 1982 (Smith et al. 1987). The funds spent in the Ohio River valley provided 94% of the residents with secondary sewage treatment facilities as of 1986. Particularly welcome were the installations of the Morris Forman treatment plant at Louisville and the Mill Creek treatment plant at Cincinnati. The Morris Forman plant began acceptable operations in 1985, and the Mill Creek plant appears to be on schedule for acceptable operation.

ORSANCO began a detection network for organic chemicals in 1978, which now consists of 13 stations where 16 halogenated compounds are monitored daily, including those compounds most often detected in river water. The stations are operated in cooperation with 11 water utilities and two industries. The purpose of the network is to provide early warnings on spills of hazardous materials, thus preventing the type of catastrophe that occurred on the Rhine River in Europe in 1986. A new monitoring system for dissolved oxygen in 1986 replaced an old system in use for 20 years (ORSANCO 1987a).

Future Prospects and Hopes

Progress has been made in improving the quality of water in the Ohio River over the past thirty years. This progress is remarkable in light of the increasing human population of the Ohio River basin. However, efforts to maintain and improve water quality must continue. Many industrial and municipal facilities still do not meet compliance standards. Even efforts to date barely bring water quality in the Ohio River up to acceptability. For example, flows fell below "critical levels" for 13 days in the summer of 1983. Critical flow is the minimum flow level used to design wastewater treatment facilities, assuming that all facilities are in compliance operation (ORSANCO 1984). Constant vigilance will be necessary to maintain levels of current water quality.

Some additional areas of concern are:

1. Identification of non-point sources of pollution, including atmospheric depositions, and reducing their effects.
2. Continuous monitoring of organic compounds, with an early-warning detection system to identify spills of hazardous materials. The system should expand to cover an increasing number of compounds, and should include spill response plans. The unprecedented spill of 2.7 million liters of diesel fuel in the Pittsburgh area in January 1988 underlines the need for better regulation of above-ground storage tanks.
3. The planned development of hydroelectric power at each navigation dam on the Ohio River may influence dissolved oxygen concentrations below each dam. The Federal Energy Regulatory Commission has promised to conduct a cluster impact assessment in the Ohio River basin to address this problem.
4. The presence of unacceptably high levels of PCB's, chlordane, mercury, and perhaps other compounds in the tissues of Ohio River fishes is a potential threat to human health and limits the use of fishery resources. The sources of these contaminants must be located and either eliminated or reduced to acceptable levels.
5. The recovering fish community of the Ohio River should be aided. Areas of clean gravels should be provided to encourage species that spawn over gravel and rubble. Selected tributaries (e.g., the Muskingum, Scioto, Salt, and Wabash rivers) could be improved to serve as refuges for fish in the event of a catastrophic chemical spill on the main-stem Ohio River.
6. As a noble and symbolic gesture, some funds now used to stock striped bass in the Ohio River might be expended on a modest program to reintroduce the magnificent lake sturgeon. A native, remnant population of this large creature

was recently found in the upper White River of Indiana; breeding stocks for hatchery rearing could be obtained there or from the upper Mississippi River. The return of the lake sturgeon would symbolize commitment to a lasting restoration of "La Belle Rivière."

Acknowledgments

One of the most difficult parts of writing this article was gathering data on fish distributions and abundances from the "grey literature" or agency reports, consultants program reports, and other sources. I appreciate the help of all who supplied me with copies of these reports. Lock chamber data from the Ohio River could be interpreted only with the generous help of H. Ronald Preston of the U. S. Environmental Protection Agency in Wheeling, West Virginia, and John Keyes and Jerry Schulte at ORSANCO. I thank them all. W. L. "Pete" Redmon of the EPA provided the initial support for undertaking the study and offered continued encouragement and assistance—thanks, Pete. Louis A. Krumholz introduced me to the Ohio River and its fishes, and I am grateful for everything he gave me.

References

Butz, B. P., D. R. Schregardus, B. A. Lewis, A. J. Policastro, and J. J. Reisa, Jr. 1974. Ohio River cooling water study. EPA-90519-74-004, U. S. Environmental Protection Agency. NTIS, Springfield, Virginia.

Cleary, E. J. 1967. The ORSANCO Story, The Johns Hopkins Press, Baltimore, Maryland.

Duquesne Light Company. 1984. 1984 Annual environmental report, Beaver Valley Power Station. Duquesne Light Company, Pittsburgh, Pennsylvania.

Duquesne Light Company. 1986. 1985 Annual environmental report, Beaver Valley Power Station. Duquesne Light Company, Pittsburgh, Pennsylvania.

ESE (Environmental Science and Engineering). 1987. Final report 1986 Ohio River ecological research program. ESE, St. Louis, Missouri.

ESE (Environmental Science and Engineering). 1988. Final report 1987 Ohio River ecological research program. ESE, St. Louis, Missouri.

Gammon, J. R. 1977. The status of Indiana streams and fish from 1800–1900. Proceedings of the Indiana Academy of Science 86:209–216.

Geo-Marine. 1984. 1983 Ohio River ecological research program. Geo-Marine, Plano, Texas.

Geo-Marine. 1986. 1985 Ohio River ecological research program. Geo-Marine, Plano, Texas.

Henley, D. T. 1987. Ohio River sport fishery investigation. Federal Aid in Fish Restoration, Project f-40 segment 8, Annual performance report. Kentucky Department of Fish and Wildlife Resources, Frankfort.

Jackson, R. V. 1986. Ohio River sport fishery investigation, Federal Aid in Fish Restoration, Project F-40 segment 9, Annual performance report. Kentucky Department of Fish and Wildlife Resources, Frankfort, Kentucky.

Janssen, R. E. 1952. The history of a river. Scientific American 186:74–80.

Miller, A. C. 1983. Report of freshwater mussels workshop. U. S. Army Engineer Waterways Experiment Station, Vicksburg, Mississippi.

Neff, S. E., W. D. Pearson, and G. C. Holdren. 1981. Aquatic and terrestrial communities on the lower Ohio River (RM 930–981). Report (Contract #DACW-27-80-C-0064) to U. S. Army Corps of Engineers, Louisville District, Louisville, Kentucky.

Ohio River Committee. 1944. Ohio River pollution control, Volumes 1–3. U. S. Government Printing Office, Washington, D.C.

ORSANCO (Ohio River Valley Water Sanitation Commission). 1949a–1987a. Annual reports. ORSANCO, Cincinnati, Ohio.

ORSANCO (Ohio River Valley Water Sanitation Commission). 1962b. Aquatic-life resources of the Ohio River. ORSANCO, Cincinnati, Ohio.

ORSANCO (Ohio River Valley Water Sanitation Commission). 1975b. Ohio River mainstem assessment of 1974 and future water quality conditions. ORSANCO, Cincinnati, Ohio.

ORSANCO (Ohio River Valley Water Sanitation Commission). 1977b. Ohio River mainstem assessment of 1976 and future water quality conditions. ORSANCO, Cincinnati, Ohio.

ORSANCO (Ohio River Valley Water Sanitation Commission). 1978b. Ohio River mainstem assessment of 1977 and future water quality conditions. ORSANCO, Cincinnati, Ohio.

ORSANCO (Ohio River Valley Water Sanitation Commission). 1986b. Assessment of water quality conditions, Ohio River 1884–85. ORSANCO, Cincinnati, Ohio.

ORSANCO (Ohio River Valley Water Sanitation Commission). 1987b. The presence of toxic substances in the Ohio River. ORSANCO, Cincinnati, Ohio.

ORSANCO (Ohio River Valley Water Sanitation Commission). 1987c. Ohio River valley fish population studies, 1968–87. ORSANCO, Cincinnati, Ohio.

Pearson, W. D., and L. A. Krumholz. 1984. Distribution and status of Ohio River fishes. Report ORNL/SUB/79-7831/1. Oak Ridge National Laboratory, Oak Ridge, Tennessee.

Pearson, W. D., and B. J. Pearson. 1989. Fishes of the Ohio River. Ohio Journal of Science 89:181–187.

Pflieger, W. L. 1971. A distributional study of Missouri fishes. University of Kansas Museum of Natural History Publication 20:225–570.

Preston, H. R. 1969. Fishery composition studies-Ohio River Basin, 1967–68. Pages G1-G21 *in* Minutes of the Engineering Committee of the Ohio River Valley Water Sanitation Commission, 10 September 1969. Cincinnati, Ohio.

Preston, H. R., and G. E. White. 1978. Summary of Ohio River fishery surveys, 1968–76. Report EPA 903/9-78-009. U. S. Environmental Protection Agency, Surveillance and Analysis Division, Region III, Philadelphia, Pennsylvania.

Rafinesque, C. S. 1820. Ichthyologia Ohiensis, or natural history of the fishes inhabiting the river and its tributary streams, preceded by a physical description of the Ohio and its branches. W. G. Hunt, Lexington, Kentucky.

Robins, C. R., et al. (six coauthors). 1980. A list of common and scientific names of fishes from the United States and Canada, 4th edition. American Fisheries Society Special Publication 12.

Ruhl, K. J. 1987. Suspended sediment transport characteristics of the Ohio River from Ashland, Kentucky to Paducah, Kentucky. Master of Engineering thesis. University of Louisville, Louisville, Kentucky.

Smith, P. W. 1979. The fishes of Illinois. University of Illinois Press, Urbana.

Smith, R. A., R. B. Alexander, and M. G. Wolman. 1987. Water-quality trends in the nation's rivers. Science (Washington D.C.) 235:1607–1615.

Taylor, R. W. 1980. A survey of the freshwater mussels of the Ohio River from Greenup Locks and Dam to Pittsburgh, PA. U. S. Army Corps of Engineers, Huntington, West Virginia.

Trautman, M. B. 1957. The fishes of Ohio. Ohio State University Press, Columbus.

Trautman, M. B. 1981. The Fishes of Ohio, revised edition. Ohio State University Press, Columbus.

USACE (U. S. Army Corps of Engineers). 1969. Ohio River basin comprehensive survey. Volumes 1–14. USACE, Ohio River Division, Cincinnati, Ohio.

USACE (U. S. Army Corps of Engineers). 1978. Ohio River navigation project, operation and maintenance. USACE, Ohio River Division, Cincinnati, Ohio.

USACE (U. S. Army Corps of Engineers). 1979. The lower Ohio River navigation project. A public information brochure. USACE, Louisville District, Louisville, Kentucky.

Williams, J. C. 1969. Mussel fishery investigations, Tennessee, Ohio and Green Rivers. Project No. 4-19-R, Final Report. Kentucky Department of Fish and Wildlife Resources, Commercial Fisheries Research and Development Act, Frankfort, Kentucky.

Wolman, M. G. 1971. The nation's rivers. Science (Washington, D.C.) 174:905–918.

Water Quality in the Cumberland River Basin

R. DON ESTES
U.S. Fish and Wildlife Service
Tennessee Cooperative Fishery
Research Unit, Tennessee Technological University
Box 5114, Cookeville, Tennessee 38505, USA

JOHN A. GORDON
Tennessee Technological University
Civil Engineering Department
Box 5015, Cookeville, Tennessee 38505, USA

WENDELL L. PENNINGTON
Tennessee Technological University Department of Biology
Box 5114, Cookeville, Tennessee 38505, USA

ABSTRACT. *The Cumberland River, which has 10 impoundments, is one of the most regulated rivers in the United States. The basin also contains some of the most scenic streams in the nation. The J. Percy Priest and Cheatham reservoirs carry a heavy nutrient load, are eutrophic, and have poor water quality whereas Center Hill, Dale Hollow, Lake Cumberland, and Cordell Hull reservoirs carry a moderate nutrient load and have good water quality. Nutrient levels are high in Martins Fork, Laurel River, Old Hickory, and Barkley reservoirs, and they are considered to be moderately polluted. The upper Cumberland Basin is severely affected by acid mine drainage, siltation, and residues from oil and gas extraction. The middle and lower basins are affected mostly by expanding urbanization and agriculture. The fauna of the Cumberland River have been altered greatly by changing hydrologic conditions, especially from hypolimnetic releases. At least 166 species of fish are known from the Cumberland River basin, of which 150 are considered endemic. The unionid mussel fauna consist of 81 species, of which eight are listed as endangered. Continued strict enforcement of state and federal environmental regulations are needed to insure that the integrity of the Cumberland River basin is maintained.*

The Cumberland River is one of the most regulated rivers in the United States as well as one of the longest (1,160 km). The length of the Cumberland River is roughly similar to that of the Green River in the western United States and the Ottawa River in eastern Canada. In many respects, the Cumberland River is representative of other regulated rivers in the southeast, such as the Chattahoochie, Ouachita, Roanoke, and White. In comparison to its better known "sister" river, the Tennessee River, the Cumberland River is longer, has a more braided channel, drains a much smaller basin (46,830 km^2 compared to 105,960 km^2), and has water of generally higher quality. Both are tributaries of the Ohio River, arise in regions of similar topography, flow through similar valleys, and run parallel courses before emptying into the Ohio River a few kilometers apart. The two rivers have been joined near their mouths by a canal built in 1966, which allows barge traffic (and fish) to cross from the Cumberland system to the Tennessee system without entering the Ohio River.

The Cumberland River basin contains some of the most beautiful and scenic streams in the country, as emphasized by the large number of state and national parks, forests, and designated scenic rivers. Two major national areas in the upper basin (above the Caney Fork River) are the Big South Fork Cumberland National River Recreational Area in Kentucky and Tennessee and the Daniel Boone National Forest in Kentucky. Nine streams or portions of them have been designated scenic rivers. More than 23 state parks and forests, nine state scenic areas, and 19 wildlife management areas are located in the Cumberland Basin (ORBC 1981). Land-between-the-Lakes is located in the lower basin and is a major national demonstration and recreation area of the Tennessee Valley Authority (TVA).

History

The Cumberland River has always been an important transportation route. Some of the first permanent settlers arrived in flatboats by way of a long journey down the

Tennessee River to the Ohio River and then up the Cumberland River to the present site of Nashville. Indians, French trappers, and "long-hunters" had used the Cumberland River for many years before the early settlers arrived.

Trade on the Cumberland River was so extensive by 1797 that Congress established a port of entry at Polymyra, Tennessee. Soon thereafter, an inspection station was established in Pulaski County, Kentucky. By 1832 improvements to navigation were begun by the Corps of Engineers (COE). Steamboats became important to the transportation needs of the region in the 1820s and 1830s, and remained so until the early 1900s. Steamboats regularly traveled upstream as far as the present site of Wolf Creek Dam, about 286 km from the mouth, and up several major tributaries. The smaller flatboats and keel boats were also important for transportation in the early days, traversing the main-stem Cumberland River upstream to Laurel River in Kentucky as well as all major tributaries.

The modern era of navigation improvements began in the 1940s with the completion of two tributary dams, Dale Hollow and Center Hill, followed by construction in the 1950s and 1960s of main-stem locks and dams. The J. Percy Priest and Wolf Creek dams were completed in the late 1960s. After Laurel River Dam was completed in 1974, almost all of the Cumberland River system downstream of Cumberland Falls was affected by dam releases, except for the Harpeth and Red rivers. A navigation channel 3 m deep is maintained for 236 km above the mouth of the Cumberland River to Celina, Tennessee.

The COE is the agency responsible for most of the development and management of the Cumberland River, not TVA, as mistakenly believed by many. However, some responsibilities are shared. Most hydroelectric power is sent through the TVA system and marketed by the Southeastern Power Association. Two steam-fired generation plants on the Cumberland River and one small hydroelectric power project (Great Falls Dam) on the Caney Fork River above Center Hill Lake are operated by TVA.

Although recreation was not an intended benefit of most of the reservoirs, it has become a major component of the management plan for the system. There were 46.5 million visitors to the 10 COE projects during 1987, most of who came to fish (34%), to boat (16%), or to camp and enjoy the scenery (33%) (USACE 1988). The J. Percy Priest Reservoir near Nashville had the largest number of visits despite its eutrophic condition. The COE has built and maintained many areas for camping, picnicking, swimming, boat launching, and other recreational activities. All projects have COE-licensed marinas for public use and many of these have lodging and dining facilities. Unlike TVA, the COE acquired extensive acreage in the watershed and maintains control of the shoreline of most projects. The Tennessee Wildlife Resources Agency and the Kentucky Department of Fish and Wildlife Resources are responsible for all fish and wildlife management on and near the reservoirs.

Morphology and Hydrology

The Cumberland River passes through five physiographic provinces. It begins at the confluence of the Poor Fork, Clover Fork, and Martins Fork in Kentucky, flows southwesterly into Tennessee, and then arches through Tennessee back into Kentucky. Its average gradient is 0.23 m/km (Figure 1). The Cumberland Basin has an average width of about 31 km, is 217 km long and covers an area of more than 15,000 km^2, of which 60% is in Tennessee. In general, the stream flows on or near bedrock from its source to near Clarksville, Tennessee, and then through alluvium-filled valleys to its mouth. The channel is well defined and has relatively stable banks that support timber and brush to the low-water line. The principal tributaries are, from source to mouth, the Laurel, Rockcastle, South Fork, Obey, Roaring, Caney Fork, Stones, Harpeth, Red, and Little rivers. Each tributary is unique in some respects, but all share a number of common characteristics. Most have well defined courses, generally flowing close to bedrock, except for the lower reaches of the Red and Little rivers, which flow in alluvial channels. Occasional sand and gravel bars or rock reefs form a series of pools and shoals.

The Cumberland River has been extensively developed by the COE, beginning in the 1800s with a series of navigation improvements (Figure 1). It is now highly regulated by 10 multipurpose tributary and main-stem dams. There are five tributary storage impoundments (Martins Fork, Laurel River, Dale Hollow Lake, Center Hill Lake, and J. Percy Priest Lake) and one main-stem storage impoundment (Lake Cumberland). The four main-stem dams with navigation locks are Cordell Hull, Old Hickory, Cheatham, and Barley (Figure 1). An additional dam (Great Falls) above Center Hill Reservoir on the Caney Fork River is operated by TVA for hydroelectric power. The reservoirs allow the COE to control flows through much of the Cumberland River basin and, thus, represent the major management system affecting water quality. Discharges from storage impoundments are mainly for hydroelectric power production and flood control, but releases are sometimes made to maintain water quality or for navigation on the Cumberland River.

The watershed of the upper Cumberland basin is generally mountainous or hilly, with low narrow valleys. The distant headwaters drain the Cumberland Mountains to the southeast and the Pine Mountain overthrust to the northwest. The

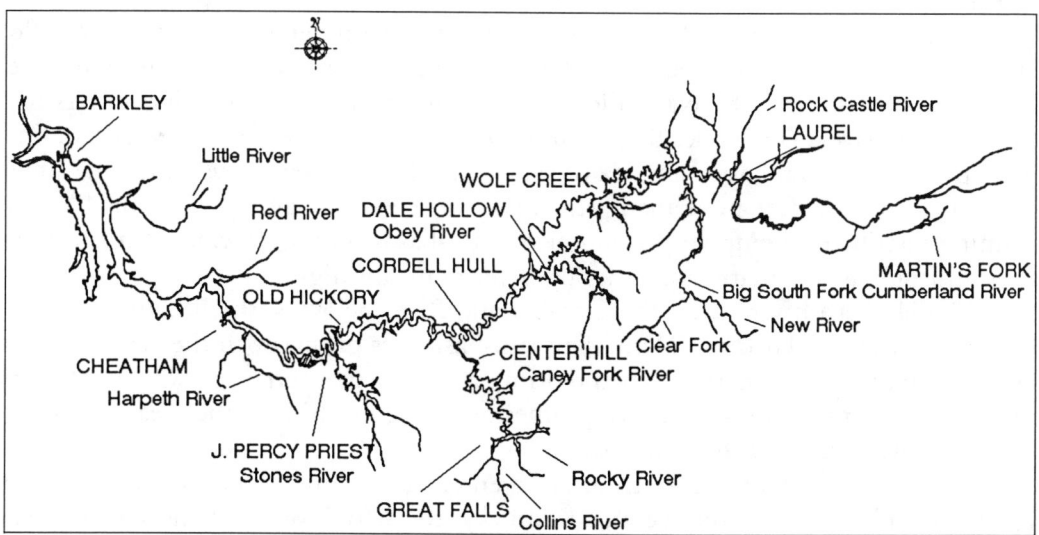

FIGURE 1. The Cumberland River basin, showing location of major tributaries and reservoirs.

eastern half of the basin lies in the mountains and coal fields of the Cumberland Plateau, and the western half lies in the Highland Rim and Nashville basin regions. The watershed extends almost to the border between Kentucky and Virginia and to Cumberland Gap, the route of the pioneers to the Cumberland region. Elevations in the upper basin generally range from 400 to 600 m above mean sea level, but the highest point in the basin is Black Mountain, Kentucky (1,263 m). The regional economy is strongly resource-based, dominated by coal production and reinforced by marginal agriculture and diminishing forest resources (ORBC 1981).

The Cumberland River flows over Cumberland Falls from the Plateau and into Lake Cumberland, the system's largest impoundment in total storage capacity. Downstream from Wolf Creek Dam (Lake Cumberland), the flow is regulated and there is no major tributary for about 50 km. The river is then joined by the Obey River, regulated by Dale Hollow Dam, and farther downstream by a smaller, unregulated stream, Roaring River. The Obey River drains the Cumberland Plateau and the eastern Highland Rim, whereas the Roaring River drains only the rim. The Obey watershed is characterized by rugged mountains and deep gorges, and is covered with extensive forests. The geology is dominated by sandstone, shale, and coal in the higher elevations, and by limestone and siltstone in the lowlands. Elevations for this region of the Cumberland Plateau range up to 700 m, compared with about 300 m on the Highland Rim. Past and present extraction of coal and oil have left scars on the landscape.

The Cumberland River flows through Cordell Hull Lake, a riverine reservoir with a hydraulic retention time of only 6–10 days in years of average rainfall, before being joined from the south by the Caney Fork River below Cordell Hull Dam. The Caney Fork, which is regulated by Center Hill Dam, drains the Cumberland Plateau and the Highland Rim. The rim is characterized by karst topography with extensive subterranean drainage, but is predominantly drained by typical surface patterns. The human population of the area is relatively high, and land use is primarily for agriculture.

From the confluence of the Caney Fork, the Cumberland River enters Old Hickory Lake, another riverine impoundment, that has a hydraulic retention time of about 14 days. Near the Nashville metropolitan area, the Stones River enters the Cumberland River. The Stones River is impounded by J. Percy Priest Dam and drains principally the central basin, an area of shallow fertile soils, extensive limestone sinks and outcrops, and few permanent streams. The elevation averages about 170 m. Farming dominates the upper portion of the sub-basin, urban development is extensive, and human populations in the lower portions are high.

The Harpeth River enters the Cumberland River below Nashville but upstream from the Cheatham Lock and Dam. This sub-basin lies primarily in the center basin, but also drains the western Highland Rim, a region of chert soils where solution cavities are common in the underlying limestone. The low hills of the area are mostly wooded, interspersed with farmland.

Farther downstream, two tributaries enter the Cumberland River from the Pennyroyal Plateau of Tennessee and Kentucky, the Red River, and the Little River. The Pennyroyal Plateau is a generally rolling upland plain with little local relief, numerous limestone sinks, and few large streams. Elevation averages 110 m. The final impoundment is Lake Barkley, the largest of the four main-stem reservoirs, which has navigation facilities.

Today, the large annual floods that once occurred near Nashville are largely controlled, but some flooding of urban areas and damage to rural properties and farmland still occurs.

Properties of the Ecosystem

Water Quality Features

In general, water quality of the Cumberland River system is good. Of the approximately 10,000 km of free-flowing streams in the Cumberland Basin, 34% were classified in 1980 as being of high quality (ORBC 1981). In a report in which the COE evaluated the water quality of all 10 reservoirs in the Cumberland Basin (USACE 1987), two reservoirs (J. Percy Priest and Cheatham) ranked as low as 4 on a scale of 1 to 5, and four (Center Hill, Dale Hollow, Lake Cumberland, and Cordell Hull) ranked as high as 2. The other four were rated moderate, or 3, in water quality. Another recent report stated that 38% of the acreage of COE reservoirs in Tennessee supported their intended use, whereas 29% was threatened by high nutrient levels and 33% was moderately impaired by high nutrient levels. A small portion (0.1%) was severely impaired by high bacterial contamination and posted against recreational water contact (TDHE 1986).

The major impact on the water quality in the upper Cumberland Basin is coal mining, which has continued since the late 1800s, typically with little regulation. In 1980, about 1,416 km of streams were severely affected by acid mine drainage and excessive siltation (ORBC 1981). For example, fauna in more than 25 km of the East Fork of the Obey River are greatly diminished because of drainage of acidic water from a few abandoned deep mines. However, some recent reclamation in the watershed has resulted in some improvement in water quality of tributary streams. Yellow Creek in Kentucky is severely impacted, in part, by mining activities. Although the Laurel River Lake has serious water quality problems, such as low dissolved oxygen (DO) and high concentrations of iron and manganese (ORBC 1981), trout fisheries are maintained in both the tailwater and reservoir. Extraction activities, including gas and oil exploration, also affect streams in the upper Cumberland Basin, including the Big South Fork and the Rockcastle rivers in Kentucky and the upper portion of the Caney Fork River in Tennessee.

The major impact on water quality of the middle and lower Cumberland River is urbanization and, to a lesser extent, agriculture and other land-use practices. The human population of the Cumberland Basin totalled 1.3 million in 1980, 80% of which lived in the lower basin. The Stones River and J. Percy Priest Reservoir have been heavily impacted by overloaded municipal wastewater treatment systems, urban development, and farming. Two reservoirs near Nashville, Old Hickory and Cheatham, are eutrophic; the DO below Old Hickory Dam failed to meet the state standard of 5 mg/L several times during 1986 (USACE 1987). The Red River at Clarksville, Tennessee, sometimes carries a heavy silt load. Despite this moderate impairment, water quality in the Cumberland River system as a whole is generally good.

Though low temperatures are not generally considered a water quality problem in the Cumberland River, depressed water temperatures caused by hypolimnetic withdrawals for hydroelectroc power production have affected a portion of the main stem and several tributaries. In more than 50 km of the main channel below Wolf Creek Dam, surface temperatures during summer are below 20°C. Temperatures before impoundment were normally 5 to 10°C higher in summer and they supported both typical warmwater and coolwater species of fish. Now the same stretch of river supports primarily a population of stocked trout and migrant cool water species (Bauer 1976). Other low-temperature tailwaters exist below Center Hill Dam (16 km), Dale Hollow Dam (4 km), Laurel River Dam (3 km), and Martins Fork Dam (2 to 3 km). These tailwaters are stocked with trout and water quality is generally acceptable; however, the endemic fish and benthic invertebrate populations have been greatly altered (Odenkirk 1987; Bauer 1976; Coopwood et al. 1987).

Aquatic Biota

Both natural and man-induced factors affect aquatic biota in the Cumberland Basin. The geographic variation in number of species of fish in different drainages illustrated in Figure 2 is, in part, due to geologic influences on stream chemistry. In general terms, the softwater streams in the upper Cumberland Basin support a less diverse fish fauna than do the moderately hardwater streams in the lower basin (Figure 2). Although surface water in the system as a whole can be classified soft (<100 mg/L $CaCO_3$), some well-buffered headwater streams have relatively high dissolved constituents, such as Sulphur Creek (Kentucky) downstream of Cumberland Falls. Likewise, some poorly buffered streams have low alkalinity, such as Martin Fork River (Kentucky) upstream of the Cumberland Falls. In comparison, specific conductance is more than a factor of 10 higher, and total alkalinity is more than 100 times higher, in Sulphur Creek than in Martins Fork. Both are high-quality streams of similar size. Only 7 species of fish and 23 taxa of benthos occur in Martin Fork, whereas 23 species of fish and 46 taxa of benthos occur in Sulphur Creek (Harker et al. 1980).

An example of how natural barriers influence the zoogeographic distribution of fish in the Cumberland system is Cumberland Falls. Downstream from the falls, 166 species of fish have been reported, of which 150 are native; above the falls, however, only 51 species of fish have been reported, of which 38 are native (Jenkins et al. 1971). Cumberland Falls has restricted the upstream dispersal of the rich fish fauna in the lower Cumberland River to the isolated upper Cumberland River. Great Falls on the Caney Fork River also isolated the upper portion of that system, to a lesser extent.

Other natural factors that influence species richness include stream size and order. In general, the larger streams or sub-basins within the Cumberland Basin

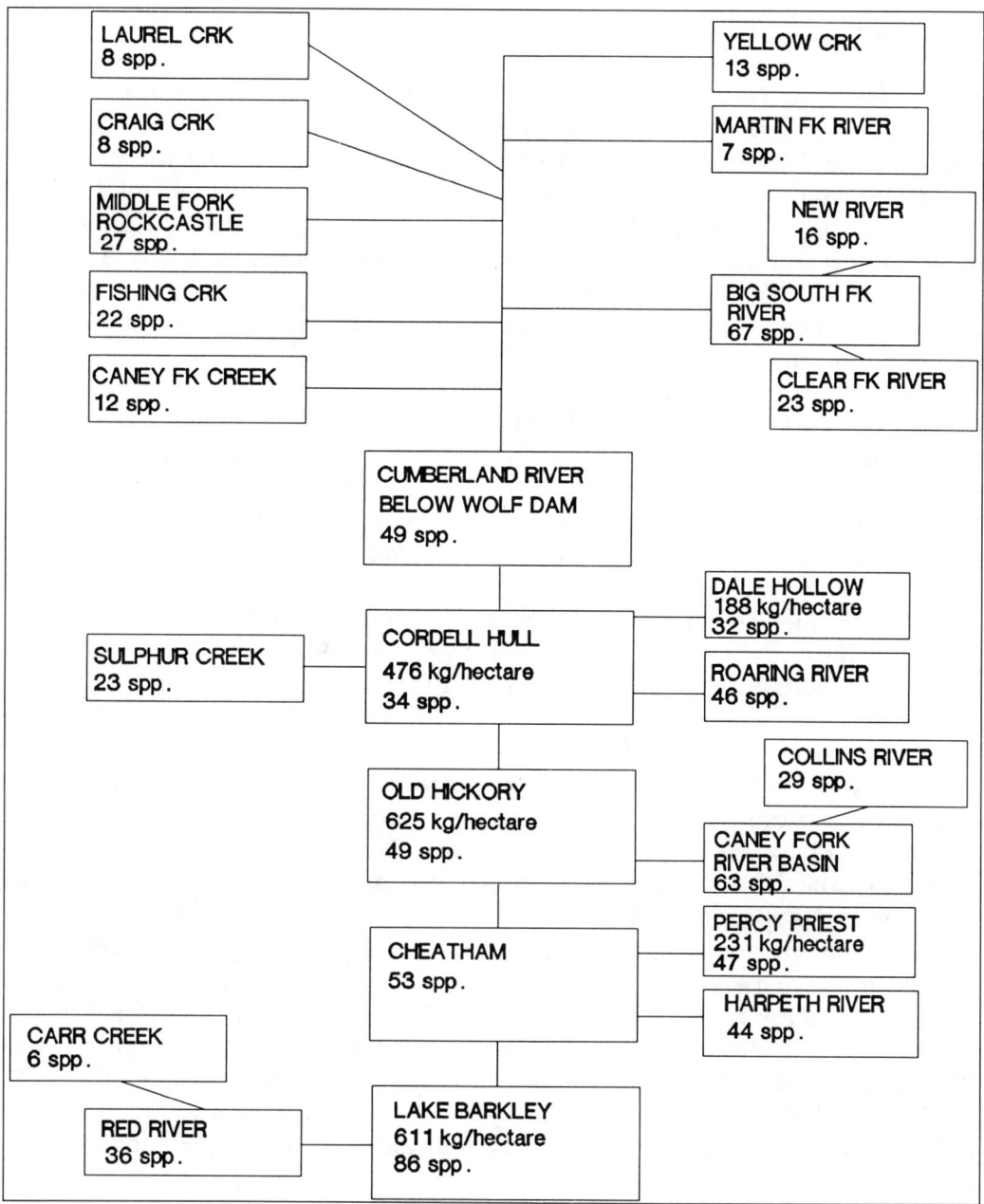

FIGURE 2. Number of fish species and density of fish in select regions of the Cumberland River basin.

have the highest number of fish species. The greatest species richness (86) has been reported from Lake Barkley, the system farthest downstream (Figure 1). Although many factors other than geology and topography (e.g., natural barriers) influence aquatic biota in the Cumberland River basin, the distribution of fish species is consistent with the concept of longitudinal gradients in fish species richness observed in smaller watersheds (Horwitz 1978).

Water Quality and Biota

In Tennessee, the New River watershed above the Big South Fork is disturbed by surface mining more than any other watershed in the state. More than 70% of the coal mined in Tennessee comes from this drainage (Larson et al. 1976). Many tributaries are either devoid of fish or have greatly reduced populations of fish and invertebrates (Winger et al. 1977). In addition, the main stream is degraded by sedimentation. However, recovery of biota is evident in the lower reaches, and impacts to New River affect the Big South Fork River only slightly. Although the impacts of mining on biota are generally widespread, several streams and even sub-basins in the New River drainage basin are not affected.

Nineteen streams of various size distributed throughout the upper Cumberland Basin in Kentucky were surveyed by Harker et al. (1980) to evaluate their water quality and aquatic biota. As judged by chemical and physical characteristics, and by existent populations of fish, macroinvertebrates, and periphyton, water quality was good in all 19 streams. About one-fourth were rated as high quality streams.

Relatively little mining occurs in the Clear Fork River watershed in Tennessee, which combines with the New River to form the Big South Fork River. This system is almost pristine and supports a diverse fauna consisting of 21 species of fish and more than 100 taxa of macroinvertebrates (Etnier et al. 1983). Additionally, 16 streams in the Big South Fork National River Recreation Area have generally good water quality and a diverse fauna, supporting at least 44 fish species and 215 macroinvertebrate taxa (O'Bara et al. 1982).

The West Fork of the Obey River is affected by acid mine drainage and by oil and natural gas extraction. These perturbations primarily affect biota near the disturbances (O'Bara 1983). In areas of the West Fork not influenced by fossil fuel production, the fish community is composed of as many as 21 species. This diversity is reduced to six species in affected areas. Despite the reduction in fish species, the richness of the macroinvertebrate community remains high and at least 117 genera have been reported (Pennington 1980; Gore et al. 1981; O'Bara 1983).

The Wolf River, also a tributary to the Obey River, is still considered a high quality system. Twenty species of fish were reported in the late 1800s from the Wolf River (Kirsch 1893) and 17 of the original species are still present in the 1980s (O'Bara and Estes 1984). The benthic community in the Wolf River is diverse and includes many species considered intolerant to most pollutants (Pennington 1980; Gore et al. 1981).

Most perturbations in the lower Cumberland River basin are unrelated to surface mining. Rather, much of the problem is due to nutrient enrichment from municipal wastewater treatment facilities and agricultural runoff. In addition, some small tributary streams receive silt from gravel removal and construction activities. The Red River at Clarksville, Tennessee, is slightly impacted (TDHE 1986), primarily by siltation. The Little River, Kentucky, is also adversely affected by siltation and nutrient enrichment but conditions have recently improved.

Another example of how aquatic fauna have been effected is the reduced number of mussel species in the Stones River upstream from J. Percy Priest Lake near Nashville.

Only 30 species were identified from this fluvial area in 1980–81 compared to about 40 species collected from the same area 15 years earlier. The disappearance of mussels was postulated to be due primarily to releases from overloaded wastewater treatment facilities, gravel removal operations, and non-point source pollution. When the impounded portion of the river was included in analysis, the reduction in mussel fauna was 40% overall (Schmidt et al. 1989).

The better quality streams in the lower Cumberland Basin include the Harpeth River, a designated scenic stream; the Roaring River, the Collins River, the Wolf River, and some smaller tributaries. Records in the Tennessee Aquatic Database System maintained by the Tennessee Wildlife Resources Agency list 53 species of fish in the Harpeth River drainage, 63 species in the upper Caney Fork basin (including the Collins River), and 46 species in the Roaring River drainage (Crumby 1987).

The water quality in the main-stem Cumberland River is generally good except near Nashville, where it is only fair, largely because of nutrient enrichment and changes brought about by the upstream reservoirs. From a fish management standpoint, moderate nutrient enrichment generally means increased productivity and a higher standing crop of fish. The biomass of fish is high in Old Hickory Reservoir (625 kg/hectare) and Lake Barkley (611 kg/hectare) downstream from the Nashville metropolitan area (Figure 2). In contrast, the reported standing crops are 188 kg/hectare in Dale Hollow reservoir, an oligotrophic body of water (TVA 1988), and 476 kg/hectare in Cordell Hull reservoir, a mainstem mesotrophic impoundment (Martinez 1980).

The exception to the rule that higher nutrient input increases standing crops of fish seems to be J. Percy Priest Reservoir, the most eutrophic impoundment in the system, where the reported standing crop is 231 kg/hectare (TWRA 1988). This reservoir is the most heavily fished in the Cumberland River system, and the standing crop estimate may simply reflect greater harvest, resulting in lower fish biomass. Of course, other factors must be considered in evaluating the effects of water quality on fish populations. The biomass of game fish may be no higher in main-stem reservoirs than in tributary reservoirs where the total biomass may be much less. For example, in Lake Barkley the standing crop of sport fish was 7.5% of the total biomass in 1983 (Swor et al. 1984) compared with 39% of the total biomass in Dale Hollow Reservoir (Tennessee Wildlife Resources Agency, Crossville, unpublished data). Although good game fish populations are generally associated with good water quality, many factors are involved and drawing definite conclusions from such data is difficult.

Nutrient enrichment usually results in greater biological oxygen demand (BOD) and reduced DO, which can significantly impact fish populations. In stratified reservoirs, low DO may not present a problem in the epilimnion but can become critical in the hypolimnion where little or no mixing occurs in summer. For example, Center Hill Reservoir supported a good "two-story" (hypolimnetic) trout fishery in the 1950s but, as eutrophication increased, the fishery was eliminated. The DO in the metalimnion now becomes too low by mid-summer to support fish over an extended period. Alternately, a modest increase of nutrients in Dale Hollow Reservoir might not adversely affect the cool water and warm water species present mostly in the epilimnion, but could conceivably produce adverse effects on lake trout (*Salvelinus namaycush*) and rainbow trout (*Oncorhynchus mykiss*) populations that depend on cold, oxygenated hypoliminic water. Similarly, low DO in release waters below some main-stem dams, such as Old Hickory Dam, are a cause for concern, especially in years with low rainfall. The tailwater fisheries below all the tributary impoundments could, at times, be affected by hypolimnetic discharges that are low in DO but this has never been documented in the Cumberland River system.

Hydrologic Effects

Perhaps the greatest adverse effect on aquatic biota of the Cumberland River is not associated with degraded water quality in the traditional sense, but rather to hydrologic features such as discharge, reservoir retention time, and temperature. Coopwood et al. (1987) studied the fishery in 128 km of the Cumberland River below Lake Cumberland, and found that indigenous fish species came either from upstream reservoir releases or from upstream migrations of potomodromous species. The cold water from Lake Cumberland was believed responsible for limiting reproduction of resident fish, especially in the upper reaches of the study site. Some recruitment occurred from tributaries farther downstream. Researchers concluded that, without stocking, the main channel would provide only a limited and sporadic sport fishery. Even with stocking, over 96 km of the river could not provide even a marginal sport fishery, except during periods when fish migrated.

Impoundments have undoubtedly had an adverse affect on the mussel population in the Cumberland River system. Before any impoundments were created, the unionid fauna along more than 608 km of the Cumberland River downstream from Cumberland Falls was commercially productive (Wilson and Clark 1914). By 1981, only a reach of about 80 km was commercially productive (Bates and Dennis 1985). Most of the original 81 species now exist only as remnant populations and eight are listed as endangered (Starnes and Bogan 1982, 1988). Loss of host fish species (for mussel larvae), depressed water temperature, scouring of the river bottom below dams, and high siltation in both tributaries and impoundments are probably the main causes for the decline in mussel populations.

Hydraulic retention time also affects the fishery of some reservoirs, both positively and negatively. A long retention time in some reservoirs is believed to improve water quality. In years with average rainfall, the retention time is about 10 days in Cordell Hull Reservoir and about 5 days in Cheatham Reservoir. The period is too short to attain maximum primary production or optimum surface temperatures for most sport fishes, except in some larger embayments. As a result, water temperatures are too low for warm water species and too high for cold water species, and the aquatic food base is depressed.

Future Prospects

It is unlikely that additional major impoundments will be built in the Cumberland River basin. The only remaining authorized project is the Celina Lock and Dam near the Kentucky-Tennessee border, but no funding is expected in the foreseeable future.

The COE is investigating the feasibility of installing additional generation capacity at some projects or modifying current operations to increase generation of hydroelectric power. Increased generation would undoubtedly affect the hydrology of the tailwaters and main channel, but the impact on aquatic biota is not known. Changes in generation patterns might also alter the magnitude and severity of water-level fluctuations in storage reservoirs, which would probably affect fish reproduction.

Water quality in the upper Cumberland Basin will continue to be affected by coal mining and other land-use practices. As oil prices declined in recent years, coal prices and production also declined. Coal production would probably increase if oil prices increased substantially. Runoff from oil and gas exploration has been minor in recent years and new regulations in Kentucky and Tennessee, if enforced, will help insure that runoff and brinewater do not contaminate surface water in the future. Strict surface mining statutes are now effective for active sites but abandoned mines still contaminate many streams. Abandoned Mine Lands funds are being used to improve some abandoned sites but much rehabilitation work remains.

The Cumberland Basin continues to attract people and industry. Care needs to be taken to insure that water quality does not decline during expansion. Nashville is growing rapidly and both point and nonpoint sources of pollution are of concern. Nashville has outdated system of combined sanitary and storm sewers and, although its main treatment plant has tertiary treatment, storm flows in excess of 375,000 m^3/day are discharged directly to Cheatham Reservoir. This combination of point and nonpoint source contamination is detrimental to water quality in the reservoir. The problem is being addressed by the U. S. Environmental Protection Agency and by state regulatory agencies by means of field surveys and water quality modeling. A satisfactory solution will not be easy or inexpensive. Both states are implementing active programs to control nonpoint pollution.

Because the drought of the 1980s had many significant impacts on the Cumberland River basin, it has heightened interest in monitoring and modeling the waterway. Two-dimensional hydrological models have been developed for the Laurel River, Center Hill, Cordell Hull, Percy Priest, Cheatham, and Barkley reservoirs, as well as for the Wolf Creek tailwater. In the future, these models should be linked to produce a basin-wide management program for effective water quality control.

Continuation of the 1980s drought may have both beneficial and adverse consequences. The present system of water storage works well when hydrological management is sound. Tributary reservoirs must be filled during winter and spring to provide flows through the main stem of the Cumberland River. In the absence of spring runoff, main-stem reservoirs are clearer and warmer, and water moves through them at lower velocities during the fish spawning season. Reproduction of some species may be adversely affected by these conditions, especially centrachids. However, primary production may be greater, resulting in faster growth and increased survival of fish. Water treatment plants operate more efficiently at lower sediment levels and higher temperatures. On the other hand, long retention times and clearer waters reduce DO, increase the production of algae that cause taste and odor problems, and allow aquatic macrophytes to spread.

Thus, the future holds both problems and opportunities for the Cumberland River and its people. There will be further population growth and industrial development in the Cumberland Basin, and perhaps greater point and nonpoint source pollution. However, professionals are finding better ways to manage water usage and the public seems willing to pay the cost for advanced treatment systems. If the drought continues, the 1980s will have provided the experience of a worst case scenario necessary for best management of the Cumberland River.

References

Bates, J. M., and S. D. Dennis. 1985. Mussel resource survey, State of Tennessee. Technical Report No. 85-3. Tennessee Wildlife Resources Agency, Nashville.

Bauer, B. H. 1976. The effects of the Cordell Hull impoundment on the tailwaters of Dale Hollow Reservoir. Master's thesis, Tennessee Technological University, Cookeville.

Coopwood, T. R. III, S. W. McGregor, T. S. Talley, and D. B. Winford. 1987. An investigation of the tailwater fishery below Wolf Creek Dam, Russellville County, Kentucky to Celina, Tennessee. Report. U. S. Fish and Wildlife Service, Cookeville.

Crumby, W. D. 1987. Growth dynamics of an introduced population of redeye bass in a north-central Tennessee stream. Master's thesis, Tennessee Technological University, Cookeville.

Etnier, D. A., D. L. Bunting, W. O. Smith, and G. A. Vaughan. 1983. Tennessee baseline stream survey. Research Report 995. Tennessee Water Resources Research Center, University of Tennessee. Knoxville.

Gore, J. A., J. D. Hughes, and W. A. Swartley. 1981. Benthic macroinvertebrates of low order streams of coal surface mining areas of the Cumberland Plateau, Kentucky, Tennessee, and Alabama. Final Report, U. S. Department of the Interior. Office of Surface Mining, Knoxville.

Harker, D. F., Jr., M. L. Warren, Jr., K. E. Camburn, S. M. Call, G. J. Fallo, and P. Wigley. 1980. Aquatic biota and water quality survey of the upper Cumberland River Basin. Technical Report, Vol. 2, Kentucky Nature Preserve Commission, Frankfort.

Horwitz, R. J. 1978. Temporal variability patterns and the distributional patterns of stream fishes. Ecological Monographs 48:307–321.

Jenkins, R. E., E. A. Lochner, and F. J. Schwartz. 1971. Fishes of the central Appalachian drainages: their distribution and dispersal. Pages 43–119 in P. C. Holt, editor. The distributional history of the biota of the southern Appalachians. Part III: Fishes. Research Division Monograph 4. Virginia Polytechnic Institute, Blacksburg.

Kirsch, P. H. 1893. Notes on a collection of fishes from the southern tributaries of the Cumberland River in Kentucky and Tennessee. Bulletin of the United States Fish Commission. 11(1891):257–265.

Larson, F. C., R. D. Minear, and B. A. Tschantz. 1976. Impact of coal strip mining on water quality and hydrology in east Tennessee. Natural Resources Research Center, Report 47. University of Tennessee, Knoxville.

Martinez, G. J. 1980. A study of age, growth, and food habits of certain game fishes in Cordell Hull Reservoir, Tennessee. Master's thesis. Tennessee Technological University, Cookville.

O'Bara, C. J., W. L. Pennington, and W. P. Bonner. 1982. A survey of water quality, benthic macroinvertebrates and fish for sixteen streams within the Big South Fork National River and Recreational Area. Final Report. U. S. Army Corps of Engineers, Nashville District, Nashville.

O'Bara, C. J. 1983. The effects of intermittent acid mine drainage and oil and natural gas runoff on the aquatic ecosystem of the West Fork Obey River, Tennessee. Master's thesis, Tennessee Technological University, Cookeville.

O'Bara, C. J., and R. D. Estes. 1984. A survey of fish communities of streams in coal surface mining areas of the Cumberland Plateau, Tennessee. Proceedings of the Southeastern Fishes Council 4:9–12.

Odenkirk, J. S. 1987. Food habits of rainbow trout and seasonal abundance of aquatic macroinvertebrates in the Center Hill tailwater. Master's thesis. Tennessee Technological University, Cookeville.

ORBC (Ohio River Basin Commission). 1981. Cumberland River Basin regional water and related land resources plan and environmental impact statement. ORBC, Cincinnati.

Pennington, W. L. 1980. Benthic populations of thirty-three stream locations draining coal reserves in Tennessee. Final Report. U. S. Geological Survey, Nashville.

Schmidt, J. E., R. D. Estes, and M. E. Gordon. 1989. Historical changes in the mussel fauna (Bivalvia: Unionoidea) of the Stones River, Tennessee. Malacological Review 1989 22:55–60.

Starnes, L. B., and A. E. Bogan. 1982. Unionid mollusca (Bivalvia) from Little South Fork Cumberland River, with ecological and nomenclatural notes. Brimleyana 8:101–119.

Starnes, L. B., and A. E. Bogan. 1988. The mussels (Mollusca: Bivalvia: Unionidae) of Tennessee. American Malacological Bulletin 6:19–37.

Swor, C. T., W. B. Wrenn, and D. R. Lowery. 1984. Results of fish community monitoring in Barkley Reservoir, 1983. Tennessee Valley Authority, Economic Development, Knoxville.

TDHE (Tennessee Department of Health and Environment). 1986. The status of water quality in Tennessee. 305(b) Report. Office of Water Management, TDHE, Nashville.

TVA (Tennessee Valley Authority). 1988. Fish standing crop data. Internal Report. TVA, Knoxville.

TWRA (Tennessee Wildlife Resources Agency). 1988. Internal Files. TWRA, Nashville.

USACE (U. S. Army Corps of Engineers). 1988. Internal Files. USACE, Nashville District, Nashville.

USACE (U. S. Army Corps of Engineers). 1987. Nashville District water quality activities for FY 87. Internal Report (EORNED-E). USACE, Nashville District, Nashville.

Wilson, C. B., and H. W. Clark. 1914. The mussels of the Cumberland River and its tributaries. U. S. Bureau of Fisheries Document No. 781:1–63.

Winger, P. V., P. W. Bettoli, M. Brazinski, and C. Lokey. 1977. Fish and benthic populations of the New River, Tennessee. Final Report. U. S. Army Corps of Engineers, Nashville.

Seasonal Patterns of Water Quality in Blackwater Rivers of the Coastal Plain, Southeastern United States

JUDY L. MEYER
Zoology Department and Institute of Ecology
University of Georgia, Athens, Georgia 30602, USA

ABSTRACT. *The Coastal Plain of the southeastern USA has predominantly sandy soils and little topographic relief. Rivers in this region have low gradients and drain extensive floodplain swamps. Their waters are low in suspended sediments and stained with humic substances leached from surrounding watersheds; hence, the name "blackwater rivers." Seasonal variation in water quality variables from two representative Georgia blackwater rivers, the Ogeechee River and the Satilla River, are discussed on basis of a 14 year record from three stations sampled by the Georgia Department of Natural Resources. These are warm water rivers that flood regularly during the winter. Water color and concentrations of dissolved organic matter are high, and alkalinity and pH are low; in these respects, the rivers are naturally of low water quality. Another water quality characteristic is low dissolved oxygen in summer due to natural loading of organic matter from floodplain swamps and, hence, high rates of ecosystem respiration. A major source of nutrients for the Ogeechee River is agricultural activity in the watershed. The Satilla River station directly below the largest city in the watershed is effected by the discharge of effluents from sanitary treatment facilities. There appears to have been little change in water quality of these rivers over the past decade. The removal of woody debris from the channel during the last century has effected aquatic biota more than any change in water quality. These rivers support productive populations of benthic invertebrates and fishes. One threat to water quality is the unregulated removal of water from the channel for agriculture. Maintaining the extensive floodplain forests is the key to protecting the integrity of both rivers. River corridors are critical components of water quality management for low-gradient blackwater rivers.*

The Coastal Plain of the southeastern United States is a region of relatively low topographic relief with mostly sandy and frequently water-logged soils. It consists of marine sediments of varying thickness deposited on a predominantly igneous and metamorphic rock complex (Hodler and Shretter 1986). The Coastal Plain is commonly divided into two physiographic regions: the upper Coastal Plain with gently rolling hills and more rapid drainage, and the lower Coastal Plain, which is flatter and more poorly drained (Bergeaux 1969). Surface sediments of the upper Coastal Plain are of Pliocene, Miocene, and Oligocene origin, whereas those of the lower Coastal Plain are younger, of Quarternary origin (Hodler and Shretter 1986).

Rivers draining the Coastal Plain have a low gradient with extensive floodplains, and are characteristically tea-colored with a low content of suspended sediments. The sandy soils have a low capacity for sorption of organic matter (St. John and Anderson 1982), and the soluble organic matter consists largely of fulvic acids (Beck et al. 1974) leached from the terrestrial environment and exported to the rivers. The fulvic acids impart a dark color to the water; hence, the rivers are called "blackwater rivers." Most Coastal Plain blackwater rivers are free of major impoundments and offer opportunities to examine unregulated lotic ecosystems. These rivers commonly have extensive floodplains where regular inundation has prevented extensive human development.

Because of the deep, porous sediments, low gradient, and extensive aquifers in many areas of the Coastal Plain, rivers in this region discharge less surface water than rivers draining similar sized basins in the Piedmont. During extended dry periods, the lower order tributaries are completely dry (GEPD 1987a).

In this paper I concentrate on two blackwater rivers in the Coastal Plain of Georgia: the Ogeechee River and the Satilla River. They are two of the best studied, large blackwater rivers in the Coastal Plain and they offer a contrast in water chemistry because of differences in their drainage basins. A small portion of the Ogeechee River drainage is in the Piedmont province, and a major portion of its drainage is in the upper Coastal Plain. As a consequence, it has somewhat higher pH and alkalinity than the Satilla River, a greater fraction of whose drainage basin lies in the lower Coastal Plain. Both rivers are typical of many blackwater rivers in the Coastal Plain without large impoundments or major urban centers in their drainage basins (Smock and Gilinsky, in press).

Morphology and Hydrology

The Ogeechee River drains a 13,500 km^2 basin, flowing for about 400 km from its headwaters in Greene County (200 m above sea level) to its mouth 24 km south of Savannah, Georgia (Figure 1). Its average gradient is less than 0.05%. A small portion of the drainage basin is in the Piedmont physiographic province. Average annual precipitation in the basin ranges from 100 to 130 cm (Hodler and Shretter 1986). Average annual discharge at the mouth of the Ogeechee River is about 115 m^3/s and the 7Q10 (the average lowest flow over a 7-d period that will occur on the average of once every 10 years) is 8.5 m^3/s (GEPD 1987a). The river floods regularly during the winter months (Figure 2). The human population is predominantly rural and only about 170,000 people live in the basin, a density of 12 persons/km^2 (GEPD 1987b). The largest city is Statesboro with a population of 15,000. The major forms of land use in the basin are agriculture (soybeans, tobacco, cotton, peanuts, corn) and livestock (cattle grazing, poultry) in the upper Coastal Plain and pulpwood production in the lower Coastal Plain (Hodler and Shretter 1986). About 25% of the drainage area is classified as agricultural (GEPD 1987b).

The major anthropogenic waste loads to the Ogeechee River come from point sources, such as municipal sewage treatment plants and treated textile wastes, and from nonpoint sources such as agricultural runoff and timber harvesting. The point source discharges to the river are relatively minor, with permitted point source discharges totaling 61,000 m^3/d (GEPD 1987b). The water quality classification for the Ogeechee River is either "fishing" or "recreation" along its entire length. Permitted consumptive water use in the basin is primarily for ground water withdrawal and it totals 121,000 m^3/d. Permitted surface water withdrawal equals 2800 m^3/d (GEPD 1987a). Non-permitted water use in the basin is largely for irrigation, which totals about 113,000 m^3/d of ground water and 76,000 m^3/d of surface water (GEPD 1987a). It is a relatively pristine river that has inspired photographers (Leigh 1986), and poets and artists (Dekle 1977). A section of the Ogeechee River was under consideration for wild and scenic river status (USNPS 1984). It has no major impoundments, although there are small ponds in its headwaters. A small dam with a water-powered woolen mill was built at the town of Shoals in the eighteenth centry, which

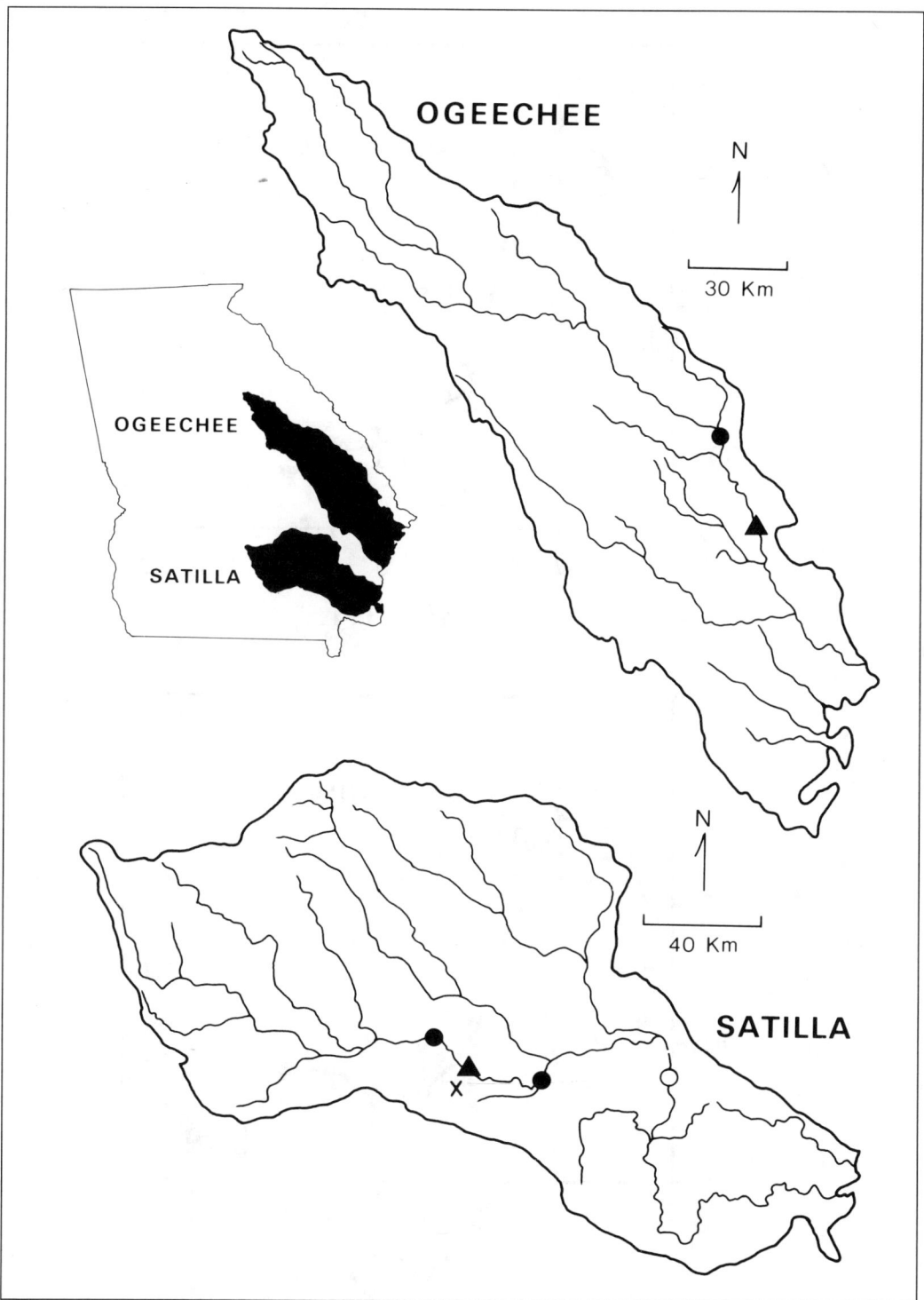

FIGURE 1. The Ogeechee and Satilla river basins, Georgia, showing locations of gaging (triangles) and water monitoring (solid circles) stations where the data were collected. In the Satilla River basin, the city of Waycross is indicated with an X and the national stream-quality accounting network (NASQAN) station with an open circle.

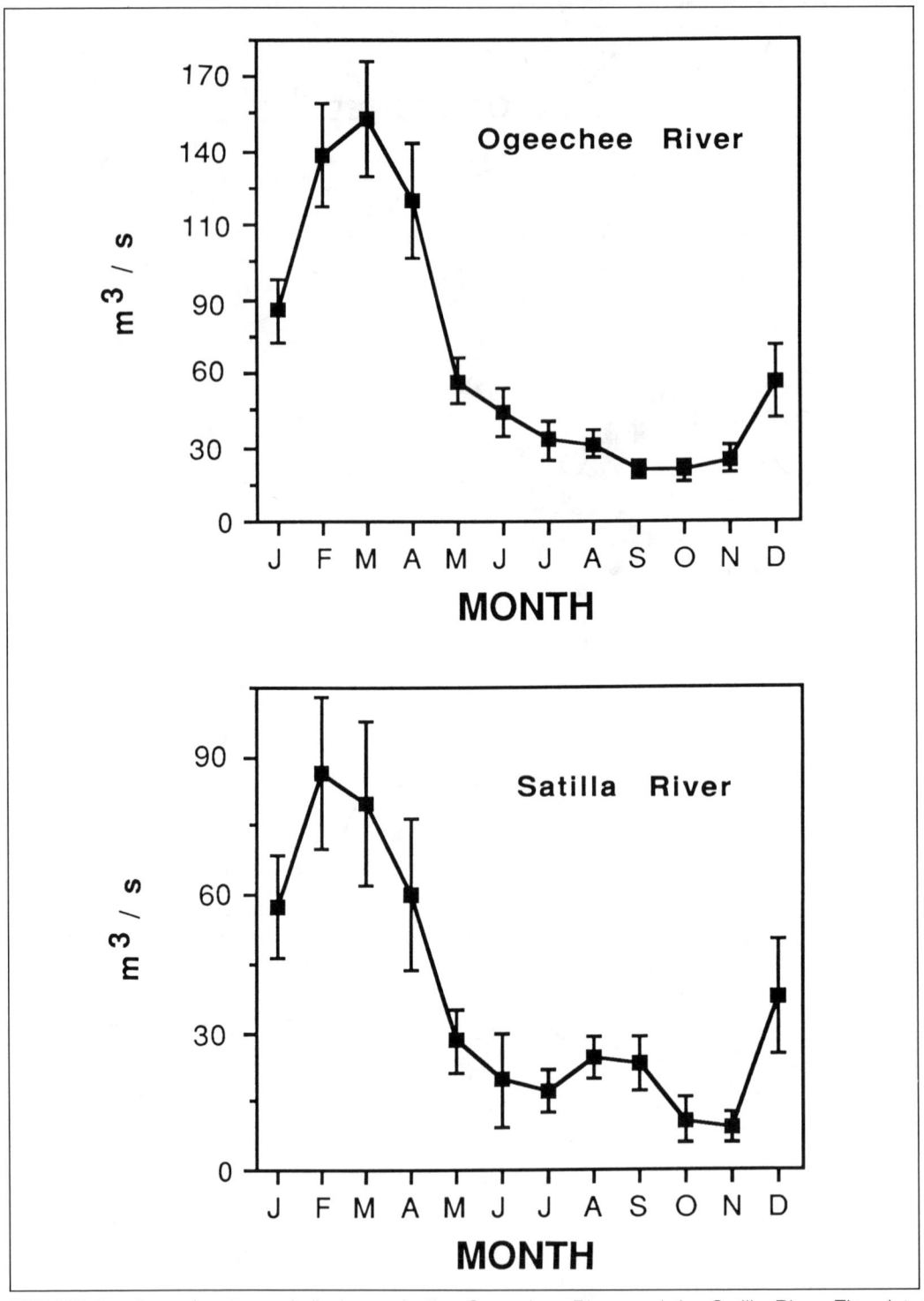

FIGURE 2. Annual pattern of discharge in the Ogeechee River and the Satilla River. The data points are means (± SE) of monthly mean daily discharges for the period 1973–1986. The Ogeechee River data are from the U.S. Geological Survey gaging station at Eden, and the Satilla River data are from the gaging station near Waycross (e.g., USGS 1973 and the following years).

was burned by Sherman (Sparks 1962). The Ogeechee River is well known locally for its good fishing (Leigh 1986).

The Satilla River lies south of the Ogeechee River (Figure 1); it drains an area of 10,200 km^2 and has a human population of about 150,000, a density of 15 persons/km^2 (GEPD 1987c). The largest population center is Waycross (19,000 people). The Satilla River also has no storage reservoirs, hydroelectric plants, or major diversions, although there are many small lakes and farm ponds on its tributaries. The river has a very low gradient of 0.03%; its headwaters are 115 m above sea level and it flows 380 km to the sea. Average annual precipitation in the Satilla River basin ranges from 110 to 130 cm (Hodler and Shretter 1986). Annual average discharge at the river's mouth is about 77 m^3/s and the 7Q10 is 1.1 m^3/s (GEPD 1987a). A winter flood is also typical of this river (Figure 2). Over 75% of the Satilla River basin is covered in woodlands and marshes; urban areas occupy less than 3%, and the remainder is cropland and pastures (GEPD 1987a). All streams in the basin are classified as "fishing" (GEPD 1987c).

The major wasteloads to the Satilla River come from municipal discharges of treated sewage and from timber harvesting operations. Large pulp mills at the town of Brunswick pump considerable amounts of ground water and discharge wastes into an estuary just north of the Satilla River. They have a major effect on surface and ground water quality in the area. I have not included this area in the material presented here because it is not part of the Satilla River watershed, although it is included in the Satilla River basin for management purposes. Permitted point source discharges to the Satilla River total 34,000 m^3 (GEPD 1987c). Permitted consumptive use of ground water totals 83,000 m^3/d. About 83,000 m^3/d of ground water and 7600 m^3/d of surface water is used for irrigation and livestock watering, uses that do not currently require a permit (GEPD 1987a).

Both the Ogeechee and Satilla rivers have low gradients and meander in sandy channels across extensive floodplains commonly 1–2 km wide, typical features of blackwater rivers on the Coastal Plain (Smock and Gilinsky, in press). Floodplain width in a 6th order section of the Ogeechee River is about 40 times the channel width (Benke and Meyer 1988). Dominant tree species in blackwater river floodplains include bald cypress *Taxodium distichum*, Ogeechee lime *Nyssa ogeechee*, sweetgum *Liquidamber styraciflua*, water oak *Quercus nigra*, swamp black gum *Nyssa biflora*, water tupelo *Nyssa aquatica*, and willow *Salix* spp. (Wharton and Brinson 1979; Wallace and Benke 1984). The forests are generally productive with an annual litterfall on the order of 600–800 g organic matter/m^2 (Mulholland 1981; Elder and Cairns 1982; Cuffney 1984).

The floodplains of these low gradient rivers are inundated annually, and there is considerable mixing between floodplain and channel waters. For example, at a sixth order site on the Ogeechee River, the floodplain is completely inundated at a discharge with a 1.2 year recurrence interval (Benke and Meyer 1988; Pernick and Roberts 1985). At this same site, 52% of total flow was through the floodplain at a discharge with a recurrence interval of 1.6 years (Benke and Meyer 1988; Pernick and Roberts 1985). Clearly the floodplain is an integral and interactive part of pristine blackwater river basins.

Another characteristic feature of blackwater rivers on the Coastal Plain is the extensive woody debris (snags) found in the channel and on the bank (Wallace and Benke 1984). The woody debris provides stable substrate in a channel of shifting sand, and supports high population densities of productive benthic macroinvertebrates (Benke et al. 1984; Smock et al. 1985; Smock and Roeding 1986). These invertebrates are a major source of food for fishes (Benke et al. 1985) and enhances their productivity.

Water Quality Characteristics

Historical Records

Water quality data were not collected regularly on the Ogeechee and Satilla rivers prior to the early 1970s, so we know little about their water quality before that time. The available data show little change in alkalinity, color, or suspended sediments in the Ogeechee River over the past fifty years. Alkalinity was measured at the national stream-quality accounting network (NASQAN) station at Eden from May 1937 through April 1938 (data retrieved from STORET, U. S. Environmental Protection Agency, Athens, Georgia). Mean alkalinity was 22.5 mg $CaCO_3$/L (range 13–31, N = 36), which is no different from the mean of 25.7 mg $CaCO_3$/L observed at the same site from October 1984 to March 1986 (range 0–38, N = 9). Color, also measured in the 1938–39 study, ranged from 26 to 108 platinum-cobalt units (mean = 50, N = 36); values measured currently are not different (see below). Chloride concentrations in 1938–39 were less than those observed today: 4.5 versus 7.4 mg/L (see below). The earliest measures of suspended sediments were taken from July 1958 through April 1959. Mean sediment concentration was 7.3 mg/L (range 3–13, N = 26), which is the same as the mean concentration of 7.4 mg/L measured between July 1985 and August 1986 (range 4–15, N = 8).

Removal of snags from river channels has had a major impact on their morphology and on the available habitat for benthic invertebrates in Coastal Plain rivers. Starting in the early 1800s, snags were removed from these rivers to facilitate navigation. This practice continued into the 1950s in some rivers, including the Satilla River (Wallace and Benke 1984). Snagging operations in the Satilla River were even more extensive than in the Ogeechee River and, as a result, it currently has less woody debris in its channel (Wallace and Benke 1984). Although snag removal had a minor long-term impact on water quality, it had a major impact on aquatic biota, many of which depend on wood as a stable substrate for attachment (Benke et al. 1984; Wallace and Benke 1984; Smock et al. 1985; Smock and Roeding 1986).

Ogeechee River

A regular water quality monitoring program was initiated on the Ogeechee River in the early 1970s by the Water Protection Branch, Environmental Protection Division, Georgia Department of Natural Resources. The sampling site (Figure 1) is located at Georgia Highway 24, downstream of the two major point source discharges (treated effluent from two textile mills). The data summarized represent mean values for a fourteen year period, 1973–1986; they were obtained either from published annual reports (e.g., GEPD 1974 and subsequently) or from the data files of that office (1984–1986). The U. S. Geological Survey has a national stream-quality accounting network (NASQAN) station at the town of Eden, about 43 river km downstream (at location of gaging station in Figure 1). Mean values for selected water quality parameters at that station are presented in Table 1 to provide comparison with another point in the river.

The discharge regime of the Ogeechee River is characterized by floods during January through April and low discharges in September and October (Figure 2). It is a warm water stream; the highest average temperatures is 27° in July and the lowest average temperature is 7.6° in January (Figure 4). Values of pH range from 6.6 to 7.2, with lows during the winter high water months and highs during the late summer when discharges are low (Figure 3). Alkalinity shows a similar seasonal pattern with average values ranging from 12 to 36 mg $CaCO_3$/L (Figure 3). Although pH and alkalinity values are less than those at 75% of the NASQAN stations (Smith et al. 1987), they are unusually high when compared with other blackwater rivers (Beck et al. 1974; Smock and Gilinsky, in press). This is because a part of the Ogeechee drainage lies in the Piedmont and because the Ogeechee River receives inflow from ground waters in the upper Coastal Plain, some of which are rich in limestone, such as Magnolia Springs near Millen, Georgia, with a daily output of 23,000 m^3.

A common feature of blackwater rivers is their extremely low levels of dissolved oxygen (DO) during the warmer months (Figure 4). Oxygen concentrations in the Ogeechee and Satilla rivers are lower than those observed at 75% of the 388 NASQAN stations, which were selected to be representative of rivers in the continental United States (Smith et al. 1987). Low DO conditions in blackwater rivers can be particularly dramatic when the discharge is low. During the 1986 drought, oxygen concentrations as low as 3.4 mg/L (41% saturation) were recorded; the annual average for 1986 was 5 mg/L, in contrast with averages ranging from 6.4 to 8.3 mg/L for the other thirteen years of record. Naturally low DO levels are common in other blackwater rivers in the southeast. In fact, many of the humic stained waters of Florida that are free

TABLE 1. Selected water quality variables for the Ogeechee River near Eden, Georgia, and the Satilla River near Atkinson, Georgia (USGS 1985). Values are the means for unfiltered water samples collected from 1974–1985 for the Ogeechee River and 1968–1985 for the Satilla River (number of samples).

Variable	Ogeechee River		Satilla River	
pH	6.50	(152)	4.90	(166)
Alkalinity (mg $CaCO_3$/L)	23.00	(129)	4.10	(145)
Nitrate (mg N/L)	0.10	(94)	0.10	(119)
Total phosphorus (mg P/L)	0.05	(117)	0.09	(135)
Total organic carbon (mg C/L)	8.80	(55)	20.00	(93)

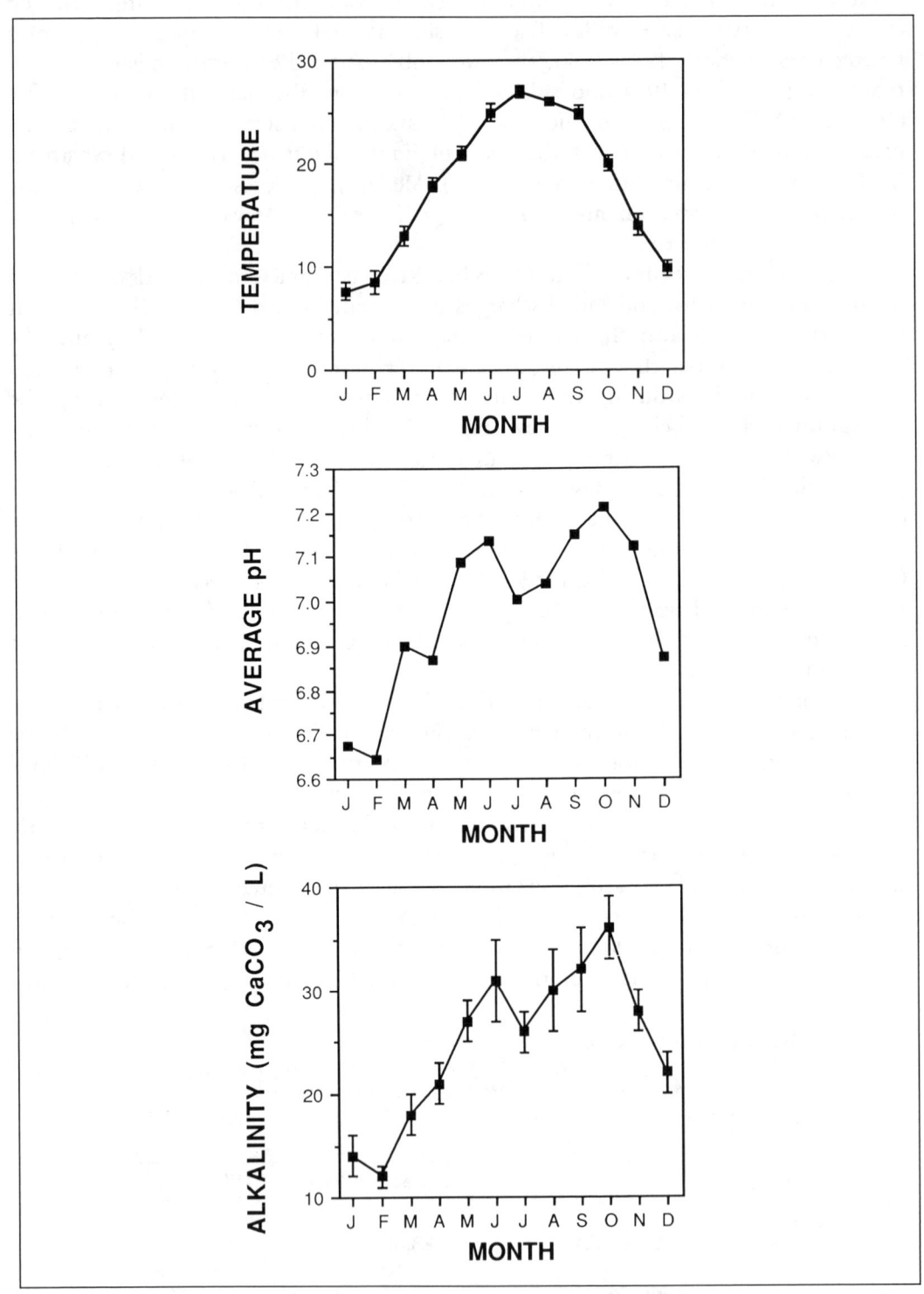

FIGURE 3. Mean (±SE) monthly water temperature, pH, and alkalinity in the Ogeechee River at Georgia Highway 24, July 1973–December 1986.

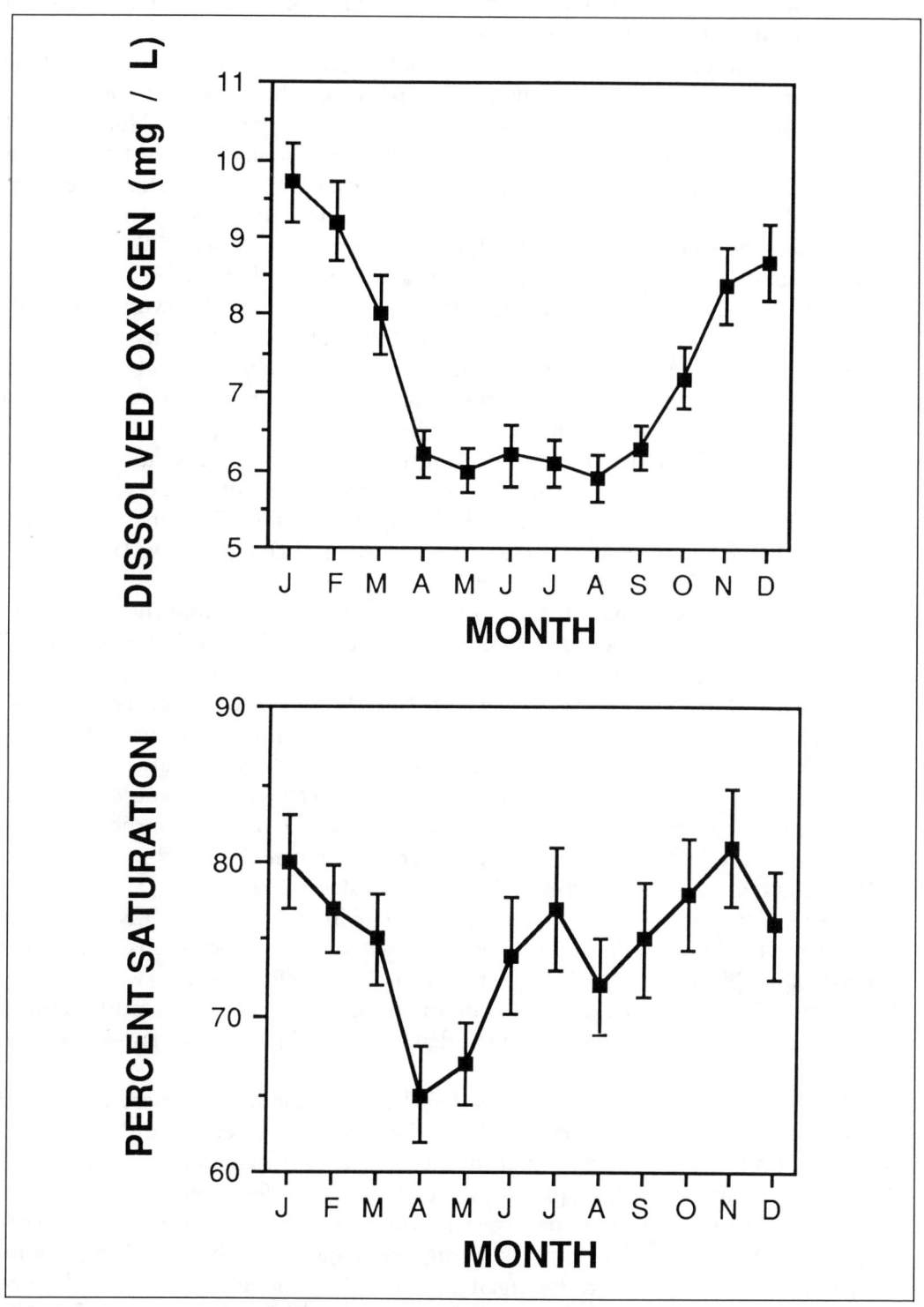

FIGURE 4. Mean (± SE) monthly dissolved oxygen concentration and percent oxygen saturation in the Ogeechee River at Georgia Highway 24, July 1973–December 1986.

from organic pollution do not meet the established 5 mg/L water quality criteria for recreational waters (Belanger et al. 1985).

There are several reasons for low levels of oxygen in blackwater systems. One is that blackwater rivers drain extensive swampy areas where there is considerable contact between organically rich bottom sediments and the water. Thus, surface waters that drain to the rivers are depleted in oxygen. In addition, although rates of photosynthesis are moderate, measures rates of respiration in the Ogeechee River are high. Much of this respiration appears to be in the sediments and is driven by inputs of organic matter from the floodplain (Edwards and Meyer 1987). The deep sandy sediments of the river channel contain active populations of bacteria to considerable depths (Meyer 1988), and respiration has been measured down to sediment depths of 60 cm (author's unpublished data). The high concentrations of dissolved organic carbon (DOC) in the water column also can support bacterial growth and respiration (Meyer et al. 1987). The DOC may become more available to bacteria after photolysis caused by exposure to sunlight.

Blackwater rivers derive their name from the humic substances that impart color to their waters, so it is not surprising that high concentrations of total organic carbon (TOC) are present (Figure 5). The highest concentrations of TOC are measured when the river is flooded and the lowest concentrations when discharge is low. The seasonal pattern for water color is the same (Figure 5).

Extensive studies of DOC at a site 50 river km downstream from the monitoring station on the Ogeechee River have demonstrated a direct relationship between DOC concentration and stream discharge (Meyer 1986). Most of the TOC is in the dissolved fraction, with DOC comprising >96% of TOC (Benke and Meyer 1988). DOC concentrations at the downstream site are considerably higher (annual average = 12 mg C/L) than upstream and, in fact, show a consistent increase as the river receives drainage from more swampy areas downstream (Meyer 1986; author's unpublished data). Floodplains appear to be an important source of DOC in the Ogeechee River. Most of the DOC in the Ogeechee River has a molecular weight in the range of 1000–10,000, which is the size of fulvic acids (Meyer 1986).

There is little indication of sewage pollution in the Ogeechee River. Average monthly fecal coliform values are generally less than 200 most probable number (MPN)/100mL (data not shown). Higher values (~2000 MPN/100 mL) were observed during low flow conditions on three dates in the early 1970s. Monthly average measures of 5-day biochemical oxygen demand (BOD) are also low—less than 1.2 mg/L (data not shown).

Nitrate concentrations in the Ogeechee River are lower than those at 75% of the NASQAN stations (Smith et al. 1987). The seasonal pattern of nitrate levels is what one would expect if agricultural inputs were a significant source. Nitrate concentrations are highest in late spring (May to June) when fertilizer is being applied to farmland (Figure 6). Ammonium concentrations are uniformly low in the Ogeechee River (monthly averages less than 0.06 mg N/L, data not shown). A significant fraction of nitrogen appears to be organic (Figure 6), although the data are limited because total nitrogen was not determined on all samples. The importance of organic N in the Ogeechee River is not surprising given the high DOC concentrations

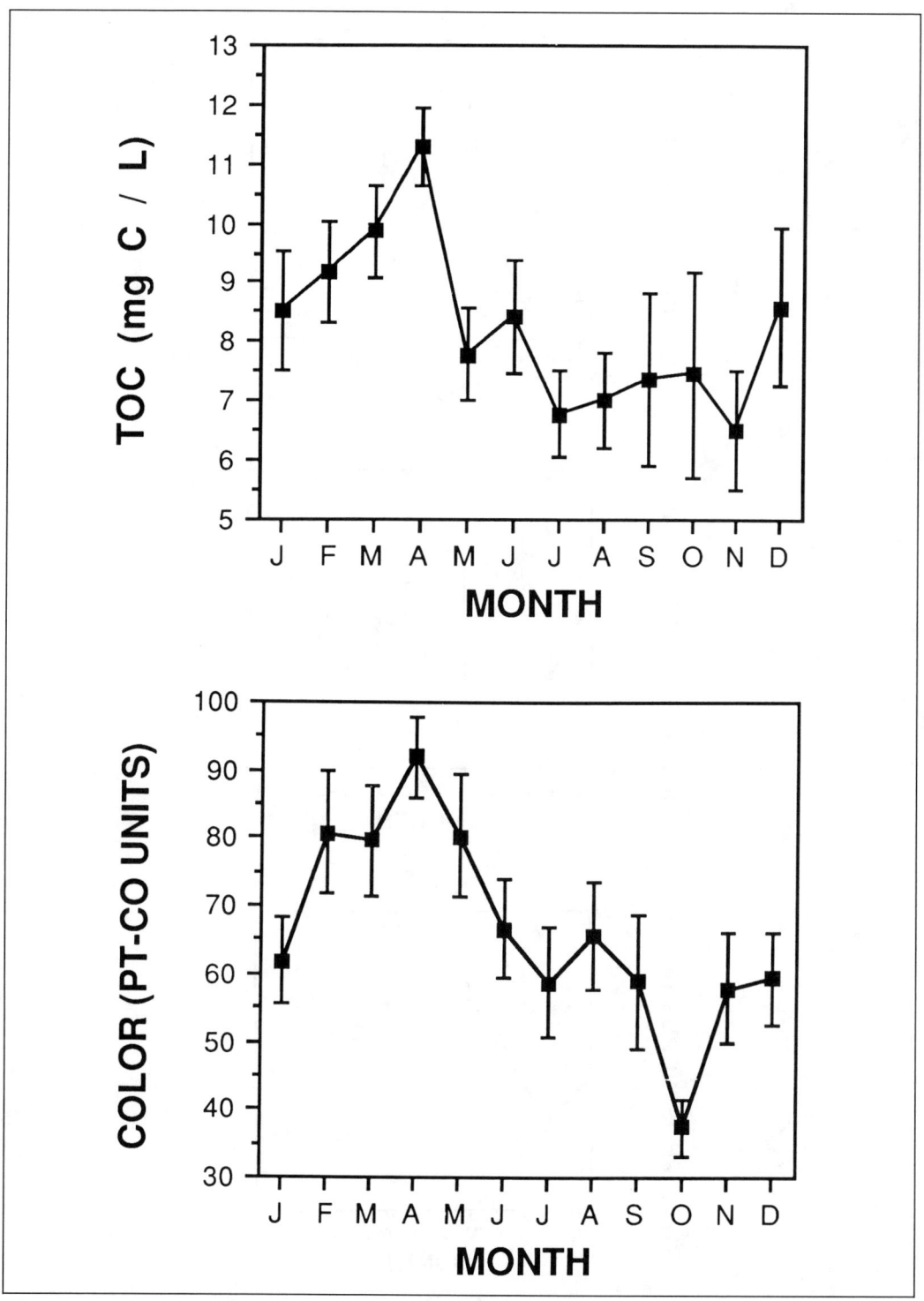

FIGURE 5. Mean (± SE) monthly total organic carbon (TOC) concentration, and color (platinum-cobalt units) in the Ogeechee River at Georgia Highway 24, July 1973–December 1986.

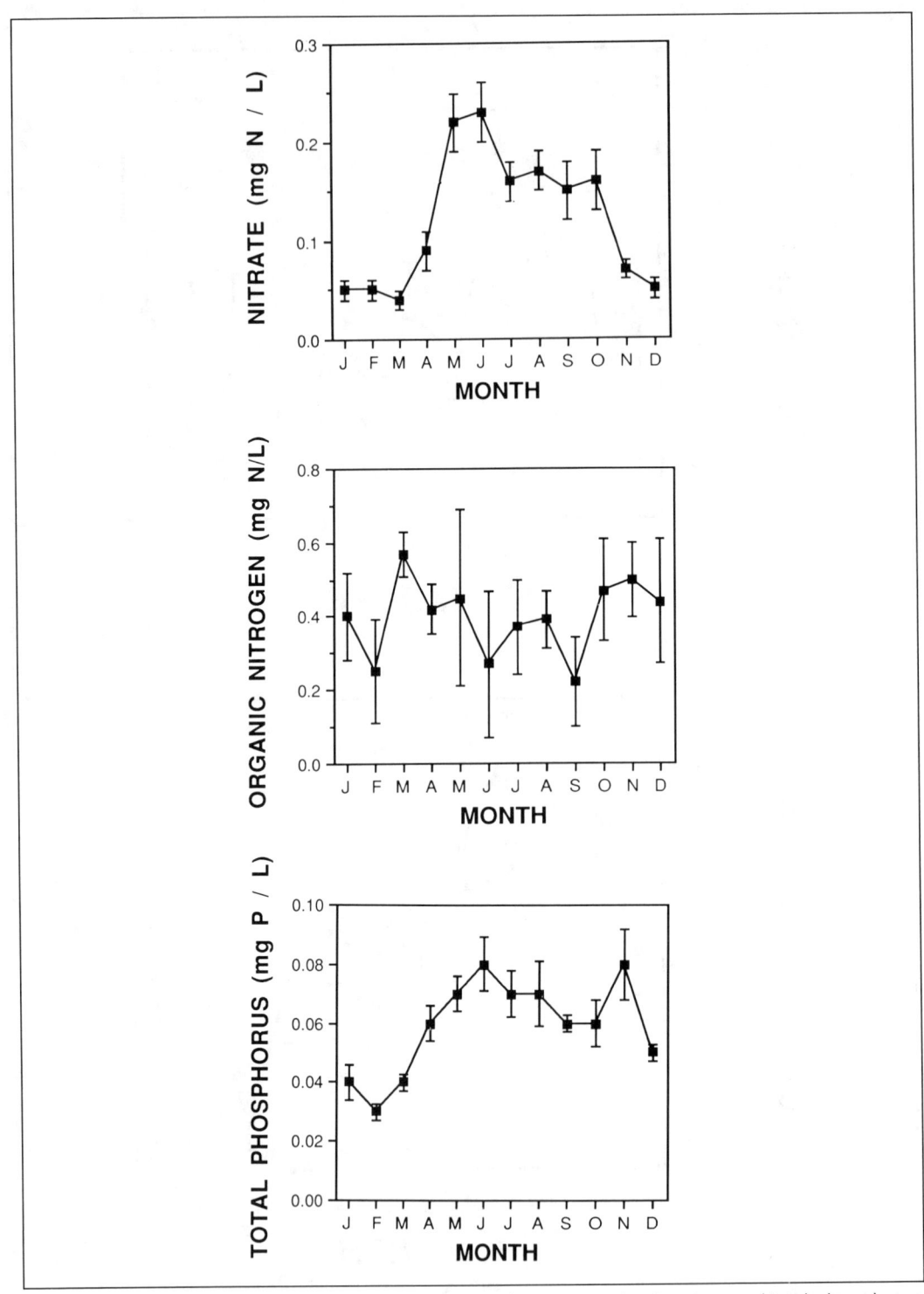

FIGURE 6. Mean (± SE) monthly concentration of nitrate, organic nitrogen, and total phosphorus in unfiltered water from the Ogeechee River at Georgia Highway 24, July 1973–December 1986.

observed. The seasonal pattern of organic N concentration is similar to that of TOC (Figure 6) and very different from the nitrate pattern, which suggests difference sources for these two forms of nitrogen. Nitrate concentrations peak during the summer when organic N concentrations are low.

Total phosphorus concentrations in the Ogeechee River show a seasonal pattern similar to that of nitrate. Concentrations are less than the EPA limit of 0.1 mg/L for controlling nuisance growths in streams (EPA 1976), and are lower than those observed at half the stations of the USGS NASQAN network (Smith et al. 1987).

Sediment and water samples for metal and organic contaminant analysis were collected during October 1979 and October 1982 from the station at Eden. Values for the organic contaminants in sediments were below detection limits and are reported in mg/kg dry weight: aldrin <1, chlordane <6, DDT <4, dieldrin <2, endrin <2, toxophene <20, methoxychlor <12, and PCB's <6. The metal concentrations reported in Table 2 are difficult to interpret because the detection limits are, in some cases, rather high.

Water quality data collected at the NASQAN station at Eden have been analyzed to detect any significant trends in water quality values from 1974 to 1981. After accounting for flow-related variation in concentration, several values showed significant trends. Conductivity, chloride, magnesium, sodium, potassium, and dissolved solids increased at a rate of about 3% (of the mean)/year; total phosphorus increased by 6%/year; ammonium increased by 32%/year; and TOC decreased by 10%/year (Smith and Alexander 1983). The increase in chloride concentration is also revealed in the historical record. Mean chloride concentration in 1938–39 was 4.5 mg/L (range 3–6, N = 37), whereas it increased to 7.4 mg/L in 1984–85 (range 5–10, N = 7) (data retrieved from STORET, USEPA). This increase could not be due to application of road salt in the basin because salt is not used, but it could indicate increased inputs of sewage to a river system where chloride levels are naturally low. Increased inputs of sewage seem likely since other values such as ammonia and phosphorus also increased over the 1974–1981 period.

Satilla River

I will present information from two water quality monitoring stations on the Satilla River established by the Water Quality Branch, Environmental Protection Division, Georgia Department of Natural Resources (Figure 1). The data span the same fourteen year period as those reported for the Ogeechee River. The upstream station is along a county road (FAS 598) north of the city of Waycross and the Satilla River receives only small amounts of effluent above this site. The downstream station is at the Highways 15 and 121 bridge about 20 km below the release of treated municipal sewage effluent from Waycross. I included this station to show the impact of municipal sewage in a blackwater stream of the Coastal Plain. Mean concentrations for some values of interest from a NASQAN station at Atkinson (Highway 84), about 45 river km below the downstream site (Figure 1) are included in Table 1 for comparison.

The Satilla River's discharge shows the same seasonal pattern as that for the Ogeechee River (Figure 2), with maximum flows from January through April and minimum flows in October and November. Its temperature regime is also similar, with little thermal difference between upstream and downstream stations (Figure 7). Alkalinity and pH are considerably lower than in the Ogeechee River, although the seasonal pattern is similar (Figure 7). Average pH and alkalinity are far below that observed at 75% of the NASQAN stations (Smith et al. 1987). Inputs of treated sewage from Waycross cause a considerable increase in alkalinity.

The Satilla River is also "blacker" than the Ogeechee River, with a darker color and higher concentrations of TOC (Figure 8). In the Satilla River, TOC is also almost exclusively DOC (Beck et al. 1974). In fact, the ratio of dissolved inorganics to organics is about 1:1, in contrast to the world average of 10:1 (Beck et al. 1974). Sewage inputs have had little impact on either TOC or color in the Satilla River (Figure 8). Because the Satilla River drains the lower Coastal Plain, a greater proportion of its drainage is swampy than is the drainage of the Ogeechee River. Since riverine swamps are a major source of the organic matter that imparts color and acidity to the blackwater rivers, differences in water quality between the Ogeechee and Satilla rivers vary accordingly.

As expected, mean DO and % oxygen saturation in the Satilla River are slightly lower below the sewage inputs, particularly during the months when river discharge is low (Figure 9). However, even at the upstream site, DO concentrations are lower than those at 75% of the NASQAN sites (Smith et al. 1987). The effects of the 1986 drought were also apparent in the Satilla River; DO fell to 3.8 mg/L during 1986, compared with annual means ranging from 6.0 to 7.9 mg/L in the other thirteen years. Coliform counts also show the impact of sewage effluents, with counts less than 400 MPN/100mL at the upstream station and counts greater than 2000 MPN/100mL common at the downstream site (data not shown). Five-day BOD was higher at the downstream site during some months, but monthly averages were <2 mg/L (data not shown).

The impact of the treated sewage is also apparent in the nitrogen and phosphorus values. Both ammonium and nitrate are considerably elevated downstream from the sewage inputs to the Satilla River (Figure 10), particularly during months when flow is low. In March 1987 the city of Waycross installed a new treatment system with higher capacity for reducing BOD and ammonia; hence, the downstream site should show some improvement in these water quality features in the future. The data on organic nitrogen concentrations (total Kjehldahl N minus ammonium) are limited and only available at the downstream site, but it is clear that organic nitrogen is an important form of nitrogen in the Satilla River (Figure 10). Nitrate concentrations at the upstream site are well below those found at 75% of the NASQAN sites and, even at the downstream station, concentrations are less than found at 50% of the NASQAN sites (Smith et al. 1987). Total phosphorus concentrations are extremely high at the downstream station; only 25% of NASQAN stations showed higher concentrations (Smith et al. 1987). The upstream station had total phosphorus concentrations similar to those observed in the Ogeechee River. In fact, phosphorus concentrations in the Satilla River are at the level that one would expect

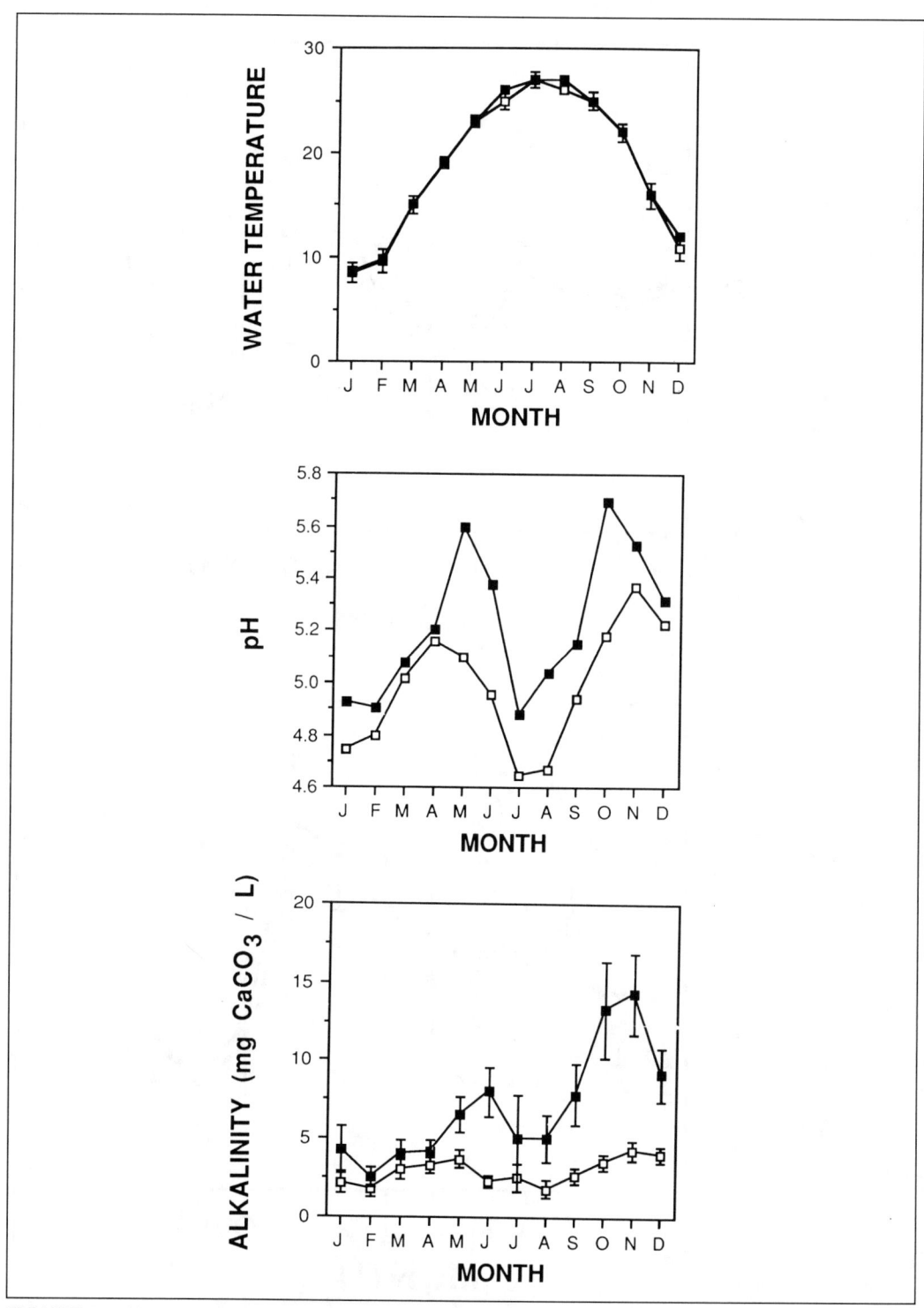

FIGURE 7. Mean (±SE) monthly water temperature, pH, and alkalinity in the Satilla River above the city of Waycross (open squares) and below the city of Waycross (closed squares), July 1973–December 1986.

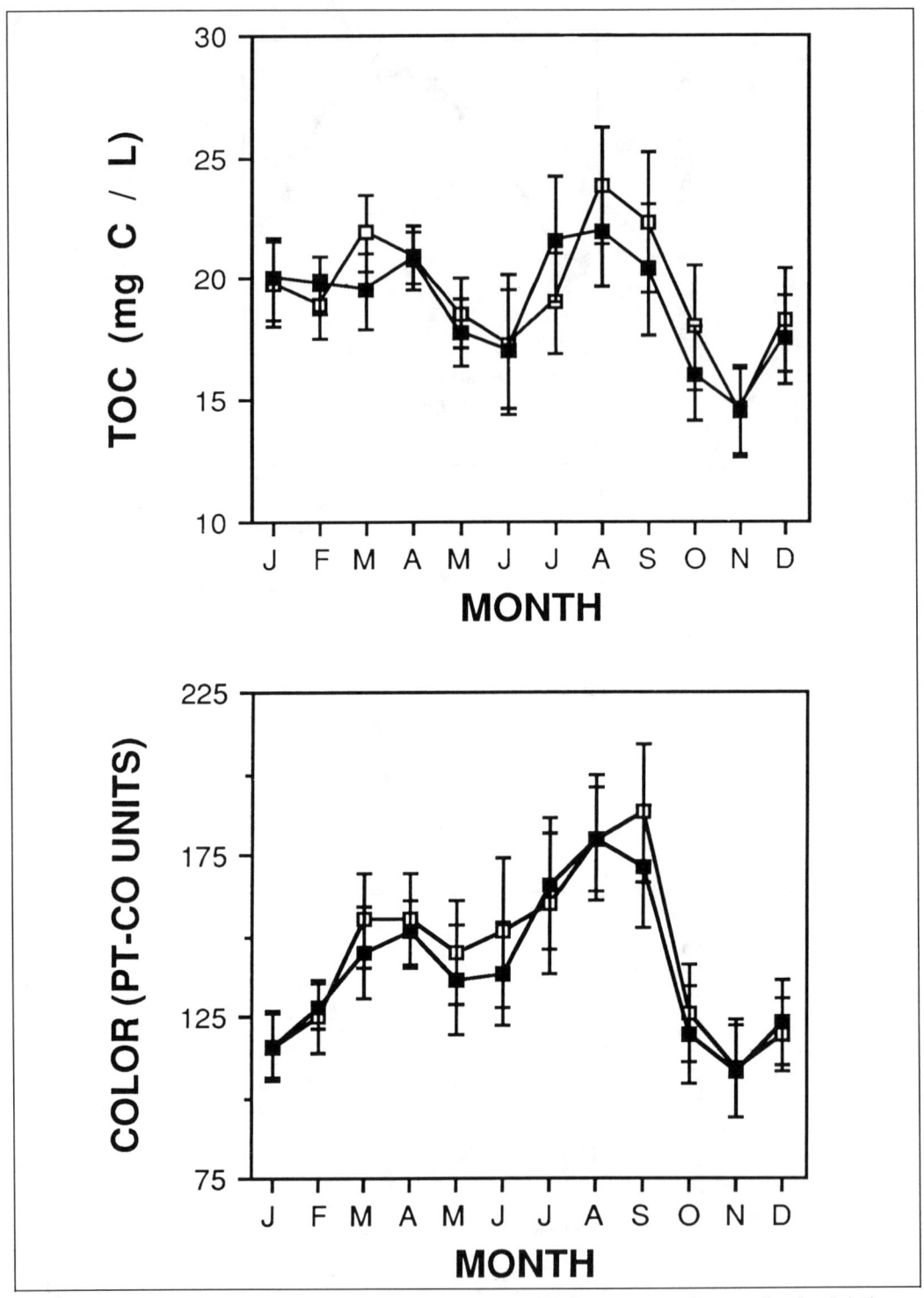

FIGURE 8. Mean (±SE) monthly total organic carbon (TOC) concentration, and color (platinum-cobalt units) in the Satilla River above the city of Waycross (open squares), and below the city of Waycross (closed squares), July 1973–December 1986.

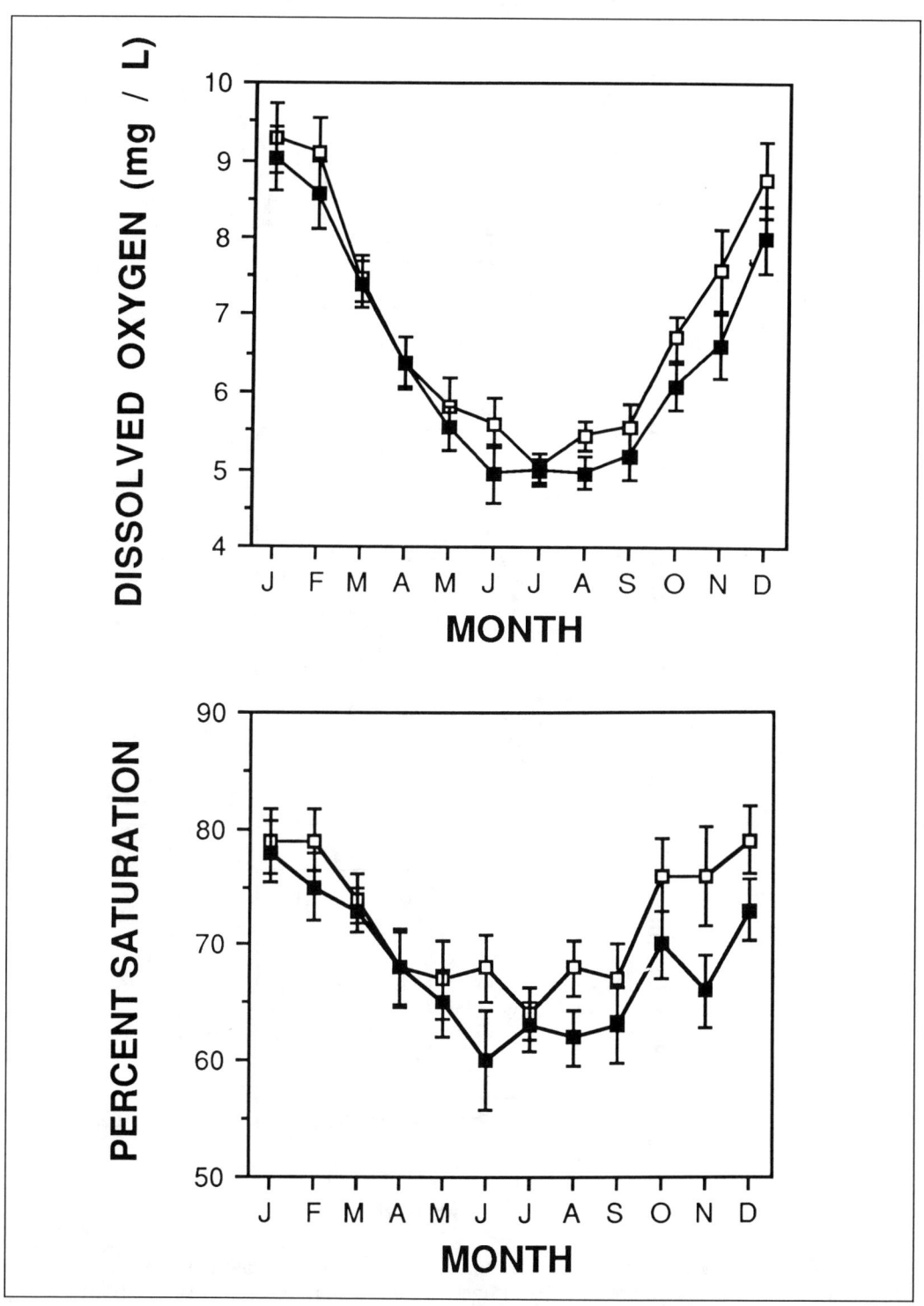

FIGURE 9. Mean (±SE) monthly dissolved oxygen concentration and percent oxygen saturation in the Satilla River above the city of Waycross (closed squares), July 1973–December 1986.

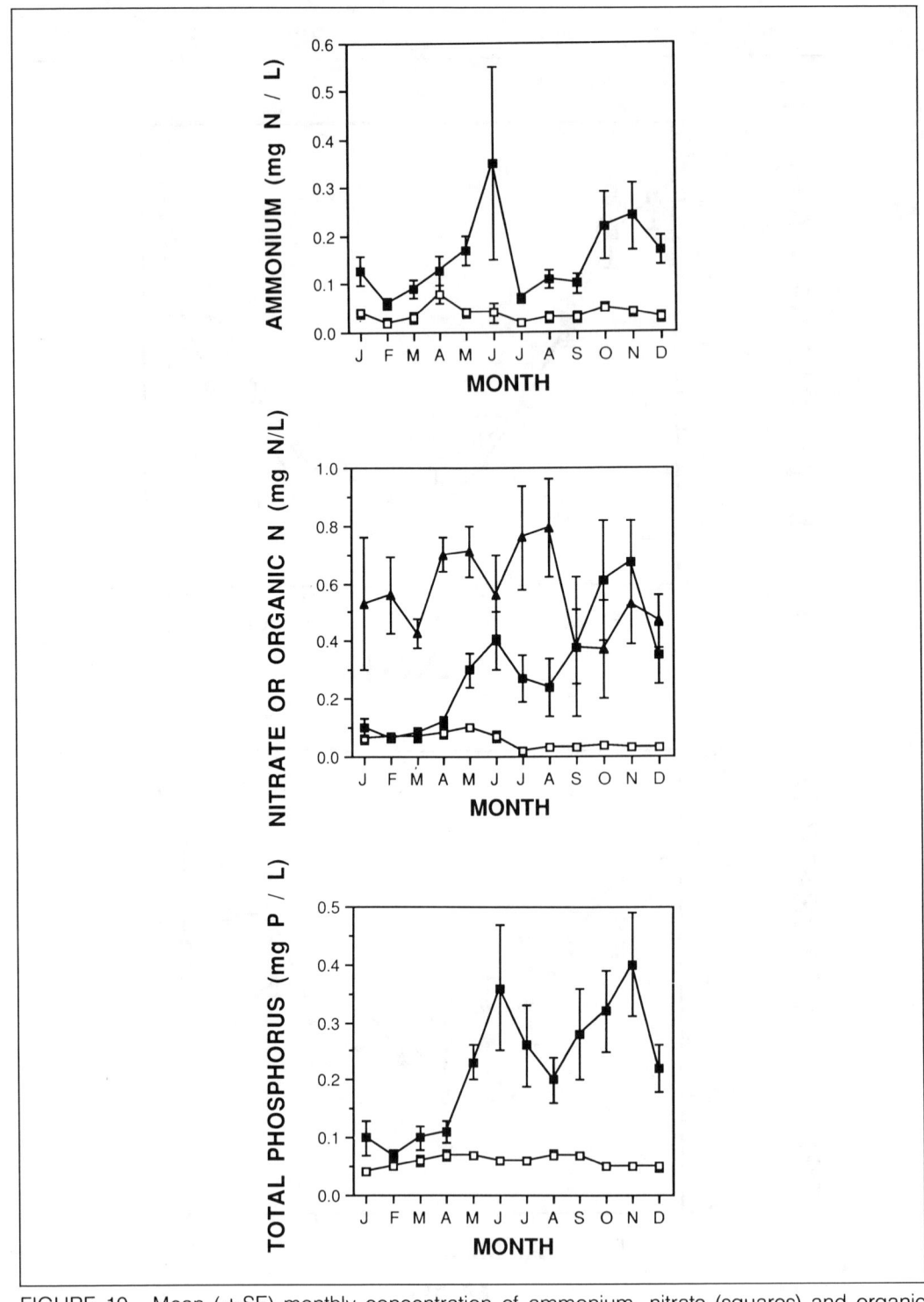

FIGURE 10. Mean (±SE) monthly concentration of ammonium, nitrate (squares) and organic nitrogen (triangles), and total phosphorus in unfiltered water from the Satilla River above the city of Waycross (open squares) and below the city of Waycross (solid symbols), July 1973–December 1986.

nuisance algal blooms (USEPA 1976). That algal blooms are generally not observed may be partly due to the limit imposed by restricted light on primary productivity in a darkly stained river.

Sediment and water samples for metal and organic contaminant analysis were collected during October 1978 and October 1981 from the downstream station on the Satilla River. Values for organic contaminants in sediments (aldrin, chlordane, DDT, dieldrin, endrin, toxophene, methoxychlor, and PCBs) were at the same levels cited for the Ogeechee River. Again, metal concentrations reported in Table 2 are difficult to interpret because the detection limits used are, in some cases, rather high.

Flow-corrected trend analysis of water quality values from 1974 to 1981 has been done with data collected at the NASQAN station at Atkinson (Figure 1), about 45 river km below the downstream station and 129 river km from the mouth of the Satilla River (Smith and Alexander 1983). Fewer significant trends were seen at this station than at the Ogeechee River station. The Satilla River showed a significant increase in conductivity and magnesium concentration (-3%/year) and a decrease in pH (-2%/hour) and fecal streptococci (66%/year) over this period (Smith and Alexander 1983).

Biota and Ecosystem Characteristics

Gross primary productivity in the Ogeechee River about 50 km downstream from the water quality station is moderate, with an annual average of 2.22 g $O_2 \cdot m^{-2} \cdot day^{-1}$ (Edwards and Meyer 1987). However, community respiration rates in this river are very high, with an annual average of 6.7 g $O_2 \cdot m^{-2} \cdot day^{-1}$ (Edwards and Meyer 1987). Because of the high respiration rates supported by inputs of organic matter from the floodplain swamps, the Ogeechee River as a whole is heterotrophic, with an annual average P/R of 0.25 (Edwards and Meyer 1987). Although most of the organic matter in the water column is DOC (Benke and Meyer 1988), the particulate organic

TABLE 2. Concentration of metals in unfiltered water and sediments collected on two dates from the Ogeechee River at Highway 24 and the Satilla River at Highway 15 and 121. Data are from Georgia Environmental Protection Division (GEPD 1979, 1981, 1982).

	Ogeechee River				Satilla River			
	October 1979		October 1982		October 1978		October 1981	
Metal	Water (μg/L)	Sediment (mg/kg)	Water (μg/L)	Sediment (mg/kg)	Water (μg/L)	Sediment (mg/kg)	Water (μg/L)	Sediment (mg/kg)
Arsenic	<25.0	<10.0	<10.0	<1.0	<50.0	13.9	<10.0	<0.5
Cadmium	<50.0	<5.0	<50.0	<5.0	<50.0	<29.0	<100.0	<5.0
Chromium	<50.0	<5.0	<50.0	<2.5	<50.0	44.0	<50.0	<2.5
Copper	<50.0	<5.0	<50.0	<2.5	<50.0	58.0	<50.0	3.0
Lead	<100.0	<10.0	<50.0	<5.0	<100.0	128.0	<100.0	5.0
Mercury	<0.2	<0.1	<0.2	<0.1	<0.2	<0.1	<0.2	<0.1
Zinc	70.0	5.0	<50.0	7.2	<50.0	255.0	<50.0	125.0

matter is an important food resource for the abundant, filter-feeding macroinvertebrates (Wallace et al. 1987). A significant fraction of the seston in the Ogeechee River is living: bacteria account for 31% (Edwards 1987), protozoa 4% (Carlough 1986), algae 6%, and drifting invertebrates 0.2% (Benke and Meyer 1988). The large bacterial populations are particularly striking (annual average, 1.5×10^{10} cells/L); they appear to come from the floodplain swamps rather than being produced in the river channel (Edwards and Meyer 1986).

Invertebrates are abundant, diverse, and productive in the Ogeechee River, as they are in most southeastern blackwater streams (Smock and Gilinsky, in press). Chrysomelid beetles are productive on the surfaces of water lilies (*Nuphar luteum*), which are found in backwater areas (Wallace and O'Hop 1985). Invertebrate production in the sandy sediments is dominated by oligochaetes (lumbriculids, tubificids, and *Barbadrilus paucisetus*), chironomids (*Dicrotendipes*, Tanypodinae, *Cryptochironomus*), ceratopogonids and the introduced Asiatic clam (*Corbicula fluminea*) (Stites 1986, 1987). The biomass and production of small benthic invertebrates such as these are similar to those in the sandy benthos of other streams (Stites 1986). Bacteria produced in the sediments may be an important source of food for aquatic insects; the productivity of bacteria and invertebrates in the sediments can be related to its organic matter content (Findlay et al. 1986; Stites 1986).

The snag habitat in the Ogeechee River is dominated by filtering and gathering collectors such as *Hydropsyche*, *Simulium*, and *Stenonema* (Wallace and Benke 1984; Benke and Meyer 1988). The snag-dwelling insects are highly productive in warm water streams (Benke and Jacobi 1986) because snags offer a stable substrate and the development time for many species is short.

The fish community of the Ogeechee River is also diverse (Mark Fisher, University of Georgia, personal communication). The most abundant piscivorous fishes in the river include the long-nosed gar *Lepisosteus osseus*, bowfin *Amia calva*, chain pickerel *Esox niger*, grass pickerel *Esox americanus*, and largemouth bass *Micropterus salmoides*. The species feeding primarily on snag dwelling insects are redbreast sunfish *Lepomis auritis*, warmouth *Lepomis gulosus*, flyer *Centrarchus macropterus*, and bluegill *Lepomis macrochirus*. Fishes feeding on insects in sandy benthic habitats include spotted sucker *Minytrema melanops*, brown bullhead *Ictalurus nebulosus*, and American eel *Anguilla rostrata*. The American shad *Alosa sapidissima* moves into the river in the spring and spawns on the floodplain in great numbers.

The invertebrate and fish communities of the Satilla River have also been extensively studied at a site close to the upstream water quality monitoring station and at the NASQAN station further downstream. Invertebrate secondary production rates are high in this river: 16.8 g dry weight/m^2 at the lower station and 33.5 g/m^2 at the upper station (Benke et al. 1984). High rates of secondary production appear to be a common feature of blackwater rivers in the Coastal Plain (Smock et al. 1985; Smock and Roeding 1986; Smock and Gilinsky, in press). The chironomid *Parakiefferiella* was the dominant species in the sandy benthos, whereas the muddy benthos contains primarily oligochaetes (*Limnodrilus*) and chironomids (Benke et al. 1984). The snag habitat had the greatest biomass, production, and diversity of benthic invertebrates. Filter-feeding caddisflies (*Hydropsyche*) and blackflies (*Simulium*) are the dominant consumers. Although the snag habitat accounts for only 6% of the area at these sites, over half of the invertebrate biomass and 16% of invertebrate production occur on them (Benke et al. 1984).

The snag habitat is particularly significant for the recreational fishery in the river. All fish species examined used snags as feeding sites to some extent. Five (spotted sunfish, redbreast sunfish, bluegill, warmouth, and pirate perch) of the eight major fish species obtained at least 50% of their prey biomass from snags (Benke et al. 1985). Probably the major significant change to affect invertebrate and fish productivity in this river has not been related to water quality but to the historical practice of removing snags to aid navigation (Wallace and Benke 1984; Benke et al. 1985).

Water Quality Control Efforts

The state of Georgia is authorized to establish water quality standards for the Ogeechee and Satilla rivers and for other Georgia rivers under the Georgia Water Quality Control Act of 1964 and its amendments. The state exercises its control on water quality by setting limits on the maximum amounts of various pollutants that can be discharged in effluents (GEPD 1987b, 1987c). The Department of Natural Resources follows the guidelines of the Federal Water Pollution Control Act Amendments of 1972 and issues NPDES permits specifying maximum allowable effluent concentrations. Each basin is divided into water quality management units to facilitate regulation. The major consumptive uses and wastewater discharges are catalogued in a basin-wide management report (e.g., GEPD 1987b). Waste load allocations for Georgia streams are developed with the aid of a computer model called GEORGIA DOSAG, which computes the dissolved oxygen sag below waste discharge points (GEPD 1987b). Water quality trends are monitored by the Water Protection Branch, Environmental Protection Division, Department of Natural Resources (e.g., GEPD 1974 and following years). This agency also inspects discharging facilities to ensure compliance with permitted discharges.

The water quality of a river is dependent not only on the materials discharged into it, but also on the consumptive use of water in the basin. Georgia has recently completed a water budget, which is published in the form of Water Availability and Use Reports for each river basin in the state (e.g., GEPD 1987a). These reports provide a comprehensive inventory of the state's water resources and their uses, and establish a basis for informed decisions on resource management. Under authorization of the Georgia Groundwater Use Act of 1972 and the 1977 Amendments to the Georgia Water Quality Control Act, the Department of Natural Resources now issues permits for surface and ground water withdrawal to non-agricultural users who require >100,000 gallons per day (GEPD 1987d). Users at the time the law was enacted were issued permits for their withdrawals at that time. The basis for permitting decisions for new or expanded withdrawals is to maintain the state's water quality standards at 7Q10 flow (GEPD 1987d).

Future Prospects

The surface water resources of the Ogeechee and Satilla river basins are not heavily used because of their naturally low quality (high color, low pH) and their very low flow during periods of limited rainfall. In contrast, ground water resources are less expensive to develop, of higher quality, and readily available (GEPD 1987a). Unlike other areas of the southeastern United States, urbanization is not expected to lead to major water quality problems in the near future (GEPD 1987b, 1987c). The extensive use of ground water by pulp processors, paper mills, and other industrial and municipal users near the coast in the Satilla basin has led to a decline in the potentiometric surfaces of the aquifer and intrusion of brackish water (GEPD 1987d). Extensive removal of ground water will remain a major water resource problem in the future unless corrective action is taken.

Because of their naturally low oxygen conditions at some times of the year, the Ogeechee and Satilla rivers are vulnerable to organic loading from anthropogenic sources. The North Newport River, a small blackwater river just south the Ogeechee River, receives wastewater from a paper mill. This site has had frequent problems with low DO and high sulfate concentrations (e.g., GEPD 1974 and following years). Because of their naturally high demand on oxygen, blackwater rivers probably cannot tolerate as much additional organic matter as clearwater rivers.

The current permitting system for water withdrawals in the state exempts agricultural users. When the permitting legislation was passed in 1972, agriculture accounted for less than 5% of water use in the state. With the adoption of center pivot irrigation systems, agricultural use of water has increased to greater than 33% of the total water withdrawals (GEPD 1987d). Legislation passed in 1982 required farmers to report estimated water usage; however, farmers have largely ignored this bill (GEPD 1987d). The 1986 drought made many farmers recognize the advantages in being allocated a quantity of water and, hence, amendments to the legislation that will require agricultural users to obtain permits are currently being considered (GEPD 1987d).

Permits for agricultural users is important if the quality of water in the Ogeechee and Satilla river basins is to be maintained. Current estimates of non-permitted withdrawal of surface water in the Ogeechee River are 27 times that of permitted surface water withdrawal. There is no permitted withdrawal of surface water in the Satilla River, but non-permitted withdrawal is an estimated 7,600 m^3/d (GEPD 1987b, 1987c). Values for use of ground water are similar. Clearly, agriculture uses a significant fraction of the water resources in these basins. It is critical that agriculture be included in the permitting process.

Agricultural activity in the watershed also has an impact on water quality due to runoff of nutrients and pesticides, soil erosion, and increased channelization of tributaries in order to drain the land. In South Carolina blackwater streams, fertilizers and lime are the major sources of cations such as calcium, magnesium, and bicarbonate (Gardner 1983). Nitrogen and phosphorus concentrations in blackwater streams increase directly with the percent of the basin in agriculture (Smock and

Gilinsky, in press). Agricultural activity in the Satilla River watershed led to increased sedimentation and higher concentrations of nitrate and fecal coliform in several small streams (GEPD 1985). Adverse effects from agriculture can be minimized if farmers follow best-management practices.

The extensive floodplains of the Ogeechee and Satilla rivers are also important in maintaining water quality. Floodplains offer a large area for interaction between elements in the water column and biologically active sediments; hence, they serve as a natural sink for pollutants from non-point sources (Wharton and Brinson 1979). Large fractions of water-borne nitrogen, phosphorus, and calcium (82%, 54%, and 42%, respectively) were removed by the riparian forest in one agricultural watershed on the Coastal Plain (Todd et al. 1983). The role of floodplains in altering water quality in these blackwater rivers becomes obvious when floodplain exchange is limited by channelization. In North Carolina blackwater streams, channelization has increased the concentrations of nitrate, total phosphorus, and specific conductance (Kuenzler et al. 1977; Yarbro et al. 1984).

Channelization of Coastal Plain streams to drain swamps for conversion to agriculture, silviculture, or extensive timber harvesting in the riparian zone is another threat to water quality in these basins. Specific steps need to be taken to preserve riparian ecosystems. For example, riparian zones in Georgia are taxed at the same rate as productive agricultural land. If they were taxed at a lower level when left undisturbed, the owner would have less incentive to exploit marginal lands (Todd et al. 1983). In blackwater rivers of the Coastal Plain, the floodplain is an essential component of the riverine ecosystem. It is critical that the river floodplain corridor be considered as a single management unit.

Acknowledgments

This research was supported by National Science Foundation grant BSR-8406631. I am particularly grateful to Jack Dozier, Ed Hall, Ted Mikalsen, and Jim Sommerville, Water Protection Branch, Georgia Environmental Protection Division, Department of Natural Resources, for providing water quality data and other information. I thank Dale Becker, Gail Cowie, Rick Edwards, Ed Hall, Kathy Hatcher, Jim Kundell, Alec Little, and Duane Neitzel, who provided useful comments on an earlier draft of the manuscript.

References

Beck, K. C., J. H. Reuter, and E. M. Perdue. 1974. Organic and inorganic geochemistry of some coastal plain rivers of the southeastern U. S. Geochimica Cosmochimica Acta 38:341–364.

Belander, T. V., F. E. Dierberg, and J. Roberts. 1985. Dissolved oxygen concentrations in Florida's humic-colored waters and water quality standard implications. Florida Scientist 48:107–119.

Benke, A. C., R. L. Henry III, D. M. Gillespie, and R. J. Hunter. 1985. Importance of snag habitat for animal production in southeastern streams. Fisheries 10:8–13.

Benke, A. C., and D. K. Jacobi. 1986. Growth rates of mayflies in a subtropical river and their implications for secondary production. Journal of the North American Benthological Society 5:107–114.

Benke, A. C., and J. L. Meyer. 1988. Structure and function of a blackwater river in the southeastern U. S. A. International Association of Theoretical and Applied Limnology, Proceedings 23:1209–1218.

Benke, A. C., T. C. Van Arsdall, Jr., and D. M. Gillelspie. 1984. Invertebrate productivity in a subtropical blackwater river: the importance of habitat and life history. Ecological Monographs 54:25–63.

Bergeaux, P. J. 1969. Soils in Georgia. Cooperative Extension Service Bulletin 662. University of Georgia, College of Agriculture, Athens.

Carlough, L. A. 1986. Protozoa in two southeastern blackwater rivers and their importance to trophic transfer. Master's thesis. University of Georgia, Athens.

Cuffney, T. F. 1984. Characteristics of riparian flooding and its impact upon the processing and exchange of organic matter in Coastal Plain streams of Georgia. Doctoral dissertation. University of Georgia, Athens.

Dekle, B. 1977. The Ogeechee River: A Poem. Ogeechee Travels Unlimited, Inc. Statesboro, GA.

Edwards, R. T. 1987. Sestonic bacteria as a food source for filtering invertebrates in two southeastern blackwater rivers. Limnology and Oceanography 32:221–234.

Edwards, R. T., and J. L. Meyer. 1986. Production and turnover of planktonic bacteria in two southeastern blackwater rivers. Applied and Environmental Microbiology 52:1317–1323.

Edwards, R. T., and J. L. Meyer. 1987. Metabolism of a subtropical low gradient blackwater river. Freshwater Biology 17:251–263.

Elder, J. F., and D. J. Cairns. 1982. Production and decomposition of forest litterfall on the Appalachicola River floodplain, Florida. Report 82-252. U. S. Geological Survey, Tallahassee, Florida.

Findlay, S., J. L. Meyer, and R. Risley. 1986. Benthic bacterial biomass and production in two blackwater rivers. Canadian Journal of Fisheries and Aquatic Sciences 43:1271–1276.

Gardner, L. R. 1983. Element mass balances for South Carolina Coastal Plain watersheds. Pages 263–279 *in* R. Lowrance, R. Todd, L. Asmussen, and R. Leonard, editors. Nutrient cycling in agricultural ecosystems. Special Publication 23. College of Agriculture Experiment Station, University of Georgia, Athens.

GEPD (Georgia Environmental Protection Division). 1974. Water quality monitoring data for Georgia streams 1973. Georgia Department of Natural Resources, Atlanta.

GEPD (Georgia Environmental Protection Division). 1979. Water quality monitoring data for Georgia streams 1978–1979. Georgia Department of Natural Resources, Atlanta.

GEPD (Georgia Environmental Protection Division). 1981. Water quality monitoring data for Georgia Streams 1981. Georgia Department of Natural Resources, Atlanta.

GEPD (Georgia Environmental Protection Division). 1982. Water quality monitoring data for Georgia Streams 1982. Georgia Department of Natural Resources, Atlanta.

GEPD (Georgia Environmental Protection Division). 1985. The Georgia non-point source impact assessment study, project summary. Georgia Department of Natural Resources, Atlanta.

GEPD (Georgia Environmental Protection Division). 1987a. Water availability and use report. Coastal plain river basins. Georgia Department of Natural Resources, Atlanta.

GEPD (Georgia Environmental Protection Division). 1987b. Ogeechee River basin. Water quality management plan. 2nd edition. Updated January 1987. Georgia Department of Natural Resources, Atlanta.

GEPD (Georgia Environmental Protection Division). 1987c. Satilla River Basin. Water quality management plan. 2nd edition. Updated January 1987. Georgia Department of Natural Resources, Atlanta.

GEPD (Georgia Environmental Protection Division). 1987d. State of Georgia water resources management strategy: A summary document. Georgia Department of Natural Resources, Atlanta.

Hodler, T. W., and H. A. Shretter. 1986. The atlas of Georgia. University of Georgia, Institute of Community and Area Development, Athens.

Kuenzler, E. J., P. J. Mulholland, L. A. Ruley, and R. P. Sniffen. 1977. Water quality in North Carolina Coastal Plain streams and effects of channelization. Report 127. Water Resources Research Institute, University of South Carolina, Raleigh.

Leigh, J. 1986. The Ogeechee: a river and its people. University of Georgia Press, Athens.

Meyer, J. L. 1986. Dissolved organic carbon dynamics in two subtropical blackwater rivers. Archive für Hydrobiologie 108:119–134.

Meyer, J. L. 1988. Benthic bacterial biomass and production in a blackwater river. International Association of Theoretical and Applied Limnology, Proceedings 23:1832–1838.

Meyer, J. L., R. T. Edwards, and R. Risley. 1987. Bacterial growth on dissolved organic carbon from a blackwater river. Microbial Ecology 13:13–29.

Mullholland, P. 1981. Organic carbon budget of a swamp-stream ecosystem. Ecological Monographs 51:307–322.

Pernick, M., and P. J. W. Roberts. 1985. Mixing in a river-floodplain system. Report SCEGIT 85-107. Georgia Institute of Technology, Atlanta.

Smith, R. A., and R. B. Alexander. 1983. A statistical summary of data from the U. S. Geological Survey's national water quality networks. Open-file report 83-533. U. S. Geological Survey, Reston, Virginia.

Smith, R. A., R. B. Alexander, and M. G. Wolman. 1987. Water quality trends in the nation's rivers. Science 235:1607–16156.

Smock, L. A., and E. Gilinsky. In press. Coastal blackwater streams. *In* W. H. Martin, editor. Biotic communities of the southeastern United States, Volume 2. Aquatic communities. Ecological Society of America, Southeastern Chapter.

Smock, L. A., E. Gilinsky, and D. L. Stoneburner. 1985. Macroinvertebrate production in a southeastern U. S. blackwater stream. Ecology 66:1491–1503.

Smock, L. A., and C. E. Roeding. 1986. The trophic basis of production of the macroinvertebrate community of a southeastern U. S. A. blackwater stream. Holarctic Ecology 9:165–174.

Sparks, Andrew. 1962. Ogeechee. Pages 13–23 *in* G. Hatcher, editor. Georgia rivers. University of Georgia Press, Athens.

Stites, D. L. 1986. Secondary production and productivity in the sediments of blackwater rivers. Doctoral dissertation, Emory University, Atlanta.

Stites, D. L. 1987. Population and production dynamics of an enchytraeid worm in a subtropical blackwater river. Canadian Journal of Fisheries and Aquatic Sciences 44:1469–1474.

St. John, T. V., and A. B. Anderson. 1982. A re-examination of plant phenolics as a source of tropical blackwater rivers. Tropical Ecology 23:151–154.

Todd, R., et al. (six coauthors). 1983. Riparian vegetation as filters of nutrients exported from a Coastal Plain agricultural watershed. Pages 485–493 *in* R. Lowrance, R. Todd, L. Asmussen, and R. Leonard, editors. Nutrient cycling in agricultural ecosystems. Special Publication 23. College of Agriculture Experiment Station, University of Georgia, Athens.

USEPA (U. S. Environmental Protection Agency). 1976. Quality Criteria for Water. Report. U. S. Environmental Protection Agency, Washington, DC.

USGS (U. S. Geological Survey). 1973–1985. Water resources data. Georgia. Reports GA-73-1 to GA-85-1. National Technical Information Service, Springfield, Virginia.

USNPA (U. S. National Park Service). 1984. Ogeechee River, Georgia. Final report, Wild and Scenic River Study. Southeast Regional Office, National Park Service, Atlanta.

Wallace, J. B., and A. C. Benke. 1984. Quantification of wood habitat in subtropical Coastal Plain streams. Canadian Journal of Fisheries and Aquatic Sciences 41:1643–1652.

Wallace, J. B., A. C. Benke, A. H. Lingle, and K. Parsons. 1987. Trophic pathways of macroinvertebrate primary consumers in subtropical blackwater systems. Archive für Hydrobiologie Supplement band 74:423–451.

Wallace, J. B., and J. O'Hop. 1985. Life on a fast pad: population dynamics, production and impact of the water lily leaf beetle on water lilies. Ecology 66:1534–1544.

Wharton, C. H., and M. M. Brinson. 1979. Characteristics of southeastern river systems. Pages 32–40 in R. R. Johnson and J. F. McCormick, editors. Strategies for protection and management of floodplain wetlands and other riparian ecosystems. U. S. Forest Service General Technical Report WO-12. National Technical Information Service, Springfield, Virginia.

Yarbro, L. A., E. J. Kuenzler, P. J. Mulholland, and R. P. Sniffin. 1984. Effects of stream channelization on exports of nitrogen and phosphorus from North Carolina Coastal Plain watersheds. Environmental Management 8:151–160.

Water Quality and Biological Communities of the Mobile River Drainage, Eastern Gulf of Mexico Region

AMELIA K. WARD AND G. MILTON WARD
Aquatic Biology Program, Department of Biology
University of Alabama, Tuscaloosa, Alabama 35487, USA

STEVEN C. HARRIS
Biological Resources Division, Geological Survey of Alabama
Tuscaloosa, Alabama 35486, USA

ABSTRACT. *The Mobile River is the largest river system draining to the Gulf of Mexico east of the Mississippi River. It has a drainage area of 110,000 km^2 and the average discharge at its mouth is 1,666 m^3/s. The Mobile River basin drains most of Alabama and portions of Mississippi and Georgia, and it includes six major rivers: the Alabama, Tallapoosa, Coosa, Cahaba, Black Warrior, and Tombigbee. It is distinguished from other river systems in North America by its geologic diversity, which has contributed to the biologic richness of its rivers and streams. Four major physiographic provinces are included in the Mobile River basin: the Appalachian Plateau, Valley and Ridge, Piedmont, and Coastal Plain provinces. The distinctive underlying parent material associated with each of these regions has affected water chemistry as well as the distribution and abundance of aquatic organisms. Dredging, channelization, and removal of snags in the large rivers of the Mobile River basin have occurred since the 1800s. Construction of impoundments began in the early 1900s and currently all of the major rivers, with the exception of the Cahaba River, are modified into a series of reservoirs. Many aquatic species once associated with the main stems are now restricted to smaller tributaries. Water quality changes associated with agricultural practices, strip-mining, and urbanization have occurred in many rivers and streams. However, the region has numerous small rivers and streams in forested watersheds, which still reflect the chemical and biological characters conferred on them by geological influences.*

The coastal region of the eastern Gulf of Mexico, which extends from western Georgia to central Mississippi, encompasses an area drained by several large river basins, bounded on the east by the Suwannee River and on the west by the Pearl River (Figure 1). The Mobile River basin, located about mid-way between the Suwanee and Pearl rivers, is the largest river system in the region with an annual discharge of 1,666 m^3s^{-1} (Morisawa 1968). The basin includes over 186,000 km of rivers and streams, the majority of which are located in Alabama, although portions extend into northwestern Georgia and eastern Mississippi.

Geological events strongly shaped the pattern of drainage of the Mobile River system and created a diversity of geological regions (Figure 2), including the Appalachian Plateau, Valley and Ridge, Piedmont, and Coastal Plain (Lineback 1973). The geological diversity, in turn, played a major role in creating aquatic environments with distinctive water chemistries and biological communities (e.g., Lay and Ward 1987).

The extensive river systems have been an important part of the life of the region since prehistoric times. Early Indian inhabitants used the rivers for fishing and transportation and the rich floodplains for farming. One of the earliest European explorers, Hernando do Soto, followed the drainages of the Coosa and Alabama rivers into the area as did many of the later immigrants from the Atlantic states (Lineback 1973). The aboriginal Indian populations were virtually eliminated by the early 1800s at about the same time that major alterations of the large river systems began.

During the last 100 years, the character of the large rivers in the Mobile River basin changed dramatically with the construction of dams for hydroelectric power, dredging and channelization for transportation, and the growth of metropolitan areas

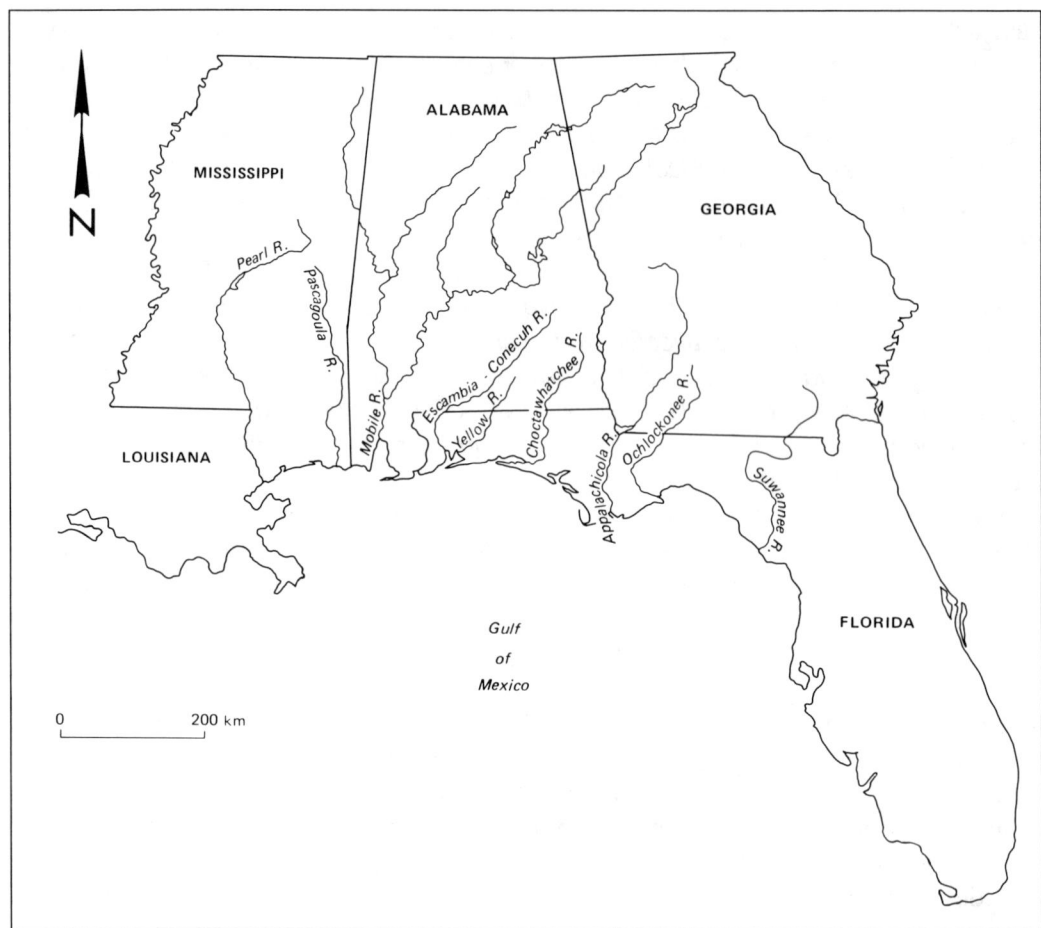

FIGURE 1. Major river basins of the eastern Gulf coast region with areas greater than 2,500 km².

as the population shifted from a primarily rural to urban one during the 20th century. Also, the acceleration of land use for silviculture and coal mining affected water quality of some small rivers and streams. For example, about 70% of Alabama is forested, and clear-cutting and other forest management practices are part of the economic base of the state. Lignite and bituminous coal mining occurs in some areas of the Mobile River basin in Mississippi and Georgia. However, the majority of coal mining associated with this drainage occurs in Alabama in the Appalachian Plateau physiographic province. About 4% of the total land has been used for bituminous coal mining since 1930. Most of this land is in the Warrior Coal Field (north central Alabama), which lies under about 20% of the surface area in the state (Tolson 1984). Strip mining often results in acidification of streams and increases in metals and other nutrients in watersheds where mines occur (Harris 1987).

The water quality and biological communities of rivers and streams in the Mobile River basin have clearly been shaped by a variety of geological and human induced events. Unperturbed rivers and streams in different physiographic provinces have different water chemistries as the result of the underlying parent material in

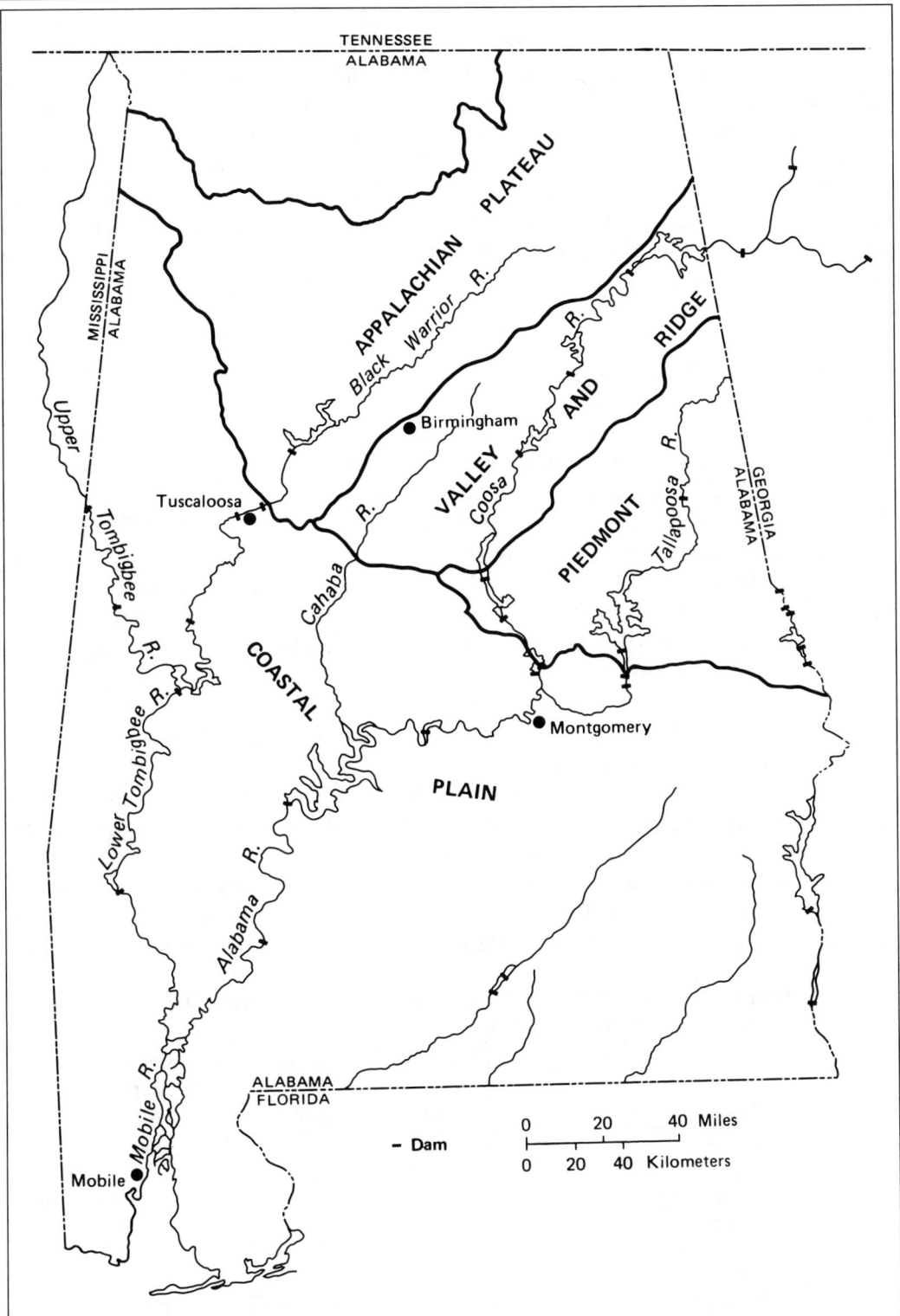

FIGURE 2. Major physiographic provinces in Alabama (Appalachian Plateau, Valley and Ridge, Piedmont, and Coastal Plain) and major subbasins of the Mobile River drainage.

those regions. Water characteristics become more diffuse in large rivers whose tributaries traverse several different geological regions. Almost all the large rivers in the Mobile River basin have been physically altered to accommodate river transportation and chemically modified by agricultural and industrial influences.

In this paper, we define the geological processes that shaped the watersheds and influenced water chemistry of rivers in the Mobile River basin, compare water chemistries of streams in relatively undisturbed watersheds with those affected by human influences, and discuss the historical perspective and current status of water quality and biological communities in some of the large rivers.

Eastern Gulf of Mexico Region

Within the eastern Gulf Region are nine major and many smaller river systems that empty into the northeastern portion of the Gulf of Mexico (Figure 1). These rivers drain almost all of Alabama, and much of Mississippi, Georgia, and northern Florida. The nine large watersheds each have an area greater than 2,500 km^2 (Table 1).

The two largest drainage basins in the eastern Gulf region are the Appalachicola and the Mobile rivers. The Appalachicola River is formed by the confluence of the Flint and the Chattahoochee rivers at Lake Seminole, on the Florida-Georgia border. The Chattahoochee River heads in northern Georgia and drains portions of eastern Alabama as well. The Flint River heads in central Georgia. South of Lake Seminole, the Appalachicola River meanders through 130 km of northwest Florida that is relatively undisturbed. Along its course the Appalachicola River forms an extensive floodplain forest of 450 km^2, portions of which may be up to 10 km wide. Near the river mouth, the Appalachicola River receives another major tributary, the Chipola River, before emptying into Appalachicola Bay, one of the most productive commercial shellfish areas in the United States (Elder and Cairns 1982).

The Mobile River basin is the largest river system in terms of basin area and discharge to the eastern Gulf of Mexico. The Mobile River is the fifth largest in terms of discharge in North America (Morisawa 1968). Formed by the confluence of

TABLE 1. Drainage basins in the eastern Gulf region greater than 2,500 km^2. (U. S. Geological Survey, Surface Water Records.)

Basin	Main stem length (km)	Basin area (km^2)	Average discharge (m^3/s)
Suwannee	386	25,000	295
Ochlockonee River	275	5,700	50
Appalachicola River	700+	50,700	873
Choctawatchee River	280	11,900	198
Yellow River	165	2,800	63
Escambia River	368	10,600	177
Mobile River	1033	110,000	1,166
Pascagoula River	359	25,000	344
Pearl River	718	16,800	277

the Alabama and the Tombigbee rivers about 125 km north of Mobile, Alabama, it drains portions of northwest Georgia, much of the eastern, northern, and western parts of Alabama, as well as northeastern Mississippi. The Mobile River system consists of 6 large tributaries: Alabama, Coosa, Tallapoosa, Cahaba, Tombigbee, and Black Warrior rivers (Figure 2, Table 2).

With the exception of the Appalachicola and Mobile rivers, most rivers draining to the eastern Gulf of Mexico arise in Coastal Plain sediments. Several of these, notably the Suwannee, Blackwater (northwest Florida), and the Escatawpa (Alabama-Mississippi) rivers may have similarities with blackwater rivers along the South Atlantic coast. However, the occurrence of blackwater systems is not universal among streams arising in the Coastal Plain. Some streams that originate from limestone springs or in upland Coastal Plain sediments do not have the same floodplain interactions, resulting in dark waters of high dissolved organic carbon content, as do other streams in this region.

River Basins of the Mobile River Drainage

Two major tributaries, the Alabama and the lower Tombigbee, form the Mobile River system. Each major tributary, in turn, begins with the confluence of two large rivers further upstream (Figure 2). The Alabama River, formed by the confluence of the Tallapoosa and Coosa rivers, flows through about 500 km of Coastal Plain sediments to its juncture with the Mobile River. The Coosa River drains the Valley and Ridge Province of northeastern Alabama and northwestern Georgia, while the Tallapoosa drains the Piedmont of eastern Alabama and western Georgia. Another major tributary, the Cahaba River, drains the central and southwestern portions of the Alabama Valley and Ridge province, plus a portion of the upper Coastal Plain, and enters the main-stem Alabama River near Selma, Alabama. The lower Tombigbee River is formed from the Black Warrior River and the upper Tombigbee River. The upper Tombigbee River drains the Coastal Plain of eastern Mississippi and western Alabama, while the Black Warrior River drains much of Alabama's Appalachian Plateau region.

TABLE 2. Major subbasins in the Mobile River drainage. (U. S. Geological Survey, Surface Water Records.)

Basin, subbasin	Physiographic province	Main stem length (km)	Basin area (km²)	Average discharge (m³/s)
Tombigbee			51,281	862
Black Warrior	Appalachian Plateau	278	16,066	287
Upper Tombigbee	Coastal Plain	428	23,288	372
Lower Tombigbee	Coastal Plain	280	11,925	
Alabama			58,240	1,018
Coosa	Valley and Ridge	457	26,401	460
Tallapoosa	Piedmont	248	11,893	136
Cahaba	Valley and Ridge	312	4,792	81

Regional Geology

The geological template, which has affected the surface topography, mineral deposition, and water chemistry characteristics of this region, is the result of complex geologic and geomorphic processes, including landmass erosion and sedimentation, volcanism, and orogeny, which have occurred over millions of years (Lineback 1973). The end result was the four major physiographic provinces: Coastal Plain, Appalachian Plateau, Valley and Ridge, and Piedmont (Figure 2). Compared with other river basins of the eastern Gulf Coast, the Mobile River basin has the greatest diversity of geology and physiography.

The Piedmont region corresponds to the distribution of crystalline rocks and includes the oldest rock formations (200 to greater than 500 million years old) in the Mobile River drainage basin. The Piedmont is underlain with igneous, crystalline, and metamorphosed sedimentary rock of Archean and Proterozoic age. The metamorphic rocks consist of slates, schists, and a variety of gneisses (Adams et al. 1926). Piedmont streams are also relatively dilute with respect to nutrients, although weathering of aluminosilicate and silicate minerals in the rocks releases cations such as K, Na, and Ca. Alkalinity values are generally higher than in streams of the Coastal Plain or Appalachian Plateau, but substantially less than in streams in the Valley and Ridge.

The Appalachian Plateau covers a large part of the region and is described as a submaturely to maturely dissected sandstone and shale plateau of moderate relief (200 to 300 m above sea level). Rocks of the Appalachian Plateau consist of largely alternating beds of Pennsylvanian age sandstones, conglomerates, shales, and siltstones of the Pottsville Formation (Lineback 1973). The pH values of undisturbed streams in most of the regions, including the Appalachian Plateau, are normally between about 6.0 and 7.5. However, the low alkalinity and generally low ionic character of stream waters in the Appalachian Plateau make them particularly susceptible to changes in pH caused by inputs from mining operations, which are also common this region.

The Valley and Ridge Province of the southern Appalachian Mountains lies west of the Piedmont and is a northeast-southwest trending series of parallel ridges and valleys extending from the Coastal Plain to the Middle Atlantic states. The ridges are topped with relatively resistant sandstone, shale, and chert, whereas limestone and dolomite formations are exposed in the valleys. Due to the solution of limestone and dolomite, carbonate alkalinity is greater in streams draining this region than in other regions.

The Coastal Plain covers more than half the basin area, and sediments of various streams exhibit considerable diversity. In general, the area is characterized by gently rolling topography from sea level to 100 m in elevation. The unconsolidated sediments are relatively young (up to 125 million years old) and consist primarily of sands, gravels, and clays. However, the existence of older, more resistant subareas breaks the homogeneity of the region, creating series of low hills and valleys and some geological diversity even within the Coastal Plain province (Lineback 1973).

Climate and Hydrology

In general, the Mobile River basin region has a humid, subtropical climate with abundant rainfall. The average annual air temperature is 18°C, ranging from 16°C in the northern portions to 20°C in the south. As a result, mean annual surface water temperatures for smaller streams range from 14–16°C in the north to more than 17°C in the south. Seasonal variations in surface water temperature range from lows of 5–7°C in the north to 7–11°C in the south, and normal highs from 25–28°C in the south to 21–25°C in the north (German and Moffett 1983).

Temperature differences among the larger rivers are less pronounced. A comparison of monthly temperature profiles between the Coosa River (northeastern Alabama) and the lower Tombigbee River (Coastal Plain) shows similar patterns (Figure 3). The average summer maxima for the Coosa and the lower Tombigbee rivers are 30°C and 31°C, respectively. Greater differences in temperature occur in winter with average winter minima of about 5°C in the Coosa River compared to 8–9°C in the lower Tombigbee (USGS 1982, 1986).

Average annual precipitation for the Mobile River basin is 137 cm, varying from 132 cm in the north to 152 cm in the lower quarter of the basin, where the Gulf of Mexico has greatest influence. For most of the basin, precipitation is entirely

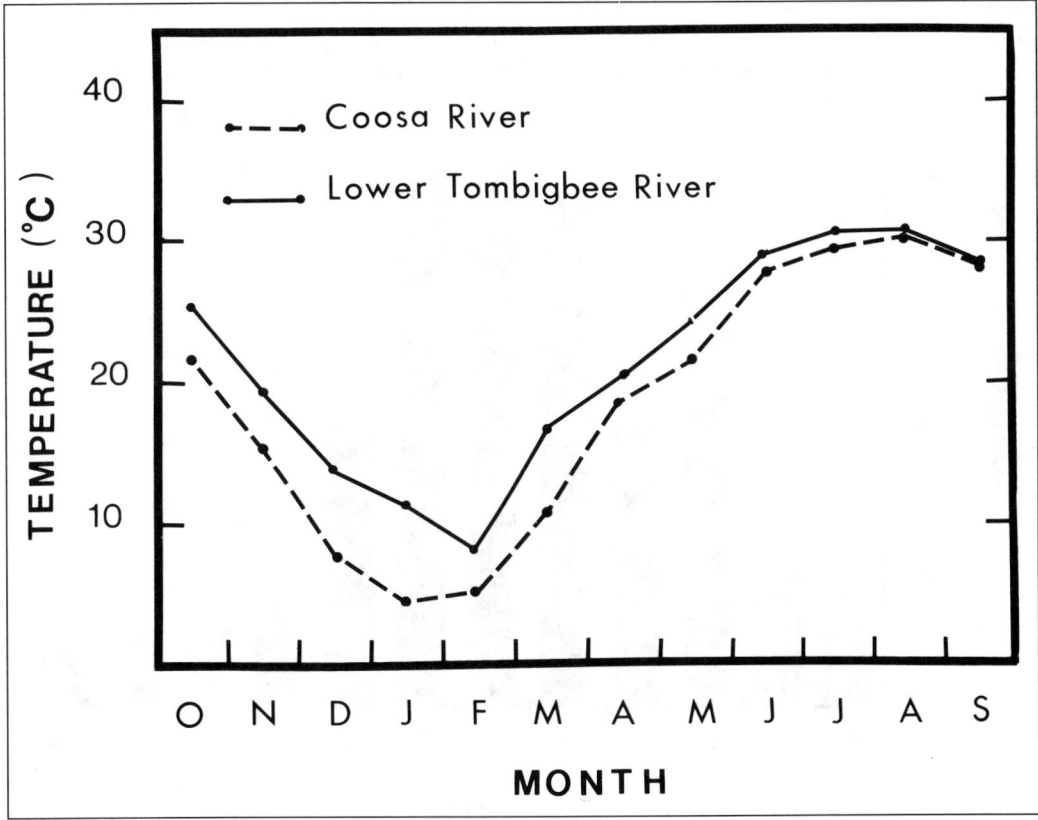

FIGURE 3. Monthly temperature averages for the Coosa (Piedmont) and lower Tombigbee (Coastal Plain) rivers.

rainfall, although the more northerly portions may receive some snowfall each year. There are no demonstrably wet and dry seasons, although late winter and spring tend to have heavier rainfall and September and October tend to be driest (USDA 1977).

Corresponding to precipitation pattern, stream flows also exhibit seasonal variations with the highest flow occurring from February to April and the lowest between August and October. Figure 4 illustrates the annual hydrograph for the Cahaba River at Centerville, Alabama, a fifth to sixth order basin (2635 km^2) draining parts of central Alabama (U. S. Geological Survey surface water records). Streams in all portions of the Mobile River basin exhibit this same basic pattern, although physiographic and geologic differences among sub-basins create differences in both peak and low flows.

One basin characteristic that influences runoff is the underlying geologic structure. The sandstone and shale (Pottsville Formation) geology of the Appalachian Plateau of Alabama has relatively less water-bearing capacities than other areas, resulting in more rapid runoff as well as rapid reductions in flow after the rainy season. Rapid runoff during the winter, particularly during storm events, produces higher peak flows in the Appalachian Plateau than in other portions of the Mobile River basin. However, because water storage capacities of the rock and soil are poor, many third to fourth order streams have very low discharge by late summer or early fall, and even streams in relatively large watersheds (25–40 km^2) may go dry.

In contrast, the geological structure of much of the Coastal Plain (unconsolidated clay, sand, and gravel) has excellent water storage capacity and streams do not normally

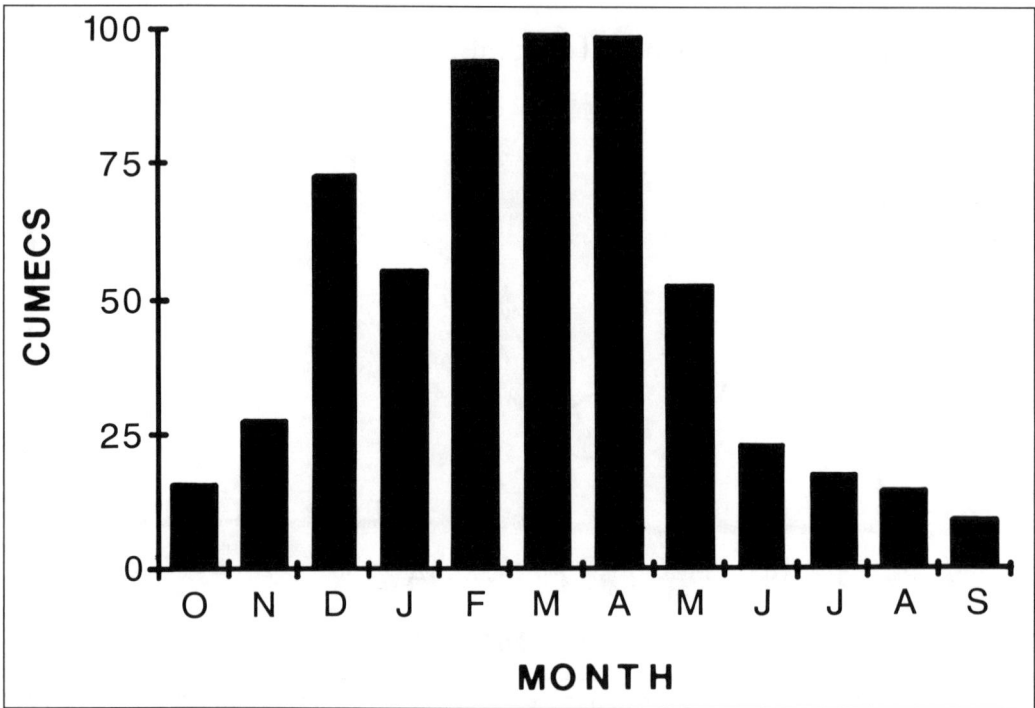

FIGURE 4. Average monthly discharge for the Cahaba River 1980–1985, at Centerville, Alabama.

exhibit extremely low flows. An exception is the chalk formations underlying the rich soils of the Black Belt region in west central Alabama. This type of geology has very low infiltration rates and water storage capacity; therefore, flows in most small streams of the Black Belt region are intermittent.

Prehistoric Perspectives

Although few data on water quality or other streams aspects were collected in the Mobile River basin prior to 1900, archaeological sites along the rivers provide clues to their importance to early human settlements and information on the productivity and biological communities associated with the rivers and their floodplains. Major river systems of the Mobile River basin have long served as a focus for human settlements and food resources. For example, as early as 2500 BC, small Indian camps were sited along the central and upper Tombigbee River drainage (Jenkins and Krause 1986). The early residents existed by hunting and gathering, and their movements were determined by the seasonal patterns of natural resources along the river. During late summer and fall, shellfish and fish were abundant in the streams and creeks draining the wooded floodplains, and waterfowl were taken as food from the bottomlands during winter.

By 600 AD, the Woodland stage Indian populations of the Southeast had expanded and their dependence on small mammals, reptiles, fish, and shellfish for food had increased. By 1000 AD, the Mississippian culture had emerged full-force with a dependency on wide-spread farming through floodplain agriculture. Many settlements flanked the meander belts of major rivers. These Indians used the nutrient-rich, well-drained natural levee soils to cultivate maize, beans, squash, and gourds and they supplemented their diet with backwater species of fish, migratory waterfowl, and other animals from backwater areas (Jenkins and Krause 1986).

The Mississippian Culture flourished from 100–1400 AD. It was already in decline by 1540, when the legendary Chief Tascalusa confronted Hernando de Soto on the banks of the Alabama River. The ensuing battle was a disastrous defeat for the Indians. However, de Soto's troops were also weakened. This battle marked the beginning of open hostilities between the Indians and the Europeans, which plagued de Soto throughout the rest of his explorations. Although Indian tribes survived in the region for several hundred more years, disease and conflicts with the European settlers took their toll. By the early 1800s, few remained. Shortly thereafter, the major rivers that had shaped the lives of the earliest inhabitants were to become major transportation arteries for resources of a different nature.

Historical Perspectives and Current Status

Physical Alteration of Rivers

Most of the major rivers of the Mobile River basin have been modified since the early part of this century. Channelization efforts to improve navigation began in the early 1800s near the port of Mobile (Mettee et al. 1987). Modifications of the river habitat, including dredging, construction of reservoirs, and cut-off of meander loops to straighten channels, have continued to the present. As a result, most large rivers (6th or greater order) have been changed from relatively shallow, fast-moving bodies of water, some with numerous shoals and outcrops, to deeper, slower-moving rivers. Large river impoundments and reservoirs range in surface area from 200 to 16,000 hectares (USACE 1981).

The Tombigbee River originates in Mississippi and flows through southwestern Alabama toward Mobile Bay. Navigation improvements on the lower Tombigbee River (below the junction with the Black Warrior River) began as early as the 1840s as part of the earliest efforts to extend navigation upstream to the north. In the ensuing years, four reservoirs were created. Studies addressing the feasibility of connecting the Tennessee River with the upper Tombigbee River for navigation purposes by the U. S. Army Corps of Engineers (COE) began as early as 1874. The early studies concluded that the cost of such a project would exceed benefits but by 1935 this conclusion had been reversed (Steward 1971). The Tennessee-Tombigbee Waterway (TTW) project was subsequently authorized by Congress in 1946, although funds were not appropriated for construction until 1971. The project was completed in 1985 at an estimated cost of 1.8 billion dollars (USACE 1987). The completion of the TTW effectively channelized the Tombigbee River north of its junction with the Black Warrior River (Figure 5) and eliminated most of the riverine habitat (Mettee et al. 1987).

The Black Warrior River has also been extensively modified since the early 1900s. One of its primary economic roles was a northern-most river link to the Warrior Coal Fields in north central Alabama (Mettee et al. 1987). Rock outcrops and a series of waterfalls upstream of the city of Tuscaloosa, the final barriers to linking the Warrior Coal Fields near Birmingham with Mobile, were inundated by 1915. At one point, 13 locks and dams were constructed on the Black Warrior River to facilitate barge traffic. The original locks and dams were replaced with the four current facilities from 1935–1957.

The Tallapoosa and Coosa rivers converge near Montgomery, Alabama, to form the Alabama River. The Tallapoosa River has been affected since the 1920s by four dams. The Coosa River has been extensively modified since 1914 with a series of seven dams and reservoirs. An extension of the Alabama River barge channel up the Coosa River from Montgomery, Alabama, to Rome, Georgia, was authorized under the River and Harbor Act of 1945. The first part of this project would provide for construction of navigation locks at dams between Montgomery and Gadsden, where channelization would also occur. The COE initiated engineering designs on the proposed waterway in 1978, and the environmental impact statement and design memorandum were completed in 1982. However, no money for construction has yet been appropriated. Development of the waterway from Gadsden, Alabama, to Rome, Georgia was deferred until the waterway from Montgomery to Gadsden was assured (USACE 1987).

The Alabama River joins the Tombigbee River in southwestern Alabama. About 75% of the Alabama River was impounded for navigation between 1963 and 1972.

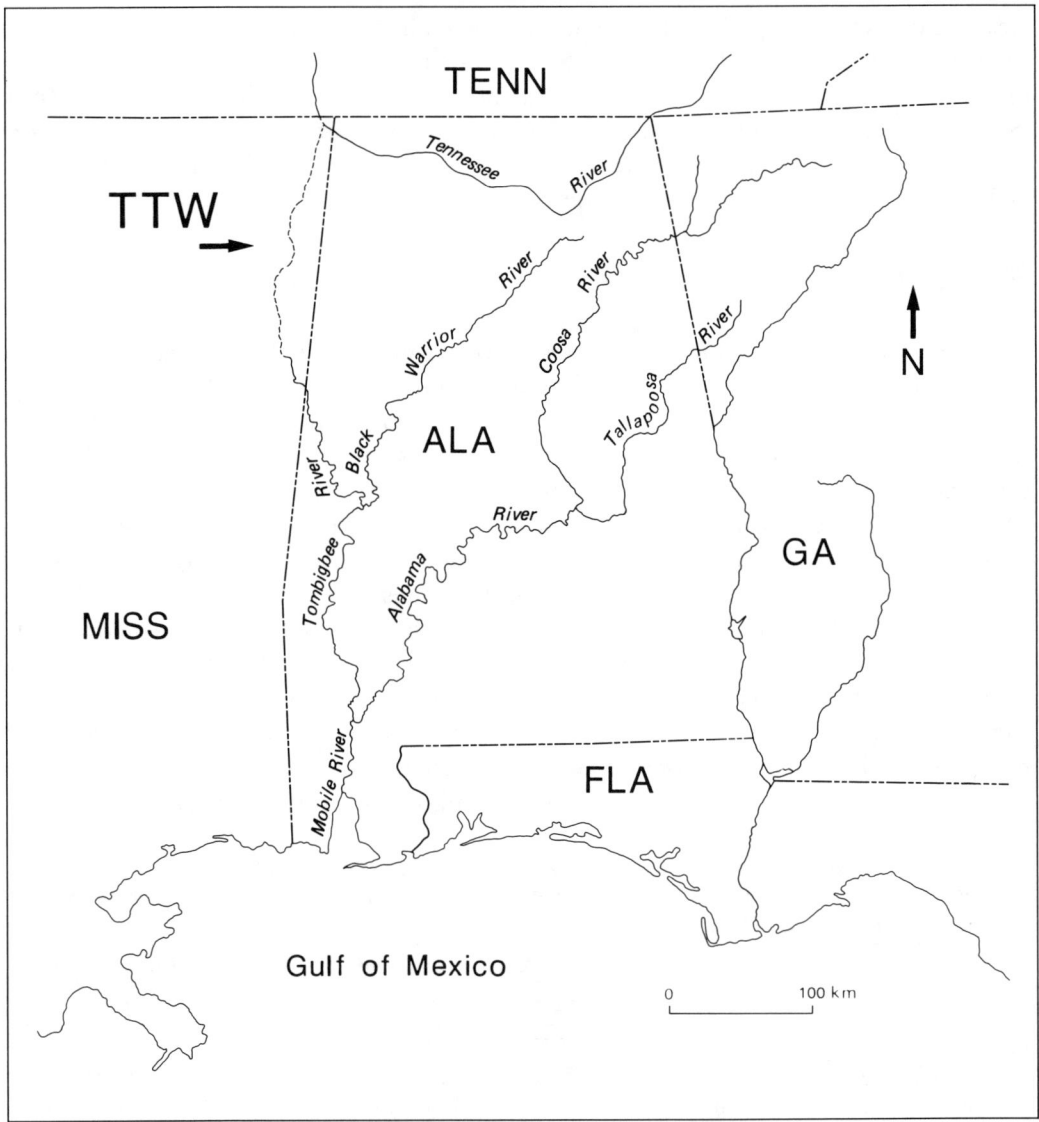

FIGURE 5. Mobile River drainage map showing the location of the Tennessee-Tombigbee Waterway (TTW).

This was accomplished by the construction of three locks and dams at a cost of nearly 1.78 billion dollars (USACE 1987).

Although the largest rivers in the Mobile River basin have been profoundly altered in the last 100 years, other medium-sized river systems remain less affected by human activity. For example, the Sipsey River (fifth order stream) originates in the northwestern Coastal Plain of Alabama and drains to the upper Tombigbee River. Because of its relatively small size, variable flow, and remote location, the Sipsey River is not impounded and its watershed is mostly undisturbed. Water quality of the Sipsey River is good (Mettee 1978). The Cahaba River (sixth order stream) in the Valley and Ridge province remains mostly free-flowing. The greater Birmingham area encroaches on the northern portion of the Cahaba River and its tributaries, but the less-populated central and lower portions of the river, where limestone outcroppings are numerous, was once considered for inclusion as a Wild and Scenic River. (See the section on Conservation Efforts and Future of Area for more details). Water quality in this region is considered good (O'Neil 1984), and its large populations of bass, sunfish, and catfish form the basis of an important sports fishery.

Water Quality of Rivers and Streams

Major Rivers. Many characteristics of surface waters in the Mobile River basin are within acceptable limits for drinking water established by the U. S. Environmental Protection Agency (USEPA 1976). However, several rivers have attributes that reflect the geology of the drainage basin and human uses, either of the river itself or the watershed in which it occurs. For example, in a comparison of multiple sites in the Black Warrior and Tombigbee rivers (Table 3), the Black Warrior River had higher concentrations of manganese and sulfate probably because of industrial inputs and surface mining. The Tombigbee River sampling sites, upstream from its confluence with the Black Warrior River, had higher suspended solids, color, dissolved organic carbon, dissolved iron, and phosphates (Table 3), reflecting the Coastal Plain province in which the Tombigbee River occurs and agricultural inputs. Seasonal variation of some water quality features in both rivers was linked primarily to precipitation with maximum values for suspended material and metals occurring during late winter and early spring, the period of highest rainfall. In contrast, highest values for

TABLE 3. Average and range of water quality variables measured at multiple sites in the Black Warrior (16 sites) and Tombigbee (5 sites) rivers between July 1978 and November 1979. (Data summarized from USACE 1983b.)

Measure	Black Warrior River		Tombigbee River	
	Mean	(Range)	Mean	(Range)
pH	7.20	(6.2–8.5)	7.80	(6.2–8.9)
Color (Pt-Co units)	15.00	(5–110)	37.00	(7–221)
Specific conductance (μS/cm)	175.00	(120–215)	125.00	(35–180)
Alkalinity (as $CaCO_3$, mg/L)	24.00	(13.0–39.0)	41.00	(21.0–54.0)
Chloride (Cl, mg/L)	7.00	(3.0–11.0)	9.00	(2.0–15.0)
Sodium (NA+, mg/L)	9.30	(6.8–11.4)	6.00	(2.6–9.4)
Sulfate (SO_4^2, mg/L)	42.00	(26–65)	8.00	(2–15)
Ammonia-N (NH_4-N, mg/L)	0.14	(0.01–1.70)	0.11	(<0.01–0.50)[a]
Nitrate + nitrite-N ($NO_3 + NO_2$-N, mg/L)	0.55	(0.16–1.20)	0.21	(<0.01–0.66)
Total Kjeldahl N (mg/L)	0.49	(0.01–1.70)[a]	0.74	(0.20–1.80)
Ortho phosphate (PO_4^{3-}, mg/L)	0.03	(<0.001–0.078)	0.05	(<0.001–<0.072)
Total phosphorus (mg/L)	0.07	(0.005–0.650)	0.18	(0.020–1.320)
Dissolved organic carbon (mg/L)	4.80	(<2–13)	8.90	(2–19)
Iron, dissolved (μg/L)	87.00	(5–420)[a]	260.00	(11–882)[a]
Manganese, dissolved (μg/L)	94.00[a]	(<2–419)[a]	26.00	(<2–90)[a]
Zinc, total (μg/L)	86.00	(<10–896)	76.00	(<10–640)

[a]These variables fall outside of Environmental Protection Agency recommendations for public water supplies (USEPA 1976).

inorganic nitrogen and phosphorus generally occurred during spring and summer, coinciding with application of fertilizers in the watersheds (USACE 1983b).

The Cahaba River, which flows over carbonate bedrock along much of its length, has generally good water quality, particularly in the lower basins (O'Neil 1984). A comparison of water quality between the total drainage and just the downstream reaches reveals that the river downstream contains much lower nutrients (Table 4). However, some streams in the upper basins receive excessive organic loads from areas affected by sewage and by general metropolitan development around Birmingham. These streams also have high biological oxygen demand (BOD) and low dissolved oxygen (DO) concentrations. High concentrations of heavy metals (iron, manganese, chromium, copper, nickel, and zinc) from mining activity also occur at some upstream sites. Continued exploitation of upstream basins clearly poses a threat to the more pristine water quality features of the lower main-stem Cahaba River.

The Alabama River, below the confluence of the Tallapoosa and Coosa rivers, also has generally good water quality based on EPA standards (USACE 1983a). Its waters are affected primarily by the city of Montgomery and by agricultural practices in the Coastal Plain. Water analyses from multiple sites in the Alabama River between Montgomery, Alabama, and its confluences with the Tombigbee River frequently show high ammonia from agriculture and high concentrations of dissolved iron (Table 5). However, other metals such as arsenic, cadmium, chromium, copper, lead, mercury, nickel, and zinc are within background limits in sediment analyses (USACE 1983a).

Analysis of sediments shows patch distribution of pesticides in the Black Warrior, Tombigbeee, and Alabama rivers (USACE 1983a, 1983b). Concentrations of such compounds as BHC-Beta, BHC-Alpha, BHC-Gamma (Lindane), Heptachlor epoxide, Dieldrin, DDT, and DDE range from 0.3 to 5 µg/kg sediment in some samples collected from the Tombigbee and Black Warrior rivers. Residues of Aldrin, DDT, and two PCBs were high in some analyses of sediment from the Alabama River.

TABLE 4. Range of water quality measurements in the Cahaba River upstream of Centerville. Ranges are compiled from 7 data sets, including 125 stations, from 1948–1982. (Summarized by O'Neil 1984.)

Measure	Total drainage	Downstream reaches
pH	5.2–9.6	6.1–8.9
Specific conductance (µS/cm)	24–680	24–380
Alkalinity (as CaCo$_3$, mg/L)	5–260	9–170
Chloride (Cl$^-$, mg/L)	0–68	0–20
Sodium (Na+, mg/L)	0.7–81.0	0.7–22.0
Ammonia-N (NH4-N,mg/L)	0.01–13.0[a]	0.01–0.85[a]
Nitrate + nitrite-N (NO$_3$ + NO$_2$-N, mg/L)	0.01–17.0[a]	0.01–0.99
Total Kjeldahl N (mg/L)	0.08–6.60	0.08–0.66
Ortho phosphate (PO$_4^{3-}$, mg/L)	0.01–430	0.10–280

[a]These variables fall outside of Environmental Protection Agency recommendations for public water supplies (USEPA 1976).

TABLE 5. Median and range of water quality measurements at 46 sites in the Alabama River and lower Coosa River between August and December 1977. (Data from USACE 1983a.)

Measure	Median	(Range)
pH	—	(5.6–7.9)
Color (Pt-Co units)	102.0	(10.0–195.0)
Specific conductance (μS/cm)	128.0	(81.0–203.0)
Alkalinity (as $CaCO_3$, mg/L)	—	(3.3–120.0)
Chloride (Cl-, mg/L)	7.0	(0.0–17.0)
Ammonia N (NH_4-N, mg/L)	0.1	(0.0–1.6)[a]
Nitrate + nitrite-N ($NO_3 + NO_2$-N, mg/L)	—	(0.00–0.28)
Total Kjeldahl N (mg/L)	—	(0.0–19.0)
Ortho phosphate (PO_4^{3-}, mg/L)	—	(0.00–0.18)
Iron, dissolved (μg/L)	94.0	(0.0–570.0)[a]
Manganese, dissolved (μg/L)	16.0	(0.0–80.0)

[a] These variables fall outside of Environmental Protection Agency recommendations for public water supplies (USEPA 1976).

Small Streams. The contribution of natural geological features to chemistry of river water is frequently obscured in the large rivers of the Mobile River basin by human activities. However, the abundance of relatively undisturbed smaller streams and rivers in the basin should provide information on how geology affects water chemistry. Unfortunately, extensive studies on replicate, physically matched streams among the different geological regions are lacking. Some information exists on the differences among streams in different geological regions and implications for their biological communities.

One obvious difference among waters in different geological regions is alkalinity, which varies from very low values of less than 5 mg/L as $CaCO_3$ in Appalachian Plateau streams to greater than 150 mg/L as $CaCO_3$ during summer months in Valley and Ridge streams. Streams in the Coastal Plain and Piedmont typically have intermediate values of alkalinity.

A comprehensive comparison of two low-order, undisturbed streams, one in the Appalachian Plateau and one in the Valley and Ridge (Lay and Ward 1987), showed other differences in water chemistry and biological features. The carbonate stream had higher nitrate concentrations, whereas the sandstone/shale stream had higher dissolved organic carbon (DOC). Both streams had ortho-phosphate concentrations near detection limits and rather low nutrient concentrations overall. There was little overlap in algal species between the two streams. The carbonate stream had both higher algal biomass and higher densities of grazers, primarily as snails. These data suggested that differences in underlying parent material may be reflected through several trophic levels and that geology can be a major organizing force in stream ecosystems. More work on a larger number of undisturbed streams is needed to see if the characteristics of these two streams represent the regions in which they occur.

An understanding of water chemistry and other features of undisturbed streams in different geological regions of the Mobile River basin is necessary to help assess responses in water quality and stream biology to human perturbations. One such perturbation in some streams, primarily in the Appalachian Plateau, is surface coal mining.

Most of coal mining in the Mobile River basin occurs in three fields in Alabama located at the southern terminus of the Appalachian coal region. The Warrior Coal Field (24,115 km^2), which accounts for 90% of coal production in the state (Harris et al. 1985), is divided into two regions. These include the Warrior Coal basin and the Plateau Coal region. Smaller and less productive regions include the Coosa Coal Field (550 km^2) and the Cahaba Coal Field (920 km^2). Coal beds occur primarily in the Pottsville formation of the Appalachian Plateau. Coal is of good quality, generally with low percentages of sulfur and ash. Alabama has about 21.5 billion tonnes of coal remaining with 4.6 billion tonnes considered surface minable. In calendar year 1983, Alabama produced more than 21 million tonnes of coal with almost half coming from surface mines (Tolson 1984).

Some typical effects of strip mining on streams include increased siltation, alteration of trace metal concentrations (particularly manganese and iron), increased sulfate concentrations, and acid mine drainage, which reduces pH and decreases or eliminates many aquatic organisms (Harkins 1980; Harris et al. 1985). Many of these effects in Alabama occur in watersheds that are strip-mined (Table 6). For example, Little Tyro Creek has high specific conductance, low pH, high sulfate and manganese concentrations, and increased sediment load compared to undisturbed streams in the same region. Strip mining lowered the total number of benthic macroinvertebrates and number of species by over 50% in Little Tyro Creek (Harris et al. 1985). Effects of strip-mining can be persistent and continue for at least seven years after mining has ceased, particularly with regard to high concentrations of dissolved solids and sedimentation (Harris 1987). Sediment deposition from erosion of strip-mined areas in some streams near the Black Warrior River has severely reduced aquatic habitat, increased flooding, and lowered aesthetic value in recreation areas (Harkins 1980).

Biological Aspects and Distribution of Organisms

Fish Communities. One striking characteristic of biological communities in the Mobile River basin is the number of fish species. The basin ranks third in North America in terms of species richness because it contains 157 fish species, of which 40 are endemic (Swift et al. 1986). Several factors contribute to species richness in this region, including little extinction of freshwater fish species (Smith 1981), a

TABLE 6. Comparison of water quality variables in streams draining mined and unmined areas of the Warrior Coal Basin. (After Harris 1987; compiled from information in Harkins 1980.)

Variable	Mined areas	Unmined areas
Sediment yield (tonnes/km^2/year)	2370–711,186	48–1896
Specific conductance (μS/cm)	30–3000	20–120
pH	2.0–5.0	4.5–7.5
Dissolved iron (μg Fe/L)	0.001–40	<0.03

relatively lengthy period of geological stability, a mild climate, a large network of rivers and streams, and the geographical setting. Also, the diversity of physiographic provinces provides a variety of habitats.

Despite the number of fish species in the Mobile River basin, very little information was available on their distribution, habits, and habitats prior to the 1940s (Mettee 1978). Since several major rivers (e.g., lower Tombigbee and Black Warrior rivers) had been physically altered for some time, detailed documentation of changes in fish communities brought about by human activity in those systems cannot be provided. However, substantial research on fish distribution in all major rivers as well as modifications attributable to more recent impoundments has been completed since 1948. These include studies of the Cahaba River (May 1963; Beckham 1974; Boschung and Mettee 1974), Warrior River and its tributaries (e.g., Harima and Mundy 1974; Jandebeur 1975), Coosa River (Boschung 1961), Sipsey River (Boschung 1973), eastern Mobile basin (Tucker 1967), lower Tombigbee River (Mettee et al. 1987), and river drainages of north central Alabama (Mettee 1978).

Of the 157 fish species in the Mobile River basin, 99 are cyprinids (Swift et al. 1986). Other families such as Percidae (perches), Centrarchidae (sunfishes), Catostomidae (suckers), and Ictaluridae (freshwater catfishes) are also well represented (Mettee et al. 1987). Total species richness before human alteration of rivers will be never be known. However, a clear trend toward elimination of some fish and other animals by channelization and impoundments (physical modification) has been well documented. The ecological implications of these changes to aquatic community interactions and productivity have not been investigated.

The destruction of native species by physical modification may be exacerbated by the introduction of non-native, larger cyprinids such as the common carp (*Cyprinus carpio*) and grass carp (*Ctenopharyngodon idella*). The first was introduced to the United States in 1831 and the second in the 1960s. The extent that these introductions have spread in the Mobile River basin is not known, but they appear to do well in reservoirs (e.g., Lake Martin on the Coosa River) and in rivers such as the Black Warrior, Sipsey, and Cahaba (Tucker 1967; Mettee 1978). Native Salmonidae (trouts) are not common, although rainbow trout (*Oncorhynchus mykiss*) have been introduced into some cooler reservoirs (e.g., Dycus and Howell 1974).

The fall line has naturally limited the northern dispersal of certain species of fishes in the Mobile River basin (Tucker 1967). The rapid change in elevation associated with the juncture of the Coastal Plain with other physiographic provinces (Figure 2) has affected about 60 species of fish, restricting 17 above and 43 below this zone. However, the construction of dams throughout the Mobile River basin created additional barriers to fish movement and limited the distribution of those species sensitive to siltation and pollution, or those that require rapidly flowing water.

Before dams were built on the Alabama River in the 1970s, construction of hydroelectric dams upstream on the Coosa and Tallapoosa rivers decreased the abundance of several freshwater species (Tucker 1967). These fish included the American eel *Anguilla rostrata*, crystal darter *Ammocrypta asperalla*, goldline darter *Percina aurolineata*, and freckled darter *Percina linticula*. Of these, the goldline darter was listed as an endangered species by the late 1970s (Mettee 1978). The spawning activities of some marine species, including Atlantic needlefish *Strongylura marina*, striped mullet *Mugil cephalus*, and salt-water striped bass *Morone saxatilis* were curtailed, presumably because dams blocked their spawning migration.

Construction of the Tennessee-Tombigbee Waterway provided an opportunity to compare the effects of pre- and post-channelization on species richness and distribution of fishes in the upper Tombigbee River. Species normally present in tributary streams of the upper Tombigbee River were least influenced, but fish fauna in the main channel changed substantially with at least 17 species being eliminated (Boschung 1987). The groups most affected were the minnows and perches. Relative abundance increased for some species such as suckers (primarily smallmouth buffalo *Ictiobus bubalus*), catfish (primarily blue catfish *Ictalurus catus*), sunfish, and herring (primarily gizzard shad *Dorosoma cepedianum*). The groups most vulnerable to decimation were those that had previously inhabited swift-moving areas of sixth and seventh-order streams.

Mollusca. At one time the Mobile River system contained one of the richest molluscan faunas in North America. Based on the distributions and phylogenetic relationships among the freshwater mussels in the Tennessee-Cumberland River drainage and those of the Alabama River, the Tennessee and Alabama drainages must have been connected during the recent geological past (Van der Schalie 1938; Starnes and Etnier 1986). The Tennessee River system was an evolutionary center for freshwater mussels. But the numerous examples of close phylogenetic relationships between species in the Tennessee and Alabama drainages suggest that the Alabama drainage was a secondary center that received and modified faunal elements from the Tennessee River.

Historically, much of the mollusc fauna of the Mobile River basin was located in the Coosa and Cahaba rivers. The main stem of the Coosa River, with its many shoals and limestone outcrops, harbored an abundant and rich fauna, particularly in the families Hydrobiidae and Pleuroceridae. Impoundments created along the Coosa River during this century flooded disparate habitats, extirpating many of the taxa restricted to them. Only those species that also inhabited tributaries survived. However, even in the tributaries, siltation from the erosion of rich agricultural lands altered much of the substrate and placed increased pressure on river refugee populations. The Tombigbee River also had a substantial mussel fauna but agricultural impacts, and finally the Tennessee-Tombigbee Waterway, eliminated most species.

Ninety-nine species of gastropods from the Mobile River basin are now either extinct or threatened and endangered (Stein 1976). Of those, 75 were from the Coosa River, 16 from the Cahaba River, one from the Tallapoosa River, and seven from the Black Warrior River. Nineteen other gastropods were noted from the Tennessee River drainage, three from the Appalachicola River, and one from the Choctawhatchee River. In the Mobile River basin, at least 24 species of bivalves of the order Unionoida were either endangered, threatened, or of special concern (Stansbery 1976). Also, 14 species of Unionidae have been extirpated from the Tennessee River drainage.

The morphology of the Coosa River has been irrevocably altered and, with it, the molluscan fauna it supported. However, the Cahaba River remains mostly unimpounded. Sediment from urbanized areas around Birmingham now threatens the river, but it remains relatively free from major impacts. Because of the undisturbed nature of the Cahaba River system, a rich molluscan fauna still survives in both the main stem and its tributaries, where the majority of the remaining endemic species in the Mobile River system drainage now reside. Species of the Pleuroceridae are particularly abundant, with *Elimia* (= *Goniobasis*) the most widespread. As many as 23 species of Pleuroceridae and 16 other gastropod taxa have been reported in the Cahaba River drainage (Goodrich 1941).

Comparison of two studies documenting numbers of mussels in the Cahaba River suggests that some species were eliminated between 1938 and 1973 (Van der Schalie 1938; Baldwin 1973). Forty-five species were reported in the earlier survey and only 30 in the more recent one, which sampled the same sites. Most of the unreported species were already rare in 1938; therefore, it is unclear whether their disappearance was due to sampling limitations, ecological processes (e.g., competition, predation), or recent disturbances in the upper drainage. Only two species have been added to the fauna since 1938, one of which is the widely distributed, introduced Asiatic clam *Corbicula manilensis*.

Aquatic Insects. Knowledge of the aquatic insect fauna is still incomplete in the Mobile River basin, although some attention has been directed towards economically important species of Diptera such as the Culicidae (Dorsie and Ward 1981), Simuliidae (Stone and Snoddy 1969), and Tabanidae (Watson 1968). Recently, however, investigations of aquatic insects of value as indicators of water quality have begun, including the Plecoptera (James 1972; Stark and Harris 1986), Ephemeroptera (Berner 1977; Kondratieff and Harris 1986), and Trichoptera (Harris 1986; Lago and Harris 1987). Little is known about the Chironomidae in the Mobile River basin.

Conclusions on the overall richness of the aquatic insect fauna in eastern Gulf Coast drainages must await further studies. However, given the diversity of fishes and mollusca, one could predict that the Mobile River drainage harbors a rich insect fauna as well. For example, the Trichoptera are represented by at least 300 species in Alabama, the majority of which are in the Mobile River basin (S. C. Harris, unpublished data), and one family, the Hydroptilidae, has three to four times as many species in Alabama as in surrounding states (Harris 1986).

Understanding how differences in physiographic provinces affect the distribution of aquatic insects is important so that the effects of human perturbation can be evaluated more accurately. Aspects of riverine systems most often used to gauge the impact of human intervention on biota (community structure, composition, and activity rates) depend on the physical and chemical characteristics of the system, which vary among physiographic regions. Because the geology of the Mobile River basin varies widely, adjacent watersheds possess different characteristics. Thus, information gained from one watershed may not be comparable or transferable to an adjacent watershed. Predictions on the effects of environmental stresses on benthic communities may not be applicable across physiographic regions.

Studies of the Plecoptera, Ephemeroptera, and Trichoptera have indicated that, while some species are widespread, distributions of many others appear to be restricted to particular physiographic regions (Table 7). As with fish distribution, the species of aquatic insects restricted to the Coastal Plain seem to represent a fairly discrete

TABLE 7. Number of aquatic insect species in Alabama limited in distribution to a single physiographic province (Kondratieff and Harris 1986; Stark and Harris 1986; S. C. Harris, unpublished data).

Taxon	Appalachian Plateau	Valley and Ridge	Piedmont	Coastal Plain
Plecoptera	5	9	2	9
Ephemeroptera	9	5	2	15
Trichoptera	20	29	8	40

unit, but the lines of demarcation between physiographic regions above the fall line are less clear.

Physical surveys of many Mobile River systems were made in the late 1800s to determine the feasibility of navigational improvements. Many surveys were detailed. For example, surveys of the upper and lower Black Warrior River (U. S. House 1875; U. S. Senate 1881) recorded numerous shoals, gravel bars, snags, sunken logs, and overhanging trees as the river meandered through 216 km of the Alabama Coastal Plain. The COE cited the need to remove 66 shoals and 729 logs in order to deepen the channel to about one meter (four feet) during low water. The Black Warrior River was about 100 m wide just below Tuscaloosa, and gradually decreased to an average of 50 m wide at its confluence with the Tombigbee River. Above Tuscaloosa, the Black Warrior River was once a series of lakes 150–200 m wide, flanked by sandstone cliffs 30–60 m high, and connected by falls or rapids flowing over sandstone ledges. Twelve rapids, many 200 m wide and 300 m long (but only a few centimeters deep at low water), existed along a 75 km stretch upstream between Tuscaloosa and the junction with Locust Fork, east of Birmingham. A rich mollusc fauna once existed in the Black Warrior River, and one can presume that the shallow riffles, shoals, snags, sunken wood, and slackwater areas also harbored a wide diversity of aquatic insects. The Black Warrior River is now a continuous series of impoundments upstream to Birmingham. The aquatic invertebrate fauna is dominated by the Asiatic clam, Oligochaeta (both Naididae and Tubificidae), and many species of lentic chironomids. The burrowing mayflies *Hexagenia* and *Caenis* are also numerous in certain localized areas (USACE 1983b).

Conservation Efforts and Future of Area

Since most of the Mobile River basin lies in Alabama, the following section will deal only with agencies from that state. Four agencies in Alabama have authority for environmental regulation and conservation of the state's aquatic resources. In 1982 the state legislature established the Alabama Department of Environmental Management (ADEM), which assumed the duties of developing environmental policy for the state and of establishing applicable environmental rules, regulations, and standards. The major focus of ADEM is implementation of federal and state environmental statutes. In addition to permitting and regulatory activities, ADEM maintains

45 water quality monitoring stations on major rivers, 19 of which coincide with EPA core network stations. ADEM reports annually on the status of water quality for each of the state's major river basins, and evaluates water quality problems and management programs within the state.

Exploration and surveys of mineral, energy, water, and biological resources of the state have been the responsibility of the Geological Survey of Alabama (GSA) since it was established in 1848. The Water Resources Division of GSA continuously and systematically collects geologic and hydrologic data on the occurrence, movement, and quality of surface and ground water resources. Basic data are collected through a state-wide network of stream gauging stations, observation wells, and water quality monitoring stations, which are summarized annually in GSA publications. The GSA also collects baseline environmental data and conducts environmental assessments on major energy resource development areas of Alabama. This involves collection of biological, sediment, and water samples from streams that drain coal fields and oil and gas fields of Alabama.

The Alabama Surface Mining Reclamation Commission was established in 1981 to regulate and control surface coal mining operations. In addition to encouraging economic development of the state's coal resources and seeing that surface mined areas are reclaimed promptly, one of the commission's primary objectives is to reduce adverse effects on water quality resulting from surface mining. Alabama has one of the more restrictive surface mine reclamation laws in the United States.

The Alabama Department of Conservation and Natural Resources is concerned with water quality as it impacts state fish and wildlife refuges. Established in 1939, the objectives of the Conservation Department are to promote, conserve, and protect state parks, monuments, and historical sites; to control public lands not administered by other agencies; and to administer game and fish laws. The Game and Fish Division enforces game and fish laws, manages 25 state parks and recreation areas, conducts research, and supervises 27 state wildlife management areas.

Four national forests occur in Alabama. The Bankhead National Forest in northwest Alabama contains 72,470 hectares, including the Sipsey Wilderness Area, 5,150 hectares of undisturbed forests, sandstone cliffs, and deep gorges within Sipsey Fork of the Black Warrior River basin. Part of this area has been designated a research natural area in the U. S. Forest Service System, and Sipsey Fork is a bench-mark watershed within the U. S. Geological Survey Network. The Talladega National Forest (145,692 hectares) has districts in both the eastern and western parts of the Mobile River basin. The eastern district lies in the mountainous Piedmont of Alabama and the western district lies in the Coastal Plain. Within the latter is the Reed Brake Research Natural Area, 242 hectares of longleaf pine, scrub oak, white oak, black oak, and northern red oak. The other two national forests in Alabama are smaller. The Tuskegee National Forest (4,361 hectares) lies in the Tallapoosa River basin in eastern Alabama, and the Conecuh National Forest (34,387 hectares) lies in the Escambia-Conecuh River drainage. The latter area also encloses another U. S. Geological Survey benchmark watershed.

The Wild and Scenic Rivers Act of 1968 established a national policy to preserve free-flowing rivers of remarkable beauty or outstanding biotic value in their natural state. This legislation authorized 16 rivers in Alabama for study and potential inclusion into the Wild and Scenic River System. Three have been studied to date. Favorable recommendations were made in the case of Sipsey Fork and Escatawpa rivers. The Cahaba River was considered for inclusion in 1979 but, because the river corridor had been modified, it was judged to be ineligible (USFS 1979).

Over the years several conferences have been held to identify endangered and threatened species in Alabama (Keeler 1972). In 1976 the concept was expanded to include several groups of invertebrates and vertebrates such as mollusks, fishes, crayfishes, shrimps, amphibians, and reptiles (Boschung 1976). Recent updates of organisms considered rare or threatened have concentrated on vertebrates (Mount 1984, 1986). Aquatic plants of Alabama considered rare were compiled by Thomas (1976) and updated by Freeman et al. (1979). However, only those species listed as rare and endangered by the U. S. Department of Interior are protected by law. Aquatic species protected in Alabama are listed in Table 8.

The geologic diversity and wealth of rivers and streams in the Mobile River basin provided it with abundant natural resources. As with other regions in the sunbelt, the area has experienced recent economic growth, which contrasts with the economic struggle and near poverty during the century after the Civil War. The Mobile River basin is a region of environmental contrasts. Despite the construction of numerous dams on large rivers, small rivers and streams retain the diverse character imparted by ancient geologic processes. To maintain this character, activities designed to enhance the economy (e.g., strip-mining, timber harvesting, expansion of metropolitan areas, etc.) must be balanced with environmental vigilance.

TABLE 8. Aquatic species in Alabama currently included on the U. S. Department of Interior's Rare and Endangered Species List.

Key Cave
Alabama cave fish - *Speoplatyrhinus poulsoni* Copper and Kuehne

Tombigbee River
Marshall's mussel - *Pleurobema marshalli* (Frierson, 1857)
Curtis' mussel - *Pleurobema curtum* (Lea, 1859)
Judge Tait's mussel - *Pleurobema taitianum* (Lea, 1934)
Stirrup shell mussel - *Quadrula stapes* (Lea, 1831)
Penitent mussel - *Epioblasma penita* (Conrad, 1834)

Mobile River
Alabama red-bellied turtle - *Pseudemys alabamensis* Baur

Warrior River
Flattened musk turtle - *Sternotherus minor depressus* Tinkle and Webb
Slackwater darter - *Etheostoma boschungi* Wall and Williams
Watercress darter - *Etheostoma nuchale* Howell and Caldwell
Snail darter - *Percina tanasi* Etnier
Alabama lamp pearly mussel - *Lampsilis virescens* (Lea, 1958)
Fine-rayed pigtoe pearly mussel - *Fusconaia cuneolus* (Lea, 1840)
Pink mucket pearly mussel - *Lampsilis orbiculata* (Hildreth, 1828)
Rough pigtoe pearly mussel - *Pleurobema plenum* (Lea, 1840)
Shiny pigtoe pearly mussel - *Fusconaia cor* (Conrad, 1834)
Green pitcher plant - *Sarracenia oreophila* (Kearney) Wherry

Conclusions

The Mobile River basin is one of the most geologically diverse and biologically rich regions in the United States. The full diversity of the biological communities in streams and rivers of this region may never be known because all major rivers in the basin have been physically modified, primarily through impoundments, for many decades. Water quality of lotic ecosystems in the basin is influenced by the composition and other attributes of the underlying parent material. However, the large rivers and some small rivers and streams are also affected by human activities, including agriculture, urbanization, and strip mining. Despite these perturbations, numerous small streams and several moderately sized rivers (e.g., the Sipsey and Cahaba rivers) remain free-flowing and relatively pristine. Other than some studies on species composition and distribution, little is known about the ecology of their biological communities as compared to other regions in the United States.

Acknowledgments

We are grateful to Drs. Arthur Benke, Robert Cushman, and James Sedell for their helpful comments on a final draft of this manuscript. We also appreciate discussions with Jim Sedell about historical perspectives of river ecosystems and their relevance to current ecological concepts and practices. This paper is contribution No. 125 from the Aquatic Biology Program, Department of Biology at the University of Alabama.

References

Adams, G. I., C. Butts, L. W. Stephenson, and W. Cooke. 1926. Geology of Alabama. Special Report 14. Geological Survey of Alabama, Tuscaloosa.

Baldwin, C. S. 1973. Changes in the freshwater mussel fauna in the Cahaba River over the past 40 years. Master's thesis. Tuskegee Institute, Tuskegee, Alabama.

Beckham, E. D. 1974. A survey of the fishes inhabiting the streams draining the Oakmulgee Division of the Talladega National Forest. Open-file report. University of Alabama, University (Tuscaloosa).

Berner, L. 1977. Distributional patterns of southeastern mayflies. Bulletin of the Florida State Museum, Biological Sciences 22:1–55.

Boschung, H. T. 1961. An annotated list of fishes from the Coosa River system of Alabama. American Midland Naturalist 66:257–285.

Boschung, H. T. 1973. A report on the fishes of the upper Tombigbee River, Yellow and Indian creek systems of Alabama and Mississippi. First supplemental environmental report. Continuing environmental studies, Tennessee-Tombigbee Waterway, Alabama and Mississippi, Vol. 18. U. S. Army Corps of Engineers, Mobile District, Mobile.

Boschung, H. T., editor. 1976. Endangered and threatened plants and animals of Alabama. Bulletin 2. Alabama Museum of Natural History, Tuscaloosa.

Boschung, H. T. 1987. Physical factors and the distribution and abundance of fishes in the upper Tombigbee River system of Alabama and Mississippi with emphasis on the Tennnessee-Tombigbee Waterway. Pages 184–192 *in* W. J. Mathews and D. C. Heins, editors. Community ecology of North American stream fishes. University of Oklahoma Press, Norman.

Boschung, H. T., and M. F. Mettee. 1974. A report on the fishes of the national forests of Alabama. Open-file report. U. S. Forest Service, Atlanta, Georgia.

Dorsie, R. F., and R. A. Ward. 1981. Identification and geographical distribution of the mosquitoes of North America, north of Mexico. Mosquito Systematics Supplement 1. American Mosquito Control Association, Fresno, California.

Dycus, D. L., and W. M. Howell. 1974. Fishes of the Bankhead National Forest of Alabama. Alabama Department of Conservation and Natural Resources, Montgomery.

Elder, J. F., and D. J. Cairns. 1982. Production and decomposition of forest litter fall on the Appalachicola River flood plain, Florida. Water-Supply Paper 2196-B. U. S. Geological Survey. U. S. Government Printing Office, Washington, DC.

Freeman, J. D., A. S. Causey, J. W. Short, and R. R. Haynes. 1979. Endangered, threatened, and special concern plants of Alabama. Journal of the Alabama Academy of Sciences 50:1–26.

German, E. R., and T. B. Moffett. 1983. Temperatures of surface waters of Alabama. Map 197, Geological Survey of Alabama, University (Tuscaloosa).

Goodrich, C. 1941. Distribution of the gastropods of the Cahaba River, Alabama. Occasional Papers of the Museum of Zoology 428, University of Michigan, Ann Arbor.

Harima, H., and P. R. Mundy. 1974. Diversity indices applied to the fish biofacies of a small stream. Transactions of the American Fisheries Society 103:457–461.

Harkins, J. R. 1980. Hydrologic assessment, eastern coal province Area 23, Alabama. Open-file report 80-683. Water Resources Investigations, U. S. Geological Survey, Tuscaloosa, Alabama.

Harris, S. C. 1986. Hydroptilidae (Trichoptera) of Alabama with descriptions of three new species. Journal of the Kansas Entomological Society 59:609–619.

Harris, S. C. 1987. Aquatic invertebrates of the Warrior Coal Basin of Alabama. Bulletin Geological Survey of Alabama 127:1–303. University (Tuscaloosa).

Harris, S. C., P. E. O'Neil, M. F. Mettee, and R. V. Chandler. 1985. Impacts of surface mining on the biology and hydrology of a small watershed in west-central Alabama. Bulletin Geological Survey of Alabama 125:1–124. University (Tuscaloosa).

James, A. M. 1972. The stoneflies (Plecoptera) of Alabama. Doctoral dissertation. Auburn University, Auburn, Alabama.

Jandebeur, T. S. 1975. Fish species diversity, occurrence, and abundance in the North River drainage system of Alabama. Doctoral dissertation. University of Alabama, University (Tuscaloosa).

Jenkins, N. J., and R. A. Krause. 1986. The Tombigbee watershed in southeastern prehistory. University of Alabama Press, University (Tuscaloosa).

Keeler, J. F., editor. 1972. Rare and endangered vertebrates of Alabama. Division of Game and Fish, Alabama Department of Conservation and Natural Resources, Montgomery.

Kondratieff, B. C., and S. C. Harris. 1986. Preliminary checklist of the mayflies (Ephemeroptera) of Alabama. Entomological News 97:230–236.

Lago, P. K., and S. C. Harris. 1987. An annotated list of the Curvipalpia (Trichoptera) of Alabama. Entomological News 98:244–262.

Lay, J. A., and A. K. Ward. 1987. Algal community dynamics in two streams associated with different geological regions in the southeastern United States. Archive für Hydrobiologie 108:305–324.

Lineback, N. G. 1973. Atlas of Alabama. University of Alabama Press, University (Tuscaloosa).

May, M. M. 1963. A study of the fishes of the Cahaba River drainage system of Alabama. Master's thesis. University of Alabama, University (Tuscaloosa).

Mettee, M. F. 1978. The fishes of the Birmingham-Jefferson County region of Alabama with ecologic and taxonomic notes. Bulletin Geological Survey of Alabama. University (Tuscaloosa).

Mettee, M. F., P. E. O'Neil, R. D. Suttkus, and J. M. Pierson. 1987. Fishes of the lower Tombigbee River system in Alabama and Mississippi. Bulletin Geological Survey of Alabama 107:1–186. University (Tuscaloosa).

Morisawa, M. 1968. Streams: their dynamics and morphology. McGraw-Hill, New York.

Mount, R. H., editor. 1984. Vertebrate wildlife of Alabama. Agricultural Experiment Station, Auburn University, Auburn.

Mount, R. H., editor. 1986. Vertebrate animals of Alabama in need of special attention. Agricultural Experiment Station, Auburn University, Auburn.

O'Neil, P. E. 1984. Historical surface-water quality analysis in the Cahaba River basin north of Centerville, Alabama. Final Report, Contract 14-16-0004-82-041. U. S. Fish and Wildlife Service, Endangered Species Office, Jackson, Mississippi.

Ramsey, J. R. 1976. Freshwater fishes. Pages 53–65 *in* H. T. Boschung, editor. Endangered and threatened plants and animals of Alabama. Bulletin 2, Alabama Museum of Natural History, Tuscaloosa.

Smith, G. R. 1981. Late Cenozoic freshwater fishes of North America. Annual Review of Ecology and Systematics 12:163–193.

Stansbery, D. H. 1976. Naiad mollusks. Pages 42–52 *in* H. T. Buschung, editor. Endangered and threatened plants and animals of Alabama. Bulletin 2, Alabama Museum Natural History, Tuscaloosa.

Stark, B. P., and S. C. Harris. 1986. Records of stonesflies (Plecoptera) in Alabama. Entomological News 97:177–182.

Starnes, W. C., and D. A. Etnier. 1986. Drainage evolution and fish biogeography of the Tennessee and Cumberland rivers drainage realm. Pages 325–361 *in* C. H. Hocutt and E. O. Wiley, editors. Zoogeography of North American freshwater fishes. Wiley, New York.

Stein, C. B. 1976. Gastropods. Pages 23–41 *in* H. T. Boschung, editor. Endangered and threatened plants and animals of Alabama. Bulletin 2, Alabama Museum Natural History, Tuscaloosa.

Steward, W. H., Sr. 1971. The Tennessee-Tombigbee Waterway. A case study in the politics of water transportation. University of Alabama, Bureau of Public Administration, University (Tuscaloosa).

Stone, A., and E. L. Snoddy. 1969. The blackflies of Alabama (Diptera; Simuliidae). Alabama Agricultural Experiment Station Bulletin (Auburn University) 390:1–93.

Swift, C. C., C. R. Gilbert, S. A. Bartone, G. H. Burgess, and R. W. Yerger. 1986. Zoogeography of the freshwater fishes of the southeastern United States: Savannah River to Lake Pontchartrain. Pages 213–265 *in* C. H. Hocutt and E. O. Wiley, editors. The zoogeography of North American freshwater fishes. John Wiley, New York.

Thomas, J. L. 1976. Plants. Pages 5–12 *in* H. T. Boschung, editor. Endangered and threatened plants and animals of Alabama. Bulletin 2, Alabama Museum of Natural History, Tuscaloosa.

Tolson, J. S. 1984. Alabama coal data for 1983. Alabama Geological Survey Information Series 58E. U. S. Government Printing Office, Washington, D.C.

Tucker, C. E. 1967. A study of the fishes of the eastern Mobile River basin. Doctoral dissertation. University of Alabama, University (Tuscaloosa).

USACE (U. S. Army Corps of Engineers). 1981. Environmental data inventory, state of Alabama. USACE, Mobile District, Mobile, Alabama.

USACE (U. S. Army Corps of Engineers). 1983a. Water quality management studies Alabama River, August–December 1977. USACE, Environmental Quality Section, Mobile, Alabama.

USACE (U. S. Army Corps of Engineers). 1983b. Water quality management studies middle Black Warrior and lower Tombigbee rivers, Warrior and Demopolis lakes, July 1978–December 1979. USACE, Environmental Quality Section, Mobile, Alabama.

USACE (U. S. Army Corps of Engineers). 1987. Water resources development in Alabama 1987. USACE, Mobile District, Mobile, Alabama.

USEPA (U. S. Environmental Protection Agency). 1976. Quality criteria for water. U. S. Government Printing Office, Washington, D.C.

USDA (U. S. Department of Agriculture). 1977. Alabama River cooperative study within Alabama, volume 1. USDA, Alabama Development Office, Auburn, Alabama.

USFS (U. S. Forest Service). 1979. Cahaba River-Alabama wild and scenic river study. Southeastern Area and State of Alabama Forestry Commission, USFS, Montgomery, Alabama.

USGS (U. S. Geological Survey). 1982. Water resources data Alabama water year 1981. Water-Data Report AL-81-1, USGS, Tuscaloosa.

USGS (U. S. Geological Survey). 1986. Water resources data Alabama water year 1985. Water-Data Report AL-85-1, USGS, Tuscaloosa.

U. S. House. 1875. Examination and survey of Black Warrior River, from Locust Fork to its mouth, Alabama. (House Executive Documents 75–76). 43rd Congress, 2nd session (Part 2, Appendix T-7, 16–26). U. S. Government Printing Office, Washington, D. C.

U. S. Senate. 1881. Survey of Black Warrior River, from Tuscaloosa to Forks of Sipsey and Mulberry, Alabama. (Senate Executive Documents 42). 46th Congress, 3rd session (Appendix K 19, 1218–1221). U. S. Government Printing Office, Washington, D. C.

Van der Schalie, H. 1938. The naides (freshwater mussels) of the Cahaba River in northern Alabama. Occasional Papers of the Museum of Zoology 392, University of Michigan, Ann Arbor.

Watson, R. L. 1968. The Tabanidae of Alabama and some aspects of their ecology. Doctoral dissertation. Auburn University, Auburn, Alabama.